*Fundamentals of*

# Geometric Dimensioning and Tolerancing

*3rd Edition*

**Based on ASME Y14.5-2009**

*by Alex Krulikowski*

**Fundamentals of Geometric Dimensioning and Tolerancing, 3E**

**Alex Krulikowski**

Vice President, Editorial: Dave Garza

Director of Learning Solutions: Sandy Clark

Acquisitions Editor: Kathryn Hall

Managing Editor: Larry Main

Senior Product Manager: John Fisher

Editorial Assistant: Diane Chrysler

Vice President, Marketing: Jennifer Baker

Marketing Director: Deborah Yarnell

Marketing Manager: Erin Brennan

Marketing Coordinator: Jillian Borden

Production Director: Wendy Troeger

Production Manager: Mark Bernard

Content Project Manager: David Barnes

Production Technology Assistant: Emily Gross

Art Director: Casey Kirchmayer

Technology Project Manager: Joe Pliss

Printed in the United States of America

8 9 10 11 12     23 22 21 20 19

For product information and technology assistance, contact us at
**Cengage Learning Customer & Sales Support, 1-800-354-9706**

For permission to use material from this text or product,
submit all requests online at **www.cengage.com/permissions**.
Further permissions questions can be e-mailed to
**permissionrequest@cengage.com**

Library of Congress Control Number: 2011945071
ISBN-13: 978- 1-1111-2982-8
ISBN-10: 1-1111-2982-7

**Delmar**
5 Maxwell Drive
Clifton Park, NY 12065-2919
USA

Cengage Learning is a leading provider of customized learning solutions with office locations around the globe, including Singapore, the United Kingdom, Australia, Mexico, Brazil, and Japan. Locate your local office at: international.cengage.com/region. Cengage Learning products are represented in Canada by Nelson Education, Ltd.

To learn more about Delmar, visit **www.cengage.com/delmar**
Purchase any of our products at your local college store or at our preferred online store **www.cengagebrain.com**

**Effective Training Inc.**
2118 S. Wayne Road
Westland, MI 48186
USA

Effective Training Inc. is a global leader in the field of geometric dimensioning and tolerancing. ETI provides training and materials to thousands of companies, corporations and educational organizations around the world.

To learn more about Effective Training, or to purchase training software, visit **www.etinews.com**

To Pat

You are amazing.
You make life so grand.
You are my Angel.

# TABLE OF CONTENTS

## Introduction                                                                    Page

## GD&T Fundamentals

## Form

## Datum System

## Orientation

## Position

## Runout, Concentricity, Symmetry

## Profile

## Appendices

# ACKNOWLEDGMENTS

You know how people say they wouldn't be standing here if it weren't for the support and friendship of certain people? So I, too, am surrounded by treasured friends and colleagues who helped make this book possible. I have the greatest admiration for this highly accomplished team of experts. They were eager to contribute, found quite a few errors, and set me straight on a few things.

The technical proofreading team:

Michael Adcock, *Effective Training Inc., Westland, MI*

Jim Beary, *Benteler Automotive Corporation, Grand Rapids, MI*

Perry Betterley, *Stryker Instruments, Kalamazoo, MI*

William Caldwell, *Delphi Automotive Systems, Auburn Hills, MI*

Dan Carlson, *TRW Automotive Electronics, Farmington, MI*

Roy Cross, *Effective Training Inc., Westland, MI*

Purushuthaman Damodaran, *College of Engineering and Technology, Northern Illinois University, North Canton, OH*

Brent Davis, *Ford Motor Company, Livonia, MI*

Dan Flick, *Ivy Tech Community College, Indianapolis, IN*

Tim Graves, *Dimensional Management Systems, St. Clair Shores, MI*

Casey Guthridge, *Stark State College of Technology, Canton, OH*

Charles (Don) Holder, *GM Powertrain, Ypsilanti, MI*

Maria Hull, *Hudson Valley Community College, Troy, NY*

Evan Janeshewski, *Axymetrix Quality Engineering Inc., Vancouver, Canada*

Wayne Lee, *Johnson Controls, Plymouth, MI*

Dale MacPherson, *GM Powertrain, Pontiac, MI*

Edward McCarthy, *Raytheon Company, Tucson, AZ*

Daniel Meyers, *AAI Corporation, Baltimore, MD*

Michael Murphy, *GM Powertrain, Pontiac, MI*

Andrew Otieno, *Northern Illinois University, Dekalb, IL*

Jan Roovers, *Jan Roovers Associates, Inc., Charlotte, NC*

David Slopsema, *GM Powertrain, Pontiac, MI*

Larry Smith, *Industrial Training Services, Ontario, CA*

Harry Taylor, *GE Aviation, Cincinnati, OH*

Carl Wargula, *Quality Transformation Systems, Goodyear, AZ*

Nathan Weister, *Eaton Corporation, Moon Twp, PA*

I salute the proofreaders. What an outstanding job they did. They challenged concepts, text, illustrations, problems, and sometimes even my interpretation of Y14.5. They provided an invaluable service to the quality of this book.

Several proofreaders are close friends and coworkers, and I sincerely appreciate their honesty in their comments.

Credit is gratefully given and acknowledgement made for the use of definitions and terms from the ASME Y14.5-2009 Dimensioning and Tolerancing Standard. This standard is published by the American Society of Mechanical Engineers (ASME), New York, NY.

Last, but certainly not least, I want to express my gratitude to the outstanding team at Effective Training:

Michael Adcock – Dimensional Engineering Mentor

Brandon Billings – Website Administrator

Ken Blinn – Mechanical Designer

Roy Cross – Dimensional Engineering Mentor

Jamy Krulikowski – Multimedia Developer

Pat Krulikowski – Financial Administrator

Jim McBreen – Customer Service

Dennis Moore – Account Executive

Branny Mrljak – Account Executive

Katherine Palmer – Writer/Editor

Matthew Pride – Graphic Designer

Mark Ramsey – Product Developer

Gary Walls – Shipping Manager

Chris Wioskowski – Network Administrator

Those who didn't directly work on the book did extra duties to free up time of those who were involved. Everyone contributed enthusiastically. I couldn't ask for a better group of coworkers. Thanks everyone.

Alex Krulikowski – ETI President

# SPECIAL NOTE TO THE STUDENT

This section contains important information about how to increase your success in learning Geometric Dimensioning and Tolerancing (GD&T).

Dear Student,

Welcome to the exciting world of geometric dimensioning and tolerancing. I call it exciting, because if you master this topic, you will possess a critical skill needed to design, produce, or inspect parts. There are six major parts to the fundamentals of GD&T:

1. Vocabulary
2. Major concepts
3. Datum system
4. Symbols
5. Tolerance zone requirements
6. Design philosophy

This textbook is designed to introduce you to the fundamentals of GD&T; however, only you can ensure a successful understanding of GD&T through proper goal setting and self-discipline. I can lead you into the world of GD&T, but you must make a conscious effort to do the work required to successfully master this subject.

Because of my firm belief that success is achieved through proper goal setting and discipline, I have provided you with goals and performance objectives at the beginning of each chapter. Acquire an understanding of the goals and objectives. Work through the chapter, step by step, and you will soon have mastered the fundamentals GD&T.

However, to be successful, you should also set some goals, set aside a specific time each day to read, study and complete the questions and problems at the end of each chapter. The results of your work and discipline will be apparent as you successfully grasp each concept and master the performance objectives. Best of all, you will soon realize the larger goal: an understanding of the preciseness, the flexibility, and the power of GD&T.

Studies about how we learn have identified that several levels of thinking skills are involved in the learning process. Many students try to learn by simply memorizing facts, without a thorough understanding of the topic; however, these studies indicate that memorization alone will not properly prepare the student for tests or using the topic on the job. The following list illustrates the levels of thinking skills:

1. **Knowledge** involves remembering (memorizing ) factual material.

2. **Comprehension** involves interpreting information, changing it from one form to another, and/or making predictions.

3. **Application** involves using facts and fundamental principles to solve problems comparisons.

4. **Analysis** involves identifying and sorting out relevant and irrelevant facts to make comparisons.

5. **Synthesis** involves combining information and developing a plan or using original ideas.

6. **Evaluation** involves judging the value of observations and calculated results in order to reach a meaningful conclusion.

As you can see, memorization alone does not develop thinking skills to a level to be able to apply the fundamentals of GD&T in real-life situations. Memorization must be combined with other levels of thinking skills before a real understanding of the topic can be accomplished.

Many educators use different teaching techniques to assist students in the learning process. In order to assist you in this course, I have adopted ten commonly known learning principles.

1. Learning occurs in small steps. Begin today—not tomorrow—to study and to solve problems.

2. Study daily. Don't expect to learn a lot the night before a test.

3. First, scan the performance objectives, then carefully read the material and ask yourself relevant questions. Write down questions you cannot answer.

4. Read the material a second time, take notes, and list key points. Learning is increased with repetition.

5. Think about interconnections with what you already know, including on-the-job applications.

6. Come to class with a list of questions.

7. Study sample problems in the text. Consider the strategies used to solve these problems and how you would recognize and approach similar problems in the text, in a test, or on the job.

8. Learn the GD&T vocabulary. The terms and their definitions are critical to understanding geometric dimensioning and tolerancing. Use the proper terms when you discuss GD&T.

9. Take notes in class.

Although these principles can work for you, you alone can decide to commit the time and effort it will take to apply them. You must first commit yourself to preparing for and attending the lectures. Before each lecture, read the goals and performance objectives to be covered, read the text material, note key points and prepare questions for during the lecture.

Remember, major geometric tolerancing topics are interrelated and build upon one another, so after studying a chapter, review the performance objectives from the beginning of the chapter to be sure you understand the major points and terminology involved. Could you explain these terms and concepts to someone else? Try it! A person who understands a topic can use the correct vocabulary needed to explain that topic.

I hope you are encouraged enough to begin the hard work needed to master the fundamentals of GD&T, if so, you will be well rewarded. Geometric dimensioning and tolerancing is a comprehensive and useful topic; understanding its six major components is rewarding. In writing this text, I tried to do everything possible to assist you in your journey to learning GD&T. The rest of the work is up to you!

If you are on LinkedIn and would like to keep abreast of the latest developments in GD&T, send me an invitation and I would be happy to join your network. Also, I have a GD&T group on LinkedIn called Effective Training (ETI) / Dimensional Engineering. I invite you to join the group.

Sincerely Yours,

Alex Krulikowski

# NOTE TO THE INSTRUCTOR

This third edition of the *Fundamentals of Geometric Dimensioning and Tolerancing* preserves the best features of the first two editions while adding new material that reflects the changes in the field of geometric tolerancing and several new features to improve the understanding of GD&T.

I believe this third edition will continue to be the most practical and easy to use text on the market. Changes and new features in this edition include:

- Technical content updated to reflect the latest information from ASME Y14.5- 2009
- A two-column format
- Information is divided into smaller topics; each chapter is based on a single goal
- Several new chapters: Chapter 1: Drawing Standards; Chapter 2: Dimensions, Tolerances, and Notes; and Chapter 4: General Dimensioning Symbols
- New terms listed at the front of each chapter
- Organization of the chapters covering the GD&T symbols is divided into three sections: terms and concepts, applications, and verification principles.
- Reorganized the chapter problems into three types true/false, multiple choice, and application problems.
- The use of mnemonics to aid in the memory of concepts
- Real-world examples for each geometric tolerance
- Chapter summaries
- Expanded coverage on the verification principles and methods for each geometric tolerance
- The use of the surface interpretation for all tolerances applied at MMC
- The use of line conventions to make the illustrations easier to interpret.
- The use of color to highlight tolerance zones, boundaries, datums, parts, and gages in the figures.
- The use of a 3D model coordinate system to denote the degrees of freedom constrained
- The use of the significant seven questions for interpreting a geometric tolerance

## Organization

The *Fundamentals of Geometric Dimensioning and Tolerancing* is divided into 27 chapters. Each chapter focuses on an important goal for understanding geometric dimensioning and tolerancing. These 27 goals are the major topics that must be mastered to be fluent in the fundamentals of geometric tolerancing. Each goal is further defined and supported by a set of performance objectives.

The performance objectives describe specific, observable, measurable actions that the student must accomplish to demonstrate mastery of each goal. There are over 260 performance objectives in this text. These performance objectives are a key to success for both the student and the instructor. The text content, problems, exercises, quizzes, and teaching materials are all based on the performance objectives. Using the performance objectives will make conducting the class easier for the instructor and more meaningful for the students.

The performance objectives are based on Bloom's taxonomy of objectives. Since this is a fundamentals course, most of the objectives involve Bloom's first three levels: knowledge, comprehension, and application, and approximately 10-20% of the objectives relate the three higher levels: analysis, synthesis, and evaluation.

## Course Supplements

A complete package of course supplements to this text are available from ETI. Visit www.etinews.com to learn more about each product.

The *Instructor Answer Guide* is an answer guide to the true/false, multiple choice, and application problems at the back of each chapter. To obtain a copy, email the author at alexk@etinews.com.

The *Fundamentals of GD&T 2009 Digital Instructor's Kit*, is a complete set of teaching materials for this course. It contains over 400 digital color images, detailed lesson plans, several course syllabuses, and a complete set of quizzes and tests with answers. The software is available from Effective Training Inc. A substantial discount is available for colleges.

The *Ultimate GD&T Pocket Guide, 2nd Edition (2009)* is a 120-page information-packed reference for GD&T. It covers the definitions, rules, and major concepts — and also explains each geometric tolerance. This mini-book summarizes the major points of GD&T into a condensed easy to read pocket reference that includes index tabs, more than 100 detailed drawings, and a glossary. A substantial discount is available for colleges.

All registered owners of the *Fundamentals of GD&T Digital Instructors Kit* for this course also have access to a dedicated web page that contains the following:

- A complete answer guide for the true/false, multiple choice, and application problems
- An errata sheet for the text
- A discussion board with topics related to GD&T and the course materials

## Instructor Site

An Instruction Companion Website containing supplementary material is available. This site contains an Instructor Guide, testbank, image gallery of text figures, and chapter presentations done in PowerPoint. Contact Delmar Cengage Learning or your local sales representative to obtain an instructor account.

Accessing an Instructor Companion Website site from SSO Front Door
1. GO TO: http://login.cengage.com and login using the Instructor email address and password.
2. ENTER author, title or ISBN in the Add a title to your bookshelf search box, CLICK on Search button
3. CLICK Add to My Bookshelf to add Instructor Resources
4. At the Product page click on the Instructor Companion site link

New Users: If you're new to Cengage.com and do not have a password, contact your sales representative.

## A Few Comments From the Author

I hope you enjoy teaching this course and using the materials I have designed. If you would like to keep in touch and share ideas about GD&T or teaching, please send me an invitation on LinkedIn and I will join your network. I also invite you to join both of my GD&T groups on LinkedIn:

- Effective Training (ETI) GD&T / Dimensional Engineering
- GD&T Trainers / instructors / mentors

More than twenty proofreaders have revised this text before publication. We have made numerous improvements and corrections as a result of their efforts. However, from my past experience, I realize that a few errors have probably slipped though in the final stages of the book production. I apologize for any inconvenience this may cause. If you find an error, please send it to me. I will maintain an errata sheet and send it upon request.

If you would like to contact me with a comment or suggestion, my email address is alexk@etinews.com.

As a parting thought, I want to share a quotation from Michelangelo Buonarroti with you:

"Ancoro Imparo"

# TEXT CONVENTIONS

## Drawing Conventions

There are many engineering drawings used in this book. In order to focus on the dimensioning topic being discussed, many of the drawings are partial drawings. In some instances, figures show added detail for emphasis; in some instances, figures are incomplete by intent. Numerical values for dimensions and tolerances are illustrative only. Other drawing conventions include:

- Notes shown in capital letters on drawings are intended to appear on actual engineering drawings.

- Notes shown in lowercase letters are for explanatory purposes only and are not intended to appear on engineering drawings.

- All drawings are in accordance with ASME Y14.5-2009.

- Unless otherwise specified, all angles ± 5°.

- All units are metric. All dimensions are in millimeters.

- Basic dimensions may be omitted for clarity.

- Third angle projection is used on all figures.

- In figure interpretations, the primary datum plane will contain the X - Y plane. The primary datum axis will contain the Z axis.

## Gage Tolerances

The gages used in this text are described with dimensions; no tolerances are shown. In the product design field, gages are considered to have no tolerances; however, in industry, gages do have tolerances. The gage tolerances are usually quite small compared to part tolerances. A rule of thumb is that gage tolerances are 10% of the part tolerances. Gage tolerances are often arranged so that a (marginally) good part may be rejected, but a bad part will never be accepted. (From paragraph 2.5.4.1 MIL-HDBK-204A[AR] Design of Inspection Equipment for Dimensional Characteristics.)

## Line Conventions

A chart showing the line conventions and color use in this text is shown on the opposite page.

## Abbreviations Used in This Text

The name of the dimensioning and tolerancing standard is ASME Y14.5-2009. It is referred to in the text as Y14.5. A list of abbreviations used in the text is located on the opposite page.

| Line and Color Conventions | |
|---|---|
| Nominal workpiece surface | ——————— |
| Real workpiece surface | ∿∿∿ |
| Datum simulator | – – – – – |
| Worst-case boundary | — · — · — |
| Drawing / figure divider | — ·· — ·· — |
| Axis / center plane | — — — — |
| Datum plane edge view | – – – – – |
| Primary datum plane | |
| Secondary datum plane | |
| Tertiary datum plane | |
| External<br>• Rule #1 boundary<br>• VC (acceptance boundary) | |
| Internal<br>• Rule #1 boundary<br>• VC (acceptance boundary) | |
| Tolerance zone line / fill | |
| Datum simulator / gage element | |
| Actual part fill | |

| Abbreviation | Meaning | Page |
|---|---|---|
| ASME | American Society of Mechanical Engineers | 4 |
| CMM* | Coordinate measuring machine | 288 |
| CR | Controlled radius | 38 |
| DIA | Diameter | 406 |
| DOF | Degrees of freedom | 158 |
| ETI | Effective Training Incorporated | 415 |
| FIM | Full indicator movement | 334 |
| GD&T* | Geometric dimensioning & tolerancing | 26 |
| IB | Inner boundary | 90 |
| LMB | Least material boundary | 408 |
| LMC | Least material condition | 56 |
| MMB | Maximum material boundary | 196 |
| MMC | Maximum material condition | 55 |
| OB | Outer boundary | 90 |
| R | Radius | 38 |
| RMB | Regardless of material boundary | 184 |
| RFS | Regardless of feature size | 56 |
| SR | Spherical radius | 38 |
| VC* | Virtual condition | 91 |

* Abbreviation not in Y14.5

# ICON AND GRAPHIC CONVENTIONS

## Author's Comments

In various places throughout this book, the author provides helpful comments that:

- Discuss a dimensioning situation that is not covered in the Y14.5-2009 dimensioning standard

- Offer the reader opinions, insights, or tips about the topic being discussed

The comments are strictly advisory and are not part of the Y14.5-2009 dimensioning standard.

***Author's Comment***
When three datums are referenced on a part with all planar datums, all six degrees of freedom are restricted.

## Design Tips

The design tips from the author help designers to apply tolerancing information in a cost-effective manner. The comments are advisory and are not part of the Y14.5-2009 dimensioning standard.

***Design Tip***
When dimensioning a part, zero tolerance at MMC should be considered wherever the function of a feature of size is assembly.

## For more info...

When this appears on a page, it is accompanied by textbook page references or sources outside the textbook that contain information related to the topic.

***For more info...***
See Paragraph 4.4.1 of Y14.5.

## Technotes

Technotes contain important facts that should be noted and remembered for better understanding of the text. These notes contain technical definitions and specific rules that are applied to information within the lessons.

Each note is clearly labeled with a technote number that corresponds to the chapter where it is found and near the information where it will be of the most help.

**TECHNOTE 23-1**
**Using a Projected Tolerance Zone Modifier**

- Where a projected tolerance zone is used, the tolerance zone is projected outside the hole.

- A projected tolerance zone is used to limit the orientation deviation of a hole to ensure assembly with the mating part.

## Website Bonus Materials

When this appears on a page, it signifies that free bonus materials related to the topic are available at the Effective Training website.

***Website Bonus Materials***
Additional questions are available at our website. To access bonus materials for this textbook, please visit:
www.etinews.com/textbookbonus

# Drawing Standards

## Goal

Understand the importance of standards on engineering drawings

## Performance Objectives

Upon completing this chapter, you should be able to:

1. Describe what an engineering drawing is (p.2)
2. Explain the importance of an engineering drawing (p.3)
3. List four consequences of engineering drawing errors (p.4)
4. List the two primary dimensioning and tolerancing standards used globally (p.4)
5. Describe which ASME standards cover dimensioning and tolerancing (p.4)
6. Describe the role of dimensioning and tolerancing standards on engineering drawings (p.5)
7. Identify which dimensioning and tolerancing standards apply to an engineering drawing (p.5)

## New Terms

- ASME Y14.5-2009
- Engineering drawing

## What This Chapter Is About

This chapter introduces the topics of engineering drawings, dimensioning and tolerancing, and the importance of standards.

Engineering drawings are the documents most companies use to communicate product requirements. Dimensioning and tolerancing is typically used on engineering drawings to define the size, shape, feature relationships, and allowable variation of a workpiece. Standards are necessary to create common specifications and promote common interpretation practices.

A good engineering drawing should be able to speak for itself, it should be able to go anywhere, and competent people should have the same interpretation of the drawing without question.

# TERMS AND CONCEPTS

## Engineering Drawings

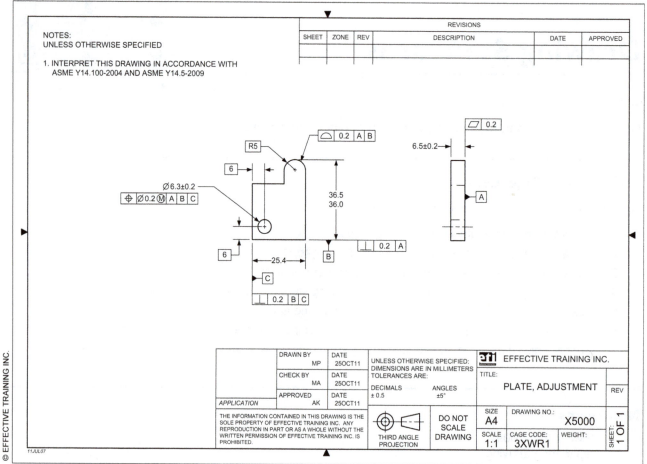

*FIGURE 1-1  Engineering Drawing*

An *engineering drawing* is a document (or digital data file) that communicates a precise description of a part. See Figure 1-1. An engineering drawing consists of pictures, words, numbers, and symbols that are used to communicate the part requirements. An engineering drawing typically includes:

- Geometry (shape, size and form of the part)

- Important functional relationships

- Tolerance (variation) permitted for proper function

- Material, heat treat, surface coatings

- Part documentation information (part number, revision level, etc.)

The engineering drawing is where the engineering requirements of the part are documented.

For more than 150 years, most engineering drawings were created using manual methods on mylar or vellum and were reproduced and distributed as blueprints. However, in the last thirty or so years, engineering drawings have been increasingly created on CAD systems. Figure 1-2 shows a comparison of "then and now" for engineering drawings.

**Author's Comment**

There are many types of engineering drawings described in ASME Y14.24. This book pertains to detail drawings.

| Engineering Drawings | | |
|---|---|---|
| | Past practices (pre 1980) | Current practices |
| Drawing medium | Vellum Mylar | Electronic files |
| Distribution methods | Bluelines Sepia prints White prints | Vendor specific CAD files CAD neutral formats IGES, STEP, JT, PDF files |
| Drawing creation tools | Compass T-Square Triangles | CAD software programs |
| Dimension methods | Coordinate tolerancing | Geometric tolerancing of CAD geometry |

*FIGURE 1-2 Engineering Drawing Practice Comparison*

## Importance of Engineering Drawings

An engineering drawing is an important document in an organization because it affects the success of both the product and the organization, as shown in Figure 1-3.

The design establishes the goal for a number of departments, so if the drawing is vague, the entire organization is less efficient. Time, money, and resources are wasted until everyone understands what the drawing is intended to convey. In a large organization, as many as 10,000 people may have to interpret an engineering drawing.

The function of a product determines the amount of variation permitted for each dimension. Therefore, the engineering drawing must communicate the allowable variation clearly and in a mathematically repeatable method. Using standards on engineering drawings is an important part of ensuring that drawings communicate clearly.

The manufacturing and inspection costs of a product are affected by the allowable tolerance of each dimension. When manufacturing a part, tolerances affect process selection, tooling, fixturing, etc. Tolerances also affect how a part is inspected, which measurement devices or gages are used, the required accuracy of the gage, and how the part is staged for measurement. To keep costs down, an engineering drawing must communicate clearly and specify the maximum allowable tolerances.

An engineering drawing is a legal document because it combines with the purchase order to form the contractual basis between a customer and supplier. It is also the basis for part acceptance to obtain payment based on the contract.

Because safety, product function, manufacturing, and inspection costs all rely on an engineering drawing, and because it is also a legal document, an engineering drawing is a crucial document in an organization.

**Author's Comment**
Engineering drawings are also used as evidence in product litigation cases.

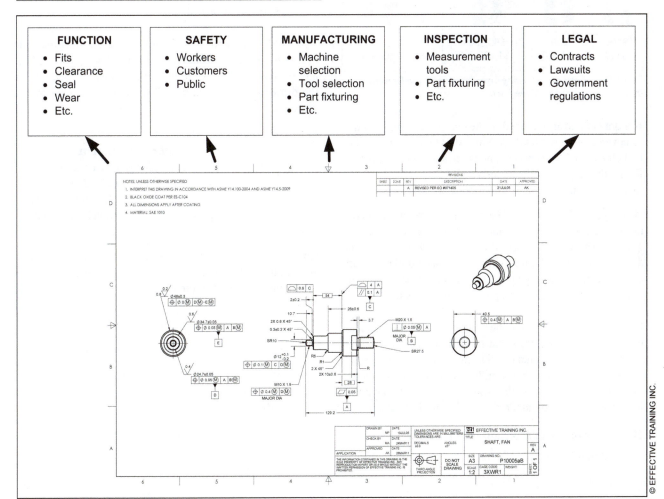

*FIGURE 1-3 Impact of an Engineering Drawing in an Organization*

## The Consequences of Vague or Flawed Engineering Drawings

Engineering drawings not only need to communicate precisely, they also need to be correct. A drawing error can be very costly to an organization. The following analysis is an example based on a medium-sized manufacturing firm.

Figure 1-4 shows typical costs that may result from a drawing error. If a drawing error is found within the design department, it can be corrected for a few dollars. The cost is simply the time required to fix the error, up to about $300.

### Cost of Drawing Error

When discovered by...

| Design | Prototyping | Production | Customer |
|---|---|---|---|
| $100 - $300 | $500 - $2,000 | $50,000 - $1,000,000 | $1,000,000 - $100,000,000 |

*FIGURE 1-4 Costs of a Drawing Error Increase as the Drawing Information is Used to Make Parts*

If a drawing error is missed in the design department and is discovered in the model shop, it may cost thousands of dollars to fix the error. This is because now—in addition to the time to fix the drawing—additional costs may be involved in loss of material and machine time, plus tooling and labor costs.

Worse yet, let's say a part described on a drawing that contains an error gets into production. Now the costs escalate quickly. The cost to process the paperwork for fixing the drawing error may be up to sixty thousands dollars. In addition, machine costs, gaging costs, tooling costs, and scrap costs can bring the total to up to a million dollars.

If a drawing error gets into the final product and it's shipped to the customer, the costs that result from the error can be much higher. If a product recall is involved, it can easily cost the organization more than a million dollars. If a product liability lawsuit is involved, the costs that result from the drawing error can run into hundreds of millions of dollars.

Drawing errors cost the organization in four ways:

1. Money
2. Time
3. Material
4. Customer dissatisfaction

# DIMENSIONING AND TOLERANCING STANDARDS

## Global Standards

There are two major dimensioning and tolerancing standards used around the world. The ASME standards are published by the American Society of Mechanical Engineers and are predominant in the U.S. Although the ASME standards are created in the U.S., they are used on engineering drawings in many countries around the world, so they are considered global standards.

The other major dimensioning and standards are the ISO standards, published by the International Organization of Standards. The ISO standards are developed by international experts and are used globally. This book is based on ASME dimensioning and tolerancing standards. Read about the History of GD&T in Appendix A.

## ASME Standards

*ASME Y14.5-2009* is the standard for dimensioning and tolerancing. At a minimum, an engineering drawing should specify this standard. An engineering drawing will often invoke several additional standards.

The ASME standards that cover engineering drawings and dimensioning and tolerancing are listed in Figure 1-5. If any of these standards are used, they must be specified on the drawing or invoked by a document referenced on the drawing.

> **Author's Comment**
> The ASME and ISO standards look similar, but are very different in application and interpretation. See Appendix E for a comparison of ASME and ISO standards

Corporate standards are another common source of dimensioning and tolerancing standards. A large corporation will often publish an addendum to ASME or ISO standards. The addendum typically covers four items. The addendum may:

- Explain a tolerancing concept in more detail
- Discourage the use of a tolerancing concept
- Select an option from a standard
- Add a tolerancing concept not in the current standards

**ASME Drawing-Related Standards***

**General Drawing**

Y14.100-2004 – Engineering Drawing Practices
Y14.1M-2005 – Metric Drawing Sheet Size and Format
Y14.3-2003 – Multiview and Sectional View Dwgs
Y14.2-2008 – Line Conventions and Lettering
Y14.38-2007 – Abbreviations and Acronyms for Use on Drawings and Related Documents
Y14.35M-1997 – Revision of Engineering Drawings and Associated Documents
Y14.34-2008 – Associated Lists
Y14.31-2008 – Undimensioned Drawings
Y14.24-1999 – Types & Applications of Engineering Dwgs

**Dimensioning & Tolerancing**

Y14.5-2009 – Dimensioning & Tolerancing of Engineering Drawings
Y14.5.1M-1994 – Mathematical Definition of Dimensioning & Tolerancing Principles
Y14.41-2003 – Digital Production Definition Data Practices (Dimensioning & Tolerancing of Solid Models)
Y14.36M-1996 – Surface Texture Symbols
Y14.43-2003 – Dimensioning and Tolerancing Principles for Gages and Fixtures

**Part Specific**

Y14.8-2009 – Casting, Forgings, and Molded Parts

* All release dates at the time of publication of this text

*FIGURE 1-5 ASME Drawing-Related Standards*

## The Role of Dimensioning and Tolerancing Standards

Dimensioning and tolerancing standards have a critical role in industry. Standards ensure:

- Common rules and conventions for specifying dimensions and tolerances

- Common interpretations of dimensions and tolerances

Without common specifications of dimensions and tolerances, engineering drawings would be more difficult to understand. Without common interpretation of dimensions and tolerances, it would be impossible to determine if parts meet their specifications. Therefore, dimensioning and tolerancing standards are used on most engineering drawings around the world.

## Identifying Which Standards Apply to an Engineering Drawing

Before interpreting an engineering drawing, you must understand which dimensioning and tolerancing standards apply to the drawing. The most common places to find this information are either in the title block or in the notes area of the drawing. Figure 1-6 shows an example of how standards are specified on an engineering drawing.

**TECHNOTE 1-1**
**Standard Identifiers**

Many ASME standards can be invoked by referencing an envelope standard, such as ASME Y14.100. However, some standards require that their actual standard identifier be indicated before their rules and practices can be applied. Some examples are Y14.41-2003, Y14.8-2003, and Y14.5-2009.

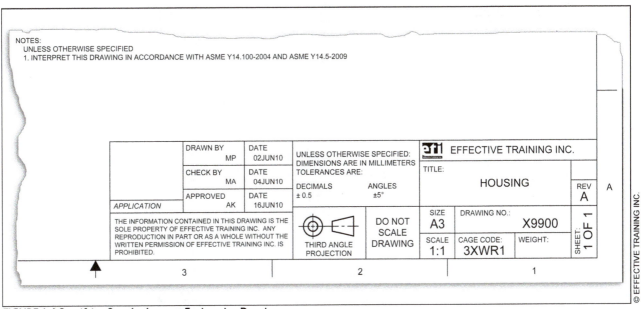

*FIGURE 1-6 Specifying Standards on an Engineering Drawing*

## SUMMARY

### Key Points

- An engineering drawing is a document (or digital data file) that communicates a precise description of a part.

- The potential consequences of flawed engineering drawings are wasted time, higher costs, wasted material, and unhappy customers.

- The least costly time to fix a drawing error is while it is still in the design department.

- The two primary dimensioning and tolerancing standards used globally are the ASME and ISO standards.

- The role of dimensioning and tolerancing standards is to establish rules and conventions for common specification and common interpretation of dimensions and tolerances.

- The primary ASME standard for dimensioning and tolerancing is ASME Y14.5-2009.

- The applicable dimensioning and tolerancing standard is typically specified in the title block or in the notes area of an engineering drawing.

### Additional Related Topics

*These topics are recommended for further study to improve your understanding of engineering drawings.*

| Topic | Source |
|-------|--------|
| ISO/ASME standard comparison | *Alex Krulikowski's ISO Geometrical Tolerancing Reference Guide* |
| Engineering drawing standards | *ASME Y14.100-2004* |
| Digital Product Definition Practices | *ASME Y14.41-2003* |

## QUESTIONS AND PROBLEMS

***Website Bonus Materials***
Additional questions are available at our website. To access bonus materials for this textbook, please visit:
www.etinews.com/textbookbonus

### True and False

*Indicate if each statement is true or false.*

T / F    1. The function of a product determines the amount of variation permitted for each dimension.

T / F    2. Dimensioning and tolerancing standards are a niche used mostly in high tech industries.

T / F    3. ASME is the only dimensioning and tolerancing standard used internationally.

T / F    4. An engineering drawing is a legal document.

T / F    5. Engineering drawings may affect product safety, tooling, and regulatory compliance.

T / F    6. At the minimum, drawings should reference ASME Y14.5-2009.

T / F    7. The applicable ASME standards may be specified in the general notes.

## Multiple Choice

*Circle the best answer to each statement.*

1. Does the ASME Y14.5-2009 standard apply if it is not indicated on the drawing?
   A. Yes, whenever ASME Y14.100-2004 is indicated, the Y14.5-2009 standard applies.
   B. Yes, all ASME standards apply to all drawings made in North America.
   C. No, Y14.5-2009 requires that it be indicated on the drawing in order for it to apply.
   D. No, all applicable ASME standards must be listed on the drawing.

2. An engineering drawing...
   A. Is a legal document that communicates a precise description of a part
   B. Includes geometry, notes, dimensions, tolerances, and material information for manufacturing purposes
   C. Defines engineering requirements for fit and function of the part
   D. All of the above

3. Engineering drawings are important because...
   A. The product specifications may affect safety and legal issues
   B. The specified tolerances affect manufacturing, inspection, and tooling costs
   C. The drawing is a legal document that is the basis for part acceptance per the purchase agreement
   D. All of the above

4. If ASME Y14.5-2009 is not specified on the drawing, what may result?
   A. The drawing interpretation may be challenged in court
   B. Ambiguous specifications may lead to nonfunctional parts being accepted and/or higher scrap costs.
   C. An incorrect specification may lead to product failure resulting in customer injury or death
   D. All of the above

5. Where does a drawing user look to identify which dimensioning and tolerancing standards apply to the drawing?
   A. A note in the drawing notes area
   B. A note in or near the title block
   C. In a specification referenced on the drawing
   D. All of the above

## Application Problems

*The application problems are designed to provide practice on applying the chapter concepts to situations that are similar to on-the-job conditions.*

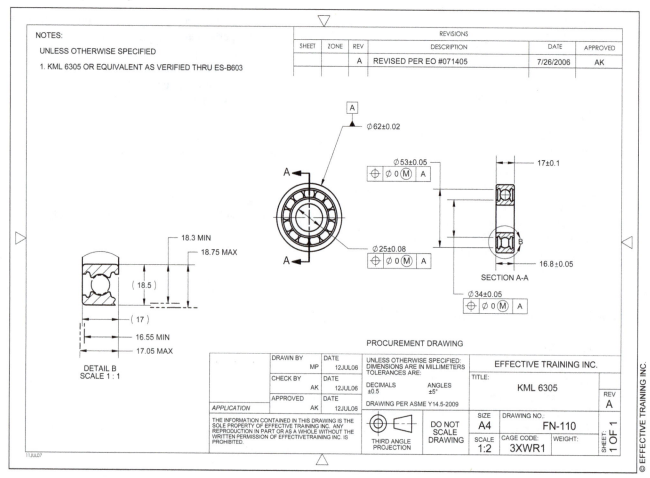

*Application questions 1–4 refer to the drawing above.*

1. Which dimensioning and tolerancing standard applies to this drawing? _____

2. Should this drawing be considered a legal document? _____ Why? _____

    _____

3. What may result if a specification on this drawing is not understood by the manufacturer / supplier? _____

    _____

4. How does ASME Y14.5-2009 affect a manufacturer or supplier (i.e., the recipient of the drawing)? _____

    _____

# Dimensions, Tolerances, and Notes Used on Drawings

## Goal

Understand the types of dimensions, tolerances, and notes

## Performance Objectives

Upon completing this chapter, you should be able to:

1. Describe three purposes of dimensions and tolerances (p.10)
2. Identify which units of linear measurement apply on a drawing (p.11)
3. Explain the options for expressing units of angular measurement (p.11)
4. Explain four conventions used when metric units apply on a drawing (p.11)
5. Describe three conventions used for angular dimensions (p.11)
6. Describe the terms, "dimension," "tolerance," and "nominal size" (p.11)
7. Describe what limit dimensioning is (p.12)
8. Describe what plus and minus tolerancing is (p.12)
9. Describe what bilateral and equal bilateral tolerances are (p.12)
10. Describe what a unilateral tolerance is (p.12)
11. Describe what an unequal bilateral tolerance is (p.12)
12. Interpret dimensional limits (p.13)
13. Define a basic dimension (p.13)
14. List two uses for a basic dimension (p.13)
15. Describe where the tolerance for a basic dimension comes from (p.13)
16. Describe general, flag, and local notes on drawings (p.15)
17. Describe why CAD models need to communicate permissible tolerances (p.16)
18. Describe how CAD models communicate permissible tolerances (p.16)

## New Terms

- Basic dimension
- Bilateral tolerance
- Dimension
- Equal bilateral tolerance
- Flag note
- General note
- Limit dimensioning
- Local note
- Nominal size
- Plus and minus tolerancing
- Tolerance
- Unequal bilateral tolerance
- Unilateral tolerance

## What This Chapter Is About

Dimensioning and tolerancing is a critical part of an engineering drawing. Dimensions and tolerances play a major role in the function, production, and inspection costs of a part. This chapter will familiarize you with some of the fundamental concepts related to dimensions and tolerances. Since most engineering drawings today are used internationally, all of the drawings in this book use metric units.

# TERMS AND CONCEPTS

## The Purpose of Dimensions and Tolerances

Dimensions and tolerances are the elements that create clarity and precision on an engineering drawing. Without them, a drawing is no more than an "artist's depiction" of a part. Three common purposes of dimensions and tolerances are shown in Figure 2-1 and described below:

- Communicate and document functional relationships

- Influence manufacturing choices

- Influence inspection choices

## Functional Relationships

Dimensions and tolerances communicate/document functional relationships by providing a toolset of symbols and rules that can be used to convert product requirements into drawing specifications. The relationships between part surfaces are defined based on how they function in the product. The amount of tolerance represents the amount of variation that will not harm the functional requirements.

The dimensions and tolerances specified on the drawing communicate and document the allowable variation of part features required for proper fit and function. Dimensions and tolerances define the shape, size, location, and orientation of tolerance zones within which part surfaces, axes, center planes, derived median planes, or derived median lines must be located.

## Manufacturing Choices

Manufacturing is influenced by the tolerances specified; they affect the choice of machines, tools, fixtures, and process sequence used to produce a part.

## Inspection Choices

Dimensions and tolerances influence inspection choices in several ways. The datum system indicates how the part is to be held during inspection, basic dimensions orient and locate tolerance zones relative to the datums, and geometric tolerances describe the shape and size of tolerance zones. The dimensions and tolerances are used to determine the inspection tools, setup requirements, acceptance criteria, and methods for verifying part acceptance.

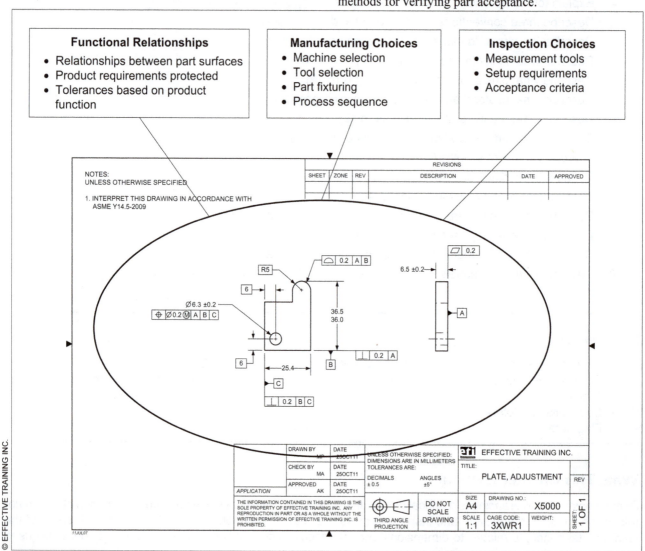

*FIGURE 2-1 Common Uses for Dimensions and Tolerances on Engineering Drawings*

## Units of Linear Measurement

English (decimal inch) or metric SI (millimeter) units are commonly used for linear dimensions on engineering drawings. The units of dimensions and tolerances must be specified on the drawing in a note or in the title block. In industry, a general note is often used to indicate units of measure. A typical general note is "UNLESS OTHERWISE SPECIFIED, ALL DIMENSIONS ARE IN MILLIMETERS."

## Units of Angular Measurement

Angular units on engineering drawings are either decimal degrees or degrees, minutes, and seconds:

- Degrees (°) and decimal parts of a degree (30.15°)

- Degrees (°), minutes ('), and seconds (") (30°25'0")

## Metric Unit Conventions

The following conventions apply for metric units:

- The use of zeros –
  - o A zero precedes the decimal point for values less than one millimeter.
  - o Trailing zeros after the last significant digit are omitted.
  - o Zeros are added to maintain uniformity between high and low tolerance values.
  - o A single zero without the plus or minus sign is shown for all nil tolerance values.

- Whole number dimensions – A zero or decimal point is not shown where the dimension is a whole number.

- The number of decimal places – A dimension may be specified with a different number of decimal places than its tolerance.

Figure 2-2 shows an example of the conventions for metric unit specifications.

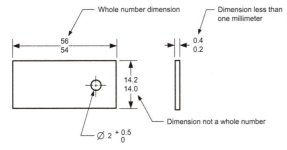

UNLESS OTHERWISE SPECIFIED,
ALL DIMENSIONS ARE IN MILLIMETERS

© EFFECTIVE TRAINING INC.

*FIGURE 2-2 Conventions for Metric Unit Specification*

## Angular Unit Conventions

Where angular dimensions are specified, the following conventions apply:

- The use of zeros – Where a dimension is less than one degree, a zero precedes the decimal point (e.g., 0.4°).

- Where only minutes or seconds are specified, a zero precedes the minutes and seconds specification (e.g., 0°30'10").

- Whole number angles – Where only degrees are specified, the value is followed by the degree symbol. A zero or decimal point is not shown after the degrees (e.g., 30°).

Examples are shown in Figure 2-3.

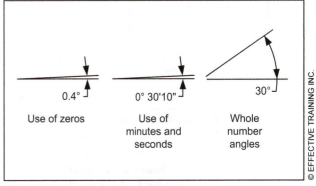

*FIGURE 2-3 Angular Dimension Specifications*

## Dimensions and Tolerances

In general, dimensions define the size, location, orientation, and form of a part. Sometimes referred to as nominal geometry. Tolerances define the acceptable deviation from the nominal geometry. Tolerances often have a greater impact on an organization than dimensions.

- *Dimension* — a numerical value(s) or mathematical expression in appropriate units of measure used to define the size, location, orientation, or form (shape) of a feature (i.e., surface) or feature of size

- *Tolerance* — the total amount that a specific dimension is permitted to vary; the difference between the maximum and minimum limits

- *Nominal size* — the designation used for the purpose of general identification

### Author's Comment

The terms "feature" and "feature of size" are important in the language of GD&T and are used extensively in the standard and in this text. Chapter 5 contains a full explanation of these terms. For now, here are simplified descriptions:

A feature is any part surface. A feature of size is the dimension (size) of a cylindrical shape or width.

Figure 2-4 shows an example with a dimension and a tolerance.

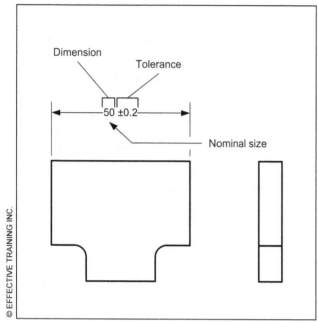

FIGURE 2-4  *Example of a Dimension and a Tolerance*

**Author's Comment**
The nominal size is not always the mean of a size dimension. Nominal size is simply the stated value.

## Methods Used to Express Tolerance on Dimensions

Two common methods used to indicate tolerances are limit tolerances and plus-minus tolerances.

*Limit dimensioning* is where a dimension has its high and low limits stated. In a limit tolerance, the high value is placed on top, and the low value is placed on the bottom. Figure 2-5A shows an example of a limit tolerance.

The high limit for this dimension is 12.5. The low limit for this dimension is 12.0. The tolerance for this dimension is the total amount of variation permitted (12.5–12.0 = 0.5). When limit tolerances are expressed in a single line, the low limit is stated first, then a dash, followed by the high limit (e.g., 12.0–12.5).

*Plus and minus tolerancing* is where the nominal of a dimension is given first, followed by a plus-minus expression of a tolerance. An example of a plus-minus dimension is shown in Figure 2-5B. For this dimension, the nominal value is 12.25. The tolerance for this dimension is 0.5.

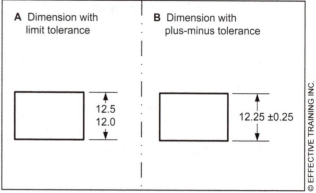

FIGURE 2-5  *Examples of Limit and Plus-Minus Tolerances*

A tolerance for a plus-minus dimension can be expressed in several ways. A *bilateral tolerance* is one that allows the dimension to vary in both directions from the specified dimension. An *equal bilateral tolerance* is where the allowable variation from the specified dimension (nominal) value is the same in both directions. Figure 2-6A shows an example.

A *unilateral tolerance* is where the allowable variation from the nominal value is all in one direction and zero in the other direction. Figure 2-6B shows an example.

An *unequal bilateral tolerance* is a tolerance in which the allowable variation from the nominal value is not the same in both directions. Figure 2-6C shows an example.

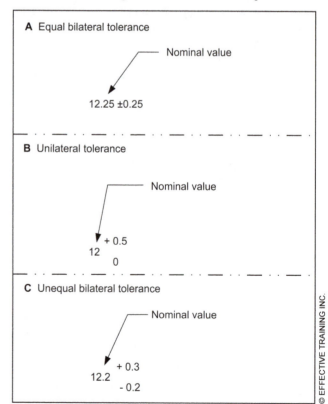

FIGURE 2-6  *Examples of Bilateral, Equal Bilateral, and Unequal Bilateral Tolerances*

## Interpreting Dimensional Limits

ASME Y14.5 states, "All limits are absolute." In other words, the last significant digit of a dimension or tolerance is considered to be followed by zeros. See Figure 2-7 for examples. To determine part acceptance, the measured value is compared directly to the drawing specification, and any deviation outside the drawing specification signifies an unacceptable part.

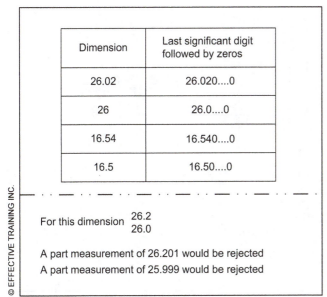

| Dimension | Last significant digit followed by zeros |
|-----------|------------------------------------------|
| 26.02     | 26.020....0                              |
| 26        | 26.0....0                                |
| 16.54     | 16.540....0                              |
| 16.5      | 16.50....0                               |

For this dimension    26.2
                      26.0

A part measurement of 26.201 would be rejected
A part measurement of 25.999 would be rejected

**FIGURE 2-7** *Interpreting Dimensional Limits*

## Basic Dimensions

A *basic dimension* is a numerical value used to describe the theoretically exact size, true profile, orientation, or true position of a feature, feature of size, or gage information (e.g., datum targets).

## Uses for Basic Dimensions

There are two uses for basic dimensions on engineering drawings. One is to define the theoretically exact location, size, orientation, or true profile of a part feature; the other use is to define gage information, such as datum targets. When a basic dimension is used to define part features or features of size, it provides the theoretical location, orientation, size, and form from which permissible variations are established by geometric tolerances.

Basic dimensions are usually specified by enclosing the numerical value in a rectangle (as shown in Figure 2-8) or in a general note, such as, "UNTOLERANCED DIMENSIONS ARE BASIC."

> **Author's Comment**
> Another way to think about basic dimensions: the basic dimension is the goal, and a geometric tolerance specifies the amount of acceptable deviation from the goal.

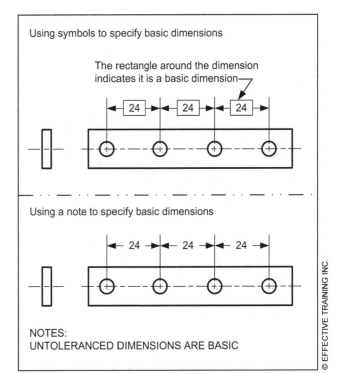

Using symbols to specify basic dimensions

The rectangle around the dimension indicates it is a basic dimension

Using a note to specify basic dimensions

NOTES:
UNTOLERANCED DIMENSIONS ARE BASIC

**FIGURE 2-8** *Methods for Specifying Basic Dimensions*

## Tolerancing Basic Dimensions

In simple terms, a basic dimension locates a geometric tolerance zone or defines gage information, such as datum targets. When basic dimensions are used to describe part features, they must be accompanied by geometric tolerances to specify how much tolerance the part feature may have. A good way to look at this is that the basic dimension only specifies half the requirement. To complete the specification, a geometric tolerance must be added to the feature associated with the basic dimension. Figure 2-9 shows examples of basic dimensions, along with their associated geometric tolerances.

Title block or general tolerances do not apply to basic dimensions. However, geometric tolerances used in general notes could apply to basic dimensions.

There are a few cases where basic dimensions are implied on engineering drawings. Some examples are a zero distance, a 90° angle, and a 180° angle. You will learn more about implied basic dimensions in Chapters 7 and 20.

© EFFECTIVE TRAINING INC.

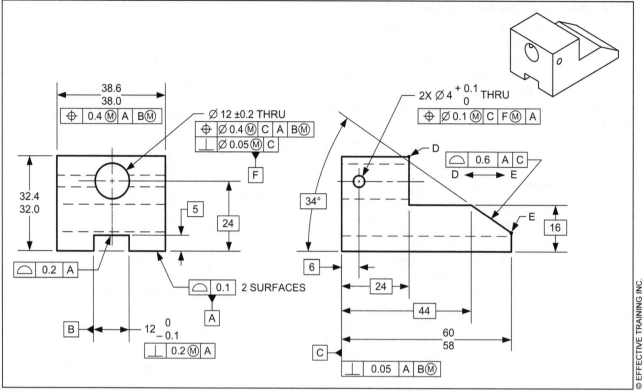

FIGURE 2-9 *Basic Dimension Examples*

### TECHNOTE 2-1
### Basic Dimensions

Basic dimensions...

- Can be used to define the theoretically exact location, orientation, or true profile of part features or gage information.

- That define part features must be accompanied by a geometric tolerance.

- That define gage information do not have a tolerance shown on the print.

- That define theoretically exact gage information are subject to gage tolerances based on the class of gage specified.

Geometric tolerances are not used on basic dimensions that specify datum targets. When basic dimensions are used to specify datum targets, they are considered gage dimensions. Gage tolerances (a very small tolerance compared to product tolerances) apply to gage dimensions. Figure 2-10 shows basic dimensions that locate datum targets. Datum targets are explained in Chapter 14.

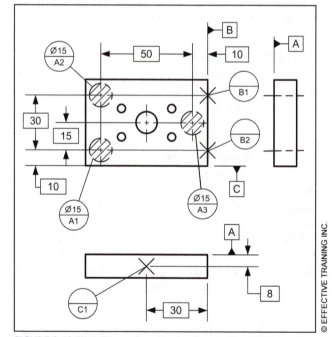

FIGURE 2-10 *Basic Dimensions Used to Locate Datum Targets*

## Engineering Drawing Notes

Drawing notes provide information that clarifies the requirements for the part shown on the drawing. Drawing notes should be unambiguous, grammatically correct statements, expressed in the present tense. The sequence of the notes does not indicate an order of importance, precedence, or sequence of manufacturing or assembly.

Notes are normally located on sheet one of a drawing. The notes area of a drawing should be identified with the heading "NOTES." There are three types of notes used on engineering drawings: general notes, flag notes, and local notes.

**Author's Comment**
Drawing notes should not include contractual requirements, such as costs, requirements for submission, etc.

### General Notes

*General notes* are notes that apply to the entire drawing. General notes (and flag notes) are always numbered as a single-numbered list in the notes area of the drawing. In Figure 2-11, notes numbered 1, 2, and 4 are general notes.

## Flag Notes

*Flag notes* are notes that are located with the general notes but apply only at specific areas or points on the drawing. A flag note should be identified with a flag note symbol, including the note number at each point of application. The flag note symbol is placed around the note number to indicate that it only applies at specific areas on the drawing. In Figure 2-11, the note numbered 3 is a flag note.

**Author's Comment**
Flag note symbols are not standardized so companies may use different symbols. Flag note symbols should not conflict with or resemble other symbols used on the drawing. In this book, we will use the flag note symbol shown in Figure 2-11.

### Local Notes

*Local notes* are notes that are located at the specific area or point of application on the drawing. Local notes are not included in the listing of general or flag notes. A requirement specified by a local note only applies to the specific area or point indicated. In Figure 2-11, zone C3, "BOTH SIDES" is an example of a local note. View titles and scales are not considered local notes.

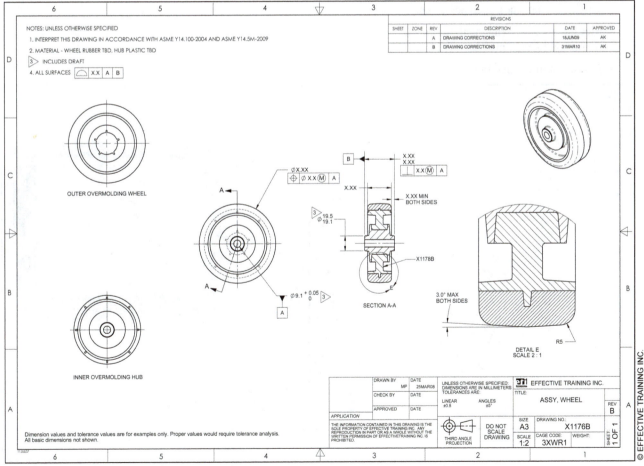

**FIGURE 2-11** *Examples of General Notes, Flag Notes, and Local Notes*

## Dimensioning and Tolerancing of CAD Models

The use of CAD models continues to increase in industry. Figure 2-12 shows an example of an engineering drawing and a CAD model.

Along with the use CAD models, a line of thinking has developed that models do not need to use dimensions or tolerances. The line of thinking is that because the CAD model is precise, and parts are directly produced from the model, dimensions and tolerances are not needed. I call this "nominal thinking" and it is flawed. The reality is the produced parts, even if produced directly from the model, will have variation that needs to be defined and measured.

First, the major source of variation of a part is not the method used to create a drawing or solid model. Part variation occurs from a variety of sources:

- Part factors – rigidity, size, cleanliness, material, heat treat, etc.

- Operator – setup of the part on the machine: clamping sequence and forces, machine adjustments, etc.

- Machine tool – speeds, feeds, wear, cleanliness, temperature, etc.

- Fixture – tolerances, wear, location, deflection, cleanliness, etc.

- Tooling – tolerances, sharpness, wear, deflection, etc.

- Measurement – gaging methods, operator error, environmental factors, measurement uncertainty, etc.

- Post machining operations – heat treat, shipping, etc.

Without dimensions and tolerances, there is no method to:

- Communicate the allowable variation

- Make engineering calculations

- Inspect the part to determine fitness for use

See the section on "the purpose of dimensions and tolerances" at the beginning of this chapter.

## Communicating Dimensions and Tolerances on CAD Models

CAD models require dimensions and tolerances. The dimensions and tolerances may be indicated:

- On a separate engineering drawing or document associated with the model

- As CAD file elements that are associated to the model, in accordance with ASME Y14.41

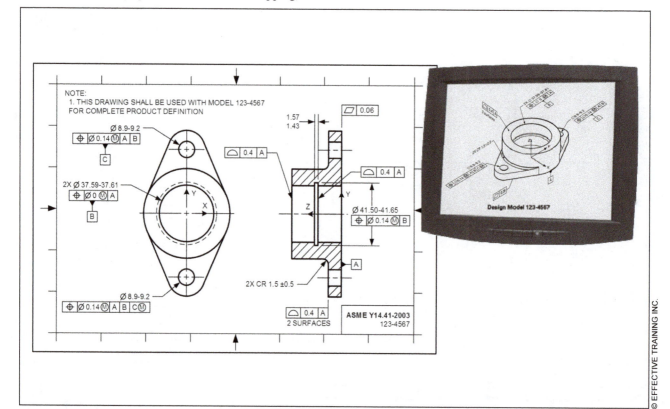

*FIGURE 2-12 Examples of an Engineering Drawing and a CAD Model*

## SUMMARY

### Key Points

- The purposes of dimensions and tolerances are to document product requirements, communicate allowable manufacturing variation, and communicate inspection requirements.

- The units of linear measurement are specified on a drawing in a note or in the title block.

- The units for angular measurement are either decimal degrees or degrees minutes and seconds.

- Where metric units apply, for dimensional values less than one millimeter, a zero precedes the decimal point.

- Where metric units apply, for whole number dimensions, a zero or decimal point is not used.

- Where metric units apply, the number of decimal places may be different for a dimension and its tolerance.

- Where an angular dimension is less than one degree, a zero is used before the decimal point.

- Where an angular dimension is a whole number of degrees, the value is followed by the degree symbol. A zero or decimal point is not shown after the degrees.

- A dimension is a numerical value(s) or mathematical expression in appropriate units of measure used to define the size, location, orientation, or form of a feature (i.e., surface) or feature of size.

- Nominal size is the designation used for the purpose of general identification.

- A tolerance is the total amount a specific dimension is permitted to vary.

- A limit tolerance is where a dimension has its high limit (maximum value) stated above the low limit (minimum value).

- A plus-minus tolerance is where the nominal of a dimension is indicated first and is followed by a plus and minus expression of a tolerance.

- An equal bilateral tolerance is where the allowable variation from the nominal dimension varies the same amount in both the plus and minus directions.

- An unequal bilateral tolerance is where the allowable variation from the nominal dimension is different in the plus and minus directions.

- A unilateral tolerance is where the allowable variation from the nominal dimension is permitted only in one direction.

- All dimensional limits are absolute.

- A basic dimension is a theoretically exact dimension.

- Basic dimensions have two main uses:
  - To define the theoretically exact size, shape, location, or orientation of features or features of size.
  - To define the theoretically exact size, shape, location, or orientation of gage features, such as datum targets.

- The tolerance for a basic dimension that describes theoretically exact part features must come from a geometric tolerance.

- Title block tolerances cannot be applied to basic dimensions.

- The tolerance for basic dimensions that describe gage information comes from the class of gage tolerance specified.

- General notes are notes that apply to the entire drawing.

- Flag notes are notes that are located with the general notes but apply only at specific areas or points on the drawing.

- Local notes are notes that are located at the specific area or point of application on the drawing.

- CAD models require tolerances for three major purposes, to:
  - Communicate allowable variation
  - Make engineering calculations
  - Inspect the part to determine fitness for use

## Additional Related Topics

*These topics are recommended for further study to improve your understanding of dimensions, tolerances, and notes.*

| Topic | Source |
|---|---|
| Tolerancing of solid models | *ASME Y14.41-2003* |
| Digital Product Definition Practices | *ASME Y14.100-2004* |

## QUESTIONS AND PROBLEMS

***Website Bonus Materials***
Additional questions are available at our website. To access bonus materials for this textbook, please visit:
www.etinews.com/textbookbonus

### True and False

*Indicate if each statement is true or false.*

T / F    1. One purpose of dimensions and tolerances is to define the manufacturing setup.

T / F    2. Relationships between part surfaces should be defined based on their function in the product.

T / F    3. The nominal size of a dimension is always the mean of the tolerance values.

T / F    4. Under ASME Y14.5, all dimensional limits are absolute.

T / F    5. A single feature control frame is required to have five or more compartments.

T / F    6. Basic dimensions are usually specified by enclosing the dimension in a rectangle.

T / F    7. The sequence of listing notes indicates the importance of notes (1 is more important than 4).

T / F    8. Using CAD models rather than an engineering drawing will eliminate most sources of variation.

T / F    9. Dimensions are used to define size, location, orientation, and filtering of a part feature or feature of size.

T / F    10. The two common linear units of measure are the decimal inch and the meter.

### Multiple Choice

*Circle the best answer to each statement.*

1. Which areas of a company are most directly impacted by dimensions and tolerances?
   A. Manufacturing (production)
   B. Assembly (function)
   C. Inspection (verification)
   D. All of the above

2. Which is the correct means of using plus-minus tolerances to permit 0.6mm of variation?

   A. $20^{+0.3}_{-0.3}$        C. $20^{+0.0}_{-0.6}$

   C. $20^{+0.4}_{-0.2}$        D. All of the above

3. In the specification $20^{+0.4}_{-0.2}$, the value 20 is known as the _____ value.
   A. Nominal        C. Theoretical
   B. Basic          D. Design

4. Dimensions and tolerance are used to ...
   A. Communicate the allowable variation
   B. Make engineering calculations
   C. Inspect the part to determine fitness for use
   D. All of the above

5. Dimensions and tolerances may be related to a CAD model by:
   A. Indicating them on a separate engineering drawing or document associated with the model
   B. CAD file elements that are associated to the model geometry in accordance with Y14.41
   C. A note stating that the CAD model is perfect, and variation from the model is not permitted
   D. Both A and B

6. Engineering drawing notes should be:
   A. Arranged in alphabetical order
   B. Expressed in the past tense
   C. Expressed in the present tense
   D. All of the above

7. The proper use of zeros for plus/minus tolerances expressed in millimeters requires:
   A. All trailing zeros are omitted for both the tolerance and nominal values
   B. A zero is shown without a plus or minus sign for nil tolerance values
   C. Zeros are used to express high and low tolerance values to the same number of digits
   D. All of the above

## Application Problems

*The application problems are designed to provide practice on applying the chapter concepts to situations that are similar to on-the-job conditions.*

1. List the numbers of the general notes: _____

2. List all local notes: _____

3. List the numbers of the flag notes: _____

4. List all the limit tolerances dimensions: _____

5. List all the bilateral plus/minus tolerance dimensions: _____

6. List all the unilateral plus/minus tolerance dimensions: _____

7. List all the limit and plus/minus tolerances that are incorrectly specified: _____

8. Show the correct specification for each of the dimensions listed in question 7. _____

9. What is the linear unit of measure for this part? _____

10. What is the angular unit of measure for this part? _____

11. Are the angular units correctly specified?_____ Why or why not? _____

12. How are the tolerances related to the CAD model? _____

# Coordinate Tolerancing and Geometric Dimensioning and Tolerancing (GD&T)

## Goal

Understand why geometric tolerancing is superior to coordinate tolerancing

## Performance Objectives

Upon completing this chapter, you should be able to:

1. Describe what the coordinate tolerancing method is (p.22)
2. Explain why coordinate tolerancing is UNSAFE (p.22)
3. Explain three potential consequences of using coordinate tolerances (p.25)
4. Describe three legitimate uses for coordinate tolerancing (p.26)
5. Describe what the geometric dimensioning and tolerancing (GD&T) system is (p.26)
6. Describe the design philosophy used with GD&T (p.26)
7. List the six major components in the GD&T language (p.31)
8. Describe where GD&T should be used (p.31)
9. List four benefits of GD&T (p.31)
10. Explain how GD&T eliminates the UNSAFE conditions of coordinate tolerancing (p.32)
11. Explain the "great myth" of GD&T (p.33)

## New Terms

- Coordinate tolerancing
- Functional dimensioning
- Geometric dimensioning and tolerancing (GD&T)

## What This Chapter Is About

This chapter explains why GD&T is replacing coordinate tolerancing, and why it is the preferred method for specifying tolerances on engineering drawings.

This chapter introduces and compares two dimensioning and tolerancing methods: coordinate tolerancing and geometric dimensioning and tolerancing. The coordinate tolerancing method is still used on many drawings. It does a good job defining size features, such as the size of holes, pins, tabs, slots, etc. However, it does a poor job locating or orienting these size features. GD&T should be used to locate and orient size features and surfaces.

This chapter is important because it highlights the shortcomings of coordinate tolerances and explains the ways in which GD&T is superior to coordinate tolerancing.

## TERMS AND CONCEPTS

### The Coordinate Tolerancing Method

For about 150 years, a tolerancing approach called "coordinate tolerancing" was the predominant tolerancing system used on engineering drawings. *Coordinate tolerancing* is a dimensioning system where the coordinates (X,Y,Z) of the centers of features of size and surfaces are located (or defined) by means of linear dimensions with plus-minus tolerances. An example of coordinate tolerancing is shown in Figure 3-1.

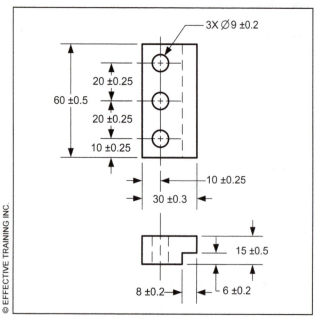

3X Ø9 ±0.2
20 ±0.25
60 ±0.5
20 ±0.25
10 ±0.25
10 ±0.25
30 ±0.3
15 ±0.5
8 ±0.2
6 ±0.2

© EFFECTIVE TRAINING INC.

*FIGURE 3-1  Coordinate Tolerancing*

**Author's Comment**
I use the term "coordinate tolerancing" in this book; however, this term has several different names throughout industry. A few common terms used to refer to coordinate tolerancing are rectangular coordinate tolerancing, plus-minus tolerancing, direct tolerancing, and others.

### Coordinate Tolerancing Is UNSAFE

Coordinate tolerancing was popular when companies were small, and it was easy for the designer to talk to the machinist or inspector to explain the drawing intent. Over the years, as companies have grown in size, many parts are no longer produced in the same building where they were designed. Parts are made in plants located in other cities, or purchased from another company that may be located in another country.

This has led to several problems. The designer can no longer explain the drawing intent to the manufacturing group and inspection departments, and the shortcomings of coordinate tolerancing have become more obvious. Coordinate tolerancing is an oversimplification of part definition and can actually raise costs.

Because of its shortcomings, the use of coordinate tolerancing is diminishing in industry. However, coordinate tolerancing is not totally obsolete; it does have some legitimate applications on engineering drawings.

Coordinate tolerancing leaves out several important pieces of information about how to inspect the part. It also increases costs and causes safety or functional problems.

Figure 3-2 shows the six major shortcomings of coordinate tolerancing. An easy way to remember them is to think of the word, "UNSAFE," a reminder that parts dimensioned with coordinate tolerancing can result in non-functional or unsafe conditions for the customer.

**U**ndefined measurement setup

**N**o indication of measurement origin

**S**quare or rectangular tolerance zones

**A**ccumulation of tolerances

**F**ixed-size tolerance zones

**E**xact start point of dimension undefined

© EFFECTIVE TRAINING INC.

*FIGURE 3-2  Coordinate Tolerancing Shortcomings*

Each letter of UNSAFE relates to one of the shortcomings of coordinate tolerancing. Let's look at each of these shortcomings in more detail.

### Undefined Measurement Setup

Coordinate tolerancing results in ambiguous instructions for inspection. Figure 3-3 shows two logical methods an inspector could use to set up the part from Figure 3-1 for inspecting the holes. The inspector could rest the part on the face first, long side second, and the short side third, or on the face first, the short side second and the long side third.

Because there are different ways to hold the part for inspection, two inspectors could get different measurements from the same part. This can result in two problems: good parts may be rejected or, worse yet, bad parts could be accepted as good parts.

The problem is that the drawing does not communicate to the inspector which surfaces should touch the gaging equipment first, second, and third. When using coordinate tolerancing, additional notes would be required to communicate this important information to the inspector. See Figure 3-3.

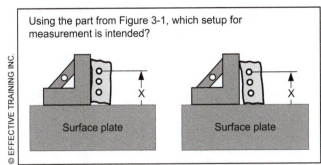

FIGURE 3-3 *Undefined Measurement Setup Using Coordinate Tolerancing*

### No Indication of Measurement Origin

With coordinate tolerancing, there is no standard for interpretation. Therefore, when a dimension exists between two surfaces, one inspector may choose to measure from the points of the surface, and another may choose to measure from a reference plane established from the surface.

If a measurement is taken from a reference plane, the form error of the surface is not part of the measurement result. If a measurement is taken from a surface, the form error will affect the measurement result. Each measurement provides a different measure. Since this is not covered in any standard, there is no way to determine which measurement is correct. An example is shown in Figure 3-4.

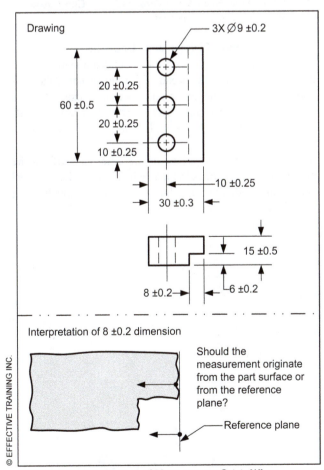

FIGURE 3-4 *No Indication of Measurement Origin Where Coordinate Tolerance is Used*

### Square or Rectangular Tolerance Zones

Since coordinate tolerances use a tolerance in the Y (vertical) direction and a tolerance in the X (horizontal) direction, the resulting tolerance zone is square or rectangular. The tolerance zone for the hole location is formed by the max and min of the X and Y location dimensions. An example is shown in Figure 3-5.

Since this is an assembly application, and there are round pins fitting into the round holes, there is equal clearance in all directions. The drawing specification allows the holes to be located at the corners of the square tolerance zone, and the part will still assemble. The distance from the nominal location of the hole center to the corner of the square tolerance zone is one-half of 0.707, or 0.3535. If a hole can have a 0.3535 deviation in four directions and still assemble, then the hole should be allowed to have a 0.3535 deviation in all directions. The square tolerance only permits a 0.25 deviation in the X & Y directions, resulting in the rejection of functional parts.

Therefore, the tolerance zones should be cylindrical to allow equal tolerance in all directions corresponding to the amount of clearance between the pin and holes.

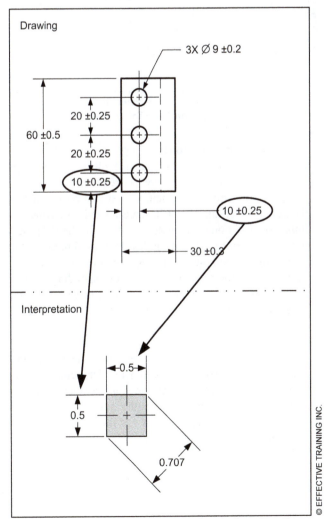

FIGURE 3-5 *Square Tolerance Zone That Results From Coordinate Dimensions*

### Accumulation of Tolerances

Coordinate tolerancing can result in unwanted tolerance accumulation. In Figure 3-6, the total deviation in the vertical direction for the lower hole is 0.5. The allowable deviation in the vertical direction for the top hole is 1.5 because it includes the tolerance accumulation from the lower two holes.

Where coordinate dimensioning includes the use of chained dimensions, the tolerances from each dimension in the chain accumulate. The tolerance accumulation can cause problems for the function of the workpiece.

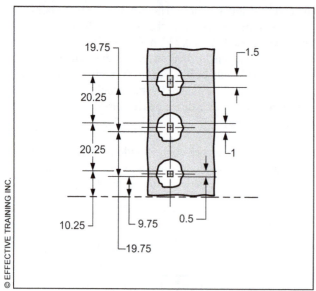

**Figure 3-6**   *How Unwanted Tolerance Accumulation Can Occur With Chained Coordinate Tolerances*

### Fixed-Size Tolerance Zones

Coordinate tolerancing uses fixed-size tolerance zones. The drawing specification in Figure 3-1 requires the center of the hole to be within a 0.5 square tolerance zone, whether the hole is at its smallest or its largest size limit. When the important function of the holes is assembly, the hole location is most critical when the hole is at its minimum limit of size. If the actual hole size is larger than its minimum size limit, its location tolerance can be correspondingly larger without affecting the part function. This is a result of the additional clearance between the mating part and larger hole.

Square and fixed-size tolerance zones can cause functional parts to be scrapped. Since coordinate tolerancing does not allow for cylindrical tolerance zones or tolerance zones that increase with the hole size, lengthy notes would have to be added to a drawing to allow for these conditions. An example is shown in Figure 3-7.

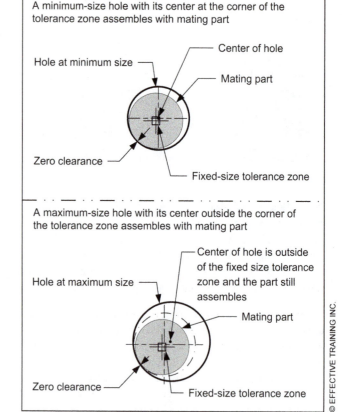

**Figure 3-7**   *How Fixed-Size Tolerance Zones Can Cause Functional Parts to be Rejected*

### Exact Start Point of Dimensions Undefined

A coordinate tolerance uses a double-ended dimension line. It does not indicate which end of the dimension is the start point and which is the end point. There are many instances where measuring from one surface or the other results in different measurements. An example is shown in Figure 3-8.

Measuring from one surface or the other also affects the orientation of the measurement which can make a difference as to whether or not a dimension is within the drawing specification.

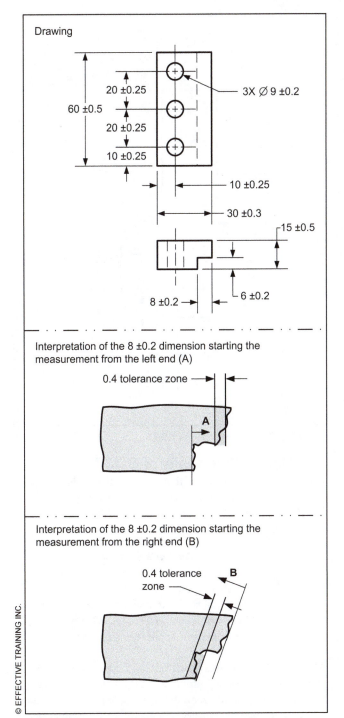

*Figure 3-8 How the Use of Coordinate Tolerances Results in the Exact Start Point of a Dimension Being Undefined*

## Potential Consequences of Coordinate Tolerances

Where the location or orientation of features of size and surfaces are dimensioned with coordinate tolerances, there are several ways to interpret the dimensions. This can result in undesirable conditions and potential consequences, including. . .

1. **Increased Manufacturing Costs**
   Due to square (or rectangular) and fixed-size tolerance zones, the part manufacturing costs may be higher. The rejection of usable parts may also increase manufacturing costs.

2. **Part Acceptance Disputes**
   As a result of several different possible interpretations where coordinate tolerances are used to locate or orient a feature of size or a surface, there is a significant chance of two inspectors getting different measurements for a dimension on the same part. This can result in waste, delays, and additional costs.

3. **Assembly Problems**
   Where coordinate tolerances are used to locate or orient features of size or surfaces, there is a risk of non-functional parts. For example, parts measured with different setups may not assemble.

---

### TECHNOTE 3-1
### Coordinate Tolerancing Caution

Specifying coordinate tolerances (direct dimensioning) does not communicate a single set of measurement requirements. Specifying coordinate tolerances to locate part surfaces or features of size should be avoided. The preferred methods for locating surfaces and features of size are:

- Geometric tolerances

- The dimension origin symbol

If coordinate tolerancing is used, it should only be used on nonfunctional part surfaces. Where coordinate tolerancing is used to locate part surfaces, the part is considered to be compliant if it meets the dimensional requirements under any of the following conditions:

- Using a measurement setup from either end of the dimension to establish the direction of measurement

- The measurement origin may be from either the part surface or a reference plane which allows the measurement process to either include or exclude the flatness of the origin surface.

## Legitimate Uses for Coordinate Tolerancing

Where coordinate tolerances are used to locate or orient features of size and surfaces, several assumptions must be made during the manufacture and inspection of the part. The part manufacturing costs are higher, disputes over part acceptance occur, and the part function may be at risk. Therefore, it is recommended that the use of coordinate tolerances be limited to size dimensions, tangent radii and non-critical chamfers. Figure 3-9 shows examples of appropriate uses of coordinate tolerances.

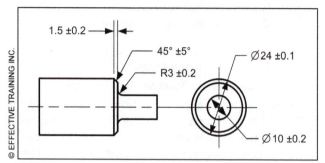

FIGURE 3-9  *Appropriate Uses of Coordinate Tolerances*

### Author's Comment

Although coordinate tolerancing may be used to define the size of the features of size, it is not capable of defining coaxial relationships. The relationship between the 24 ±0.1 and the 10 ±0.2 diameters is undefined in Figure 3-9.

### Author's Comment

If a dimension can't be measured with the jaws of a caliper, radius gage, it needs a datum reference frame.

## The Geometric Dimensioning and Tolerancing System

*Geometric dimensioning and tolerancing (GD&T)* is a symbolic language used on engineering drawings and CAD models to define part geometry and communicate allowable variation. GD&T is a design tool. The language of GD&T consists of a set of well-defined symbols, rules, definitions, and conventions. GD&T is a mathematical language that can be used to define the size, form, orientation, and location of part features.

A part defined with GD&T may reflect the function and assembly conditions of the part. Using GD&T to properly define a part provides the best conditions for consistent interpretation, proper function and cost-effective manufacturing. Figure 3-10 shows an example of a part defined with GD&T.

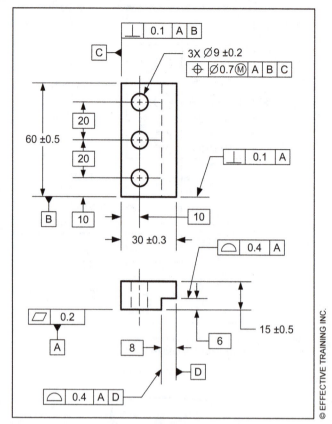

FIGURE 3-10  *Part Defined With GD&T*

## The Design Philosophy of GD&T

The design philosophy of GD&T is functional dimensioning. *Functional dimensioning* is a dimensioning approach that defines a part based on the product (fit and function) requirements.

### Author's Comment

Learning how to apply the functional dimension philosophy is not part of the scope of this text. Functional dimensioning is explained in ETI's GD&T advanced concepts and applications courses.

Although dimensioning based on part functional requirements is the philosophy promoted in the standard, it doesn't mean that the designer should dimension a part without understanding or taking into account manufacturing and inspection needs. Often, a drawing contains a mixture of functional dimensioning and dimensions based on manufacturing capabilities and inspection in non-critical areas of the part.

### Author's Comment

The design philosophy is a controversial area of engineering drawings. Many organizations struggle with dimensioning a part based on manufacturing methods or functional requirements. However, in more than 30 places, the Y14.5 Standard states that functional requirements are the basis for dimensions.

## Six Major Components of the GD&T Language

Because GD&T is a comprehensive language, understanding it can seem overwhelming. It helps to divide the GD&T language into six major components.

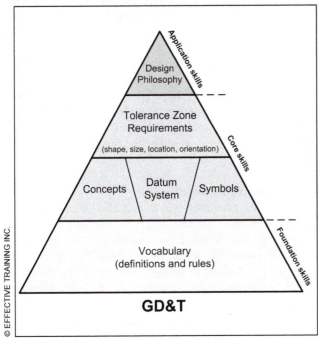

Figure 3-11 The Six Major Components of GD&T

As you study the GD&T language, think of how each topic fits into this structure. It will help you to understand GD&T and be confident with interpreting it. Learning the vocabulary is a key to building your core skills in GD&T.

## Where GD&T Should Be Used

Designers and engineers often wonder where to use GD&T. GD&T should be used when you want to:

- Reduce manufacturing costs
- Protect an important functional relationship on a part
- Ensure that the part can be assembled
- Reduce drawing revisions
- Verify the part with a functional gage
- Improve measurement repeatability
- Outsource the part
- Analyze the design

The bottom line is that you should use GD&T wherever you want to ensure part function, ensure uniform interpretation, and reduce costs.

**Author's Comment**
Some people think GD&T is limited to machined parts. This is not accurate. GD&T may be used to define parts made from any process. There are examples of GD&T applied to machined, stamped, and cast or molded parts throughout this text.

**Author's Comment**
Some engineers use what I call "wish list" tolerances that work in this manner: "If manufacturing can hold this tolerance, then I know the part will work."

Unfortunately, these wish list tolerances are often extremely small and are very expensive to manufacture and verify. Keep in mind that tolerances should be as large as the product function will allow.

## Benefits of GD&T

GD&T provides four important benefits to an organization:

1. **Improves Communication**
   GD&T can provide uniformity in drawing specifications and interpretation, thereby reducing controversy, guesswork and assumptions. Design, production, and inspection all work in the same language.

2. **Provides Better Product Design**
   The use of GD&T can improve your product designs by providing designers with the tools to "say what they mean," and by following the functional dimensioning philosophy.

3. **Increases Tolerances for Production**
   There are two ways tolerances are increased through the use of GD&T. First, under certain conditions, GD&T provides "bonus"—or extra—tolerance for manufacturing. This additional tolerance can make a significant savings in production costs.

   Second, by the use of functional dimensioning, the tolerances are assigned to the part based upon its functional requirements. This often results in a larger tolerance for manufacturing. It eliminates the problems that result when designers copy existing tolerances, or assign tight tolerances, because they don't know how to determine a reasonable (functional) tolerance.

4. **Lowers Costs**
   The result common to each of the first three benefits is lower costs for manufacturing and inspection of the part. The lower costs result from savings in time, tooling, and gaging costs.

## How GD&T Eliminates the UNSAFE Coordinate Tolerancing Shortcomings

Let's focus on how GD&T eliminates the shortcomings of coordinate tolerancing. Some designers think that it's faster to dimension a part with coordinate tolerancing than with GD&T. This is not true. Omitting GD&T simply shifts the designer's time from dimensioning and tolerancing activities to attending meetings so the design intent can be explained to manufacturing, inspection, and suppliers.

The drawing in Figure 3-12 revises the coordinate dimensioning from Figure 3-1 to use GD&T. Each letter shows how the GD&T eliminates the shortcomings of the coordinate tolerancing.

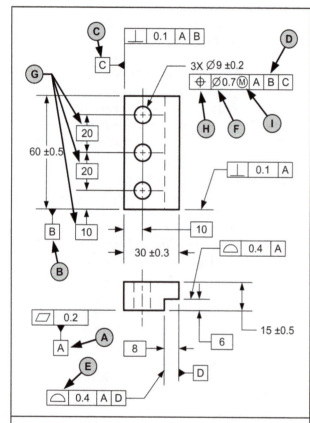

A,B,C - Symbols indicate which surfaces touch inspection equipment for measurement

D - Indicates measurement setup sequence

E - Uses profile tolerance to relate measurement to reference (or datum) plane

F - Diameter symbol indicates cylindrical tolerance zone

G - Basic dimension symbol denotes theoretical location for the tolerance zone of each hole

H - Amount of location tolerance for each hole comes from position tolerance

I - MMC modifier denotes that location tolerance for tolerance of each hole applies when hole is at smallest size

**FIGURE 3-12 Using GD&T to Eliminate Shortcomings of Coordinate Tolerancing (Drawing From Figure 3-1)**

### Defined Measurement Setup

Geometric tolerancing eliminates the shortcoming of undefined measurement setup by using a concept called "the datum system" that allows the designer to define the measurement setup for the inspector. First, symbols to indicate which part surfaces are to touch the inspection equipment are added to the drawing. See Figure 3-12, the arrows labeled "A," "B," and "C."

Then the measurement setup sequence for locating the part on the inspection equipment is indicated inside a frame (see the arrow labeled "D"). This sequence is indicated by the order of the letters, reading left to right. Using the geometric tolerance specification shown in Figure 3-12, the intended measurement would be the one shown in setup A of Figure 3-3.

### Measurement Origin Indication

Another way geometric tolerancing eliminates the shortcoming of no measurement origin indication is by using a profile tolerance to relate the measurement to a reference (or datum) plane. See the arrow labeled "E" in Figure 3-12. The interpretation for the profile tolerance is shown in Figure 3-13. Note how the measurements originate from the reference plane.

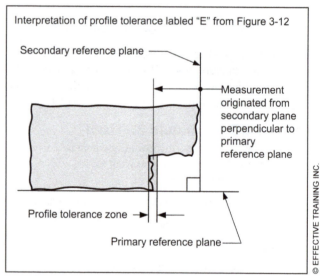

**FIGURE 3-13 Measurement Origin Indicated When Geometric Tolerances Are Used**

### Cylindrical Tolerance Zones

Geometric tolerancing also eliminates the shortcoming of square or rectangular tolerance zones by providing cylindrical tolerance zones. In Figure 3-12, the arrow labeled "F" points to one of the symbols used in GD&T. This symbol is called a diameter symbol and it denotes a cylindrical tolerance zone.

The cylindrical tolerance zone allows for 57% greater tolerance than the square tolerance zone common to coordinate tolerancing. An example is shown in Figure 3-14.

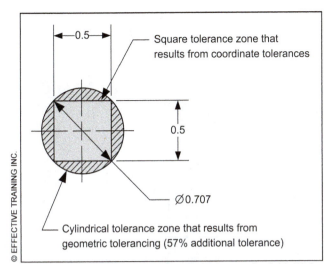

**FIGURE 3-14 Comparison of Amount of Tolerance Available With Cylindrical and Square Tolerance Zones**

### No Accumulation of Tolerances

Geometric tolerancing eliminates the shortcoming of accumulation of tolerances through the use of basic dimensions and the position tolerance. In Figure 3-12, the arrows labeled "G" point to basic dimension symbols that denotes a theoretical exact location for the tolerance zone of each hole.

The location tolerance for each hole comes from the position tolerance specified (0.7) in the tolerance portion of the feature control frame of the tolerance (labeled "H"). The interpretation for the position tolerance and basic dimensions are shown in Figure 3-15. Note how the tolerance zones originate from the theoretical location of each hole.

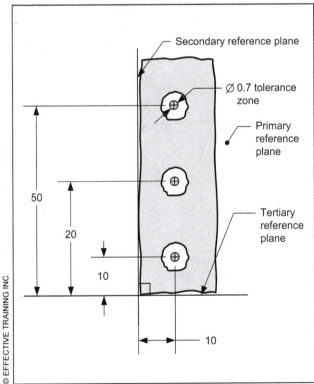

**FIGURE 3-15 No Accumulation of Tolerances With Basic Dimensions**

### No Fixed-Size Tolerance Zones

Geometric tolerancing eliminates the shortcoming of fixed-size tolerance zones through the use of the MMC modifier (see Figure 3-12, the arrow labeled "I"). This symbol, called the maximum material symbol, denotes that the location tolerance for the tolerance of each hole applies when the hole is at its smallest size. The amount of location tolerance for each hole may increase as the hole gets larger. An example is shown in Figure 3-16.

We will discuss this symbol and its effects in more detail in Chapters 6 and 8.

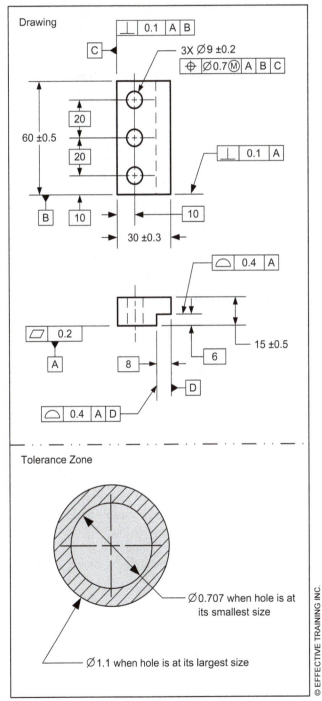

**FIGURE 3-16 No Fixed-Size Tolerance Zones With MMC Modifier**

### Exact Start Point of Dimensions Defined

Geometric tolerancing also eliminates the shortcoming of the exact start point of dimensions not being defined by using basic dimensions and profile callouts. Where coordinate tolerancing is used, there is no indication about which end of the dimension (i.e., left or right) the measurement should start from. (The 8.0+/-0.2 dimension from Figure 3-1 could be measured by starting from the left or the right end.)

In Figure 3-12, the coordinate dimension was replaced with a basic dimension and a profile tolerance (arrow labeled "E") was used to define the location of the surface. The use of the profile callout indicates that the exact start point of the dimension originates from the secondary reference plane. An example is shown in Figure 3-17.

As you can see, through the use of geometric tolerancing, the coordinate dimensioning shortcomings are eliminated. Let's take a look at what the drawing would look like if we tried to accomplish the same level of drawing completeness with coordinate tolerancing.

Figure 3-18 shows the drawing from Figure 3-1. This time the part is dimensioned with coordinate dimensions to the same level of completeness as the GD&T version, but using words instead of symbols. When comparing the drawings, ask yourself these questions:

- Which drawing do you think would be easier to create?
- Which drawing is more clear?
- Which drawing is easier to use in an international supply chain?

There are three shortcomings of using notes:

1. They are language dependent
2. Words have multiple interpretations
3. Notes use more drawing space than symbols

When the goal is to dimension both drawings to the same level of completeness, it is faster and more effective to use geometric tolerances.

The differences between coordinate tolerancing and geometric tolerancing are summarized in Figure 3-19. When comparing these tolerancing methods, it is easy to understand why geometric tolerancing is replacing coordinate tolerancing.

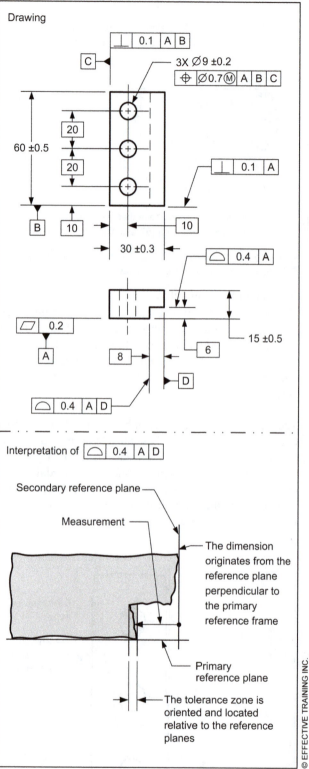

**FIGURE 3-17  Exact Start Point of Measurements Defined Where Geometric Tolerances Are Specified**

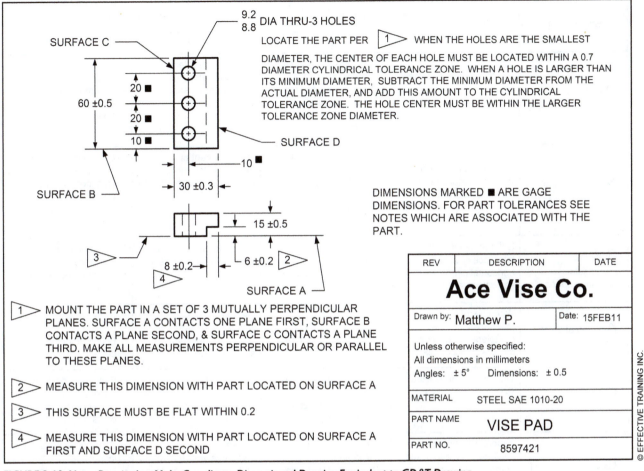

9.2 / 8.8 DIA THRU-3 HOLES

LOCATE THE PART PER [1]▷ WHEN THE HOLES ARE THE SMALLEST DIAMETER, THE CENTER OF EACH HOLE MUST BE LOCATED WITHIN A 0.7 DIAMETER CYLINDRICAL TOLERANCE ZONE. WHEN A HOLE IS LARGER THAN ITS MINIMUM DIAMETER, SUBTRACT THE MINIMUM DIAMETER FROM THE ACTUAL DIAMETER, AND ADD THIS AMOUNT TO THE CYLINDRICAL TOLERANCE ZONE. THE HOLE CENTER MUST BE WITHIN THE LARGER TOLERANCE ZONE DIAMETER.

DIMENSIONS MARKED ■ ARE GAGE DIMENSIONS. FOR PART TOLERANCES SEE NOTES WHICH ARE ASSOCIATED WITH THE PART.

| REV | DESCRIPTION | DATE |
|-----|-------------|------|

# Ace Vise Co.

| Drawn by: Matthew P. | Date: 15FEB11 |
|----------------------|---------------|

Unless otherwise specified:
All dimensions in millimeters
Angles: ±5°      Dimensions: ±0.5

| MATERIAL | STEEL SAE 1010-20 |
|----------|-------------------|

| PART NAME | **VISE PAD** |
|-----------|--------------|

| PART NO. | 8597421 |
|----------|---------|

© EFFECTIVE TRAINING INC.

[1]▷ MOUNT THE PART IN A SET OF 3 MUTUALLY PERPENDICULAR PLANES. SURFACE A CONTACTS ONE PLANE FIRST, SURFACE B CONTACTS A PLANE SECOND, & SURFACE C CONTACTS A PLANE THIRD. MAKE ALL MEASUREMENTS PERPENDICULAR OR PARALLEL TO THESE PLANES.

[2]▷ MEASURE THIS DIMENSION WITH PART LOCATED ON SURFACE A

[3]▷ THIS SURFACE MUST BE FLAT WITHIN 0.2

[4]▷ MEASURE THIS DIMENSION WITH PART LOCATED ON SURFACE A FIRST AND SURFACE D SECOND

**FIGURE 3-18** *Notes Required to Make Coordinate Dimensioned Drawing Equivalent to GD&T Drawing*

| Shortcoming | Coordinate Tolerancing | | Geometric Tolerancing | |
| --- | --- | --- | --- | --- |
| | Condition | Results | Condition | Results |
| **U**ndefined measurement setup | Implied datums allow choices for measurement setup | • Disputes over part acceptance<br>• Functional parts may be scrapped<br>• Higher manufacturing costs | The datum system communicates the measurement setup | • Clear instructions for inspection<br>• Use more functional parts<br>• Lower manufacturing costs |
| **N**o indication of measurement origin | The origin of a measurement may come from the surface or from a reference plane | • Disputes over part acceptance<br>• Functional parts may be scrapped<br>• Higher manufacturing costs | Measurement of geometric tolerances originates from the datum | • Clear instructions for inspection<br>• Use more functional parts<br>• Lower manufacturing costs |
| **S**quare or rectangular tolerance zones | Different amount of tolerance in vertical and horizontal directions than across corners | • Functional parts may be scrapped<br>• Higher manufacturing costs | Round tolerance zones have the same tolerance in all directions | • 57% additional tolerance<br>• Lower manufacturing costs |
| **A**ccumulation of tolerances | When coordinate tolerances are chained, the tolerance from each dimension accumulates | • Parts may not assemble or function<br>• Higher manufacturing costs | Geometric tolerances with basic dimensions eliminate tolerance accumulation | • Parts will assemble & function<br>• Lower manufacturing costs |
| **F**ixed-size tolerance zones | Tolerance zone is fixed in size | • Functional parts may be scrapped<br>• Higher manufacturing costs | Use of MMC modifier allows tolerance zones to grow under certain conditions | • Functional parts will be used<br>• Lower manufacturing costs |
| **E**xact start point of dimension undefined | A measurement may originate from either end of the dimension | • Disputes over part acceptance<br>• Functional parts may be scrapped<br>• Higher manufacturing costs | A measurement originates from the datum and goes to the tolerance zone | • Clear instructions for inspection<br>• Use more functional parts<br>• Lower manufacturing costs |

**FIGURE 3-19** *Comparison Between Coordinate Tolerancing and Geometric Tolerancing*

## The Great Myth of GD&T

Even though geometric tolerancing has been accepted by many companies and individuals, it is still associated with a myth. The "Great Myth of GD&T" is the misconception that geometric tolerancing raises part costs. See Figure 3-20.

The myth stems from two factors:

1. **Fear of the unknown** — It is common to be skeptical of things that are not well understood. When a part dimensioned with GD&T is sent out for a cost estimate, people tend to inflate their assessment of how much the part will cost simply because they are fearful that the drawing contains requirements they may not be able to easily meet. Geometric tolerancing gets the blame for the higher cost, but in reality, GD&T probably allowed the part more tolerance, and the drawing user did not understand how to read the drawing.

2. **Poor tolerancing practices** — Many drawings contain tolerances that are very difficult to achieve in production, regardless of what dimensioning system is used. This stems from unskilled designers who assign tolerances with best guess values, copy from similar designs, or assign tolerance thinking, "It would be great if manufacturing could hold..." Somehow the language of GD&T gets the blame. It's not the fault of the language; it's the fault of the unskilled designers.

The fact is, that when properly used, **GD&T SAVES MONEY**.

The great myth about geometric tolerancing can be eliminated with a better understanding of geometric tolerancing by both drawing makers and drawing users. Simply put, knowledge is the key to eliminating the myth.

Let's review a few FACTS about geometric tolerancing:

- GD&T increases tolerances with cylindrical tolerance zones.
- GD&T allows additional (bonus) tolerances.
- GD&T allows the designer to communicate more clearly.
- GD&T eliminates confusion at inspection.

*FIGURE 3-20 The Great Myth of GD&T*

© EFFECTIVE TRAINING INC.

## SUMMARY
### Key Points

- Coordinate tolerancing is a dimensioning method where a part surface (or feature of size) is located or (defined by) means of directly toleranced linear dimensions.

- Coordinate tolerancing should not be used to locate or orient features of size or surfaces.

- Where coordinate tolerances are used to locate or orient features of size or surfaces, it is not clear which surface to use to establish the direction of the measurements.

- Where coordinate tolerances are used to locate or orient features of size or surfaces, it is not clear if the measurements originate from the surface or from a reference plane.

- Where coordinate tolerances are used to locate or orient features of size, it results in a square or rectangular tolerance zone.

- Where coordinate tolerances are used to locate or orient features of size, it results in a fixed-size tolerance zone.

- Where coordinate tolerances are used to locate or orient features of size, the measurement setup for the part is undefined.

- Three appropriate uses for coordinate tolerances are size dimensions, tangent radii, and non-critical chamfers.

- GD&T is a symbolic language used on engineering drawings and CAD models to define part geometry and allowable variation.

- The ASME Y14.5-2009 GD&T standard encourages the use of functional dimensioning.

- The six major components of the GD&T language are: definitions and rules, symbols, concepts, the datum system, tolerance zone requirements, and the design philosophy.

- GD&T should be used wherever you want to ensure part function, ensure uniform interpretation, and reduce costs.

- GD&T eliminates the six shortcomings of coordinate tolerancing.

- The great myth that "using GD&T raises part costs" is not true.

- Two common tolerancing methods are direct (coordinate) tolerancing and geometric tolerancing.

- Coordinate tolerancing is a tolerancing method where variation between two part features or features of size is controlled by linear dimension and its stated tolerance.

- Geometric tolerancing is a tolerancing method where the allowable variation of size, location, orientation, and form of part features or features of size is defined with a set of symbols and supporting rules.

## Additional Related Topics

*These topics are recommended for further study to improve your understanding of geometric tolerancing.*

| Topic | Source |
|---|---|
| Functional dimensioning | *ASME Y14.5-2009* |

## QUESTIONS AND PROBLEMS

***Website Bonus Materials***
Additional questions are available at our website. To access bonus materials for this textbook, please visit: www.etinews.com/textbookbonus

## True and False

*Indicate if each statement is true or false.*

T / F    1. Coordinate tolerancing refers to tolerances verified with a coordinate measuring machine (CMM).

T / F    2. There are four major shortcomings of using coordinate tolerancing.

T / F    3. An undefined measurement setup is one of the major shortcomings of coordinate tolerancing.

T / F    4. Part acceptance disputes are one consequence of coordinate tolerancing.

T / F    5. The design philosophy of GD&T is to assign tolerances so that manufacturing is economical.

T / F    6. One benefit of GD&T is that it improves communication.

T / F    7. GD&T eliminates the problem of tolerance accumulation through the use of basic dimensions.

T / F    8. There are three major components of GD&T.

T / F    9. Geometric tolerances should be used whenever you want to increase manufacturing costs.

## Multiple Choice

*Circle the best answer to each statement.*

1.  When inspecting a coordinate toleranced dimension, the measurement may originate from _____.
    A.  High points of a part surface
    B.  Low point of a part surface
    C.  A datum plane (reference plane)
    D.  All of the above

2.  The use of coordinate tolerances for cylindrical holes results in _____ tolerance zones.
    A.  Cylindrical
    B.  Rectangular
    C.  Parallel
    D.  Uniform

3.  The fact that functional parts may be scrapped is the result of using ...
    A.  Coordinate tolerancing
    B.  Square-shaped tolerance zones
    C.  Cylindrical tolerance zones
    D.  Flexible tolerance zones

4.  The "Great Myth of GD&T" is that...
    A.  GD&T improves quality
    B.  GD&T reduces costs
    C.  GD&T is easy to measure
    D.  GD&T raises part costs

5.  Where coordinate tolerances are used to locate or orient features of size or surfaces, the result may be...
    A.  Reduced manufacturing costs
    B.  A better understanding of functional relationships
    C.  Build (assembly) problems
    D.  Fewer drawing revisions

6.  GD&T uses a design philosophy that defines the...
    A.  Method of manufacturing a part
    B.  Method of inspection
    C.  Functional requirements of a part
    D.  All of the above

7.  The term, "UNSAFE" applies to:
    A.  Geometric dimensioning and tolerancing
    B.  Coordinate tolerancing
    C.  Functional dimensioning
    D.  The Great Myth of GD&T

8.  GD&T eliminates the shortcoming of undefined measurement setup by using:
    A.  The MMC modifier
    B.  The datum system
    C.  Cylindrical tolerance zones
    D.  Notes

## Application Problems

*The application problems are designed to provide practice on applying the chapter concepts to situations that are similar to on-the-job conditions.*

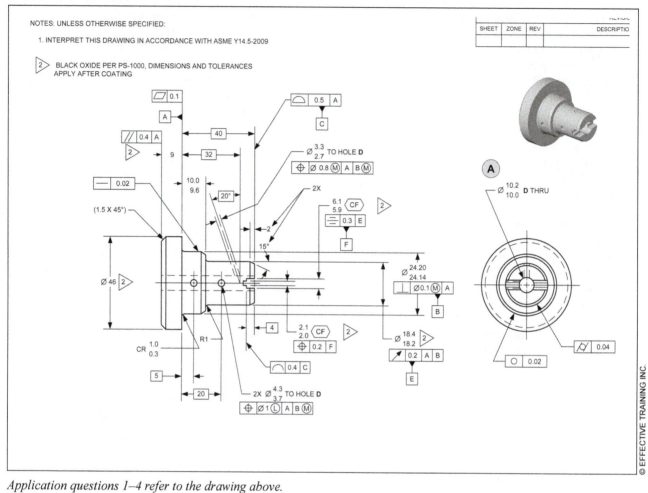

*Application questions 1–4 refer to the drawing above.*

1.  What system of dimensioning and tolerancing is used on this drawing? _____

    _____

2.  When inspecting the location and orientation of the hole labeled "A," which features should the inspector use to orient and locate the part? _____

    _____

    _____

3.  What is the tolerance for the location of hole labeled "A"? _____

    How did you determine this? _____

    _____

    _____

4.  Would this drawing be easier or more difficult to work with if only coordinate tolerances where used and no geometric tolerances were specified? Explain your answer. _____

    _____

    _____

# General Dimensioning Symbols

## Goal

Understand the general dimensioning symbols

## Performance Objectives

Upon completing this chapter, you should be able to:

1. Interpret radius, controlled radius, and spherical radius (p.38)
2. Interpret diameter, spherical diameter, and square dimensions (p.39)
3. Interpret counterbore, spotface, and depth dimensions (p.40)
4. Interpret countersink, number of places, and "by" dimensions (p.40)
5. Interpret maximum, minimum, and reference dimensions (p.41)
6. Interpret dimension origin dimensions (p.41)
7. Interpret the model coordinate system symbol (p.42)

## New Terms (or Symbols)

- By
- Controlled radius
- Counterbore
- Countersink
- Countersink symbol
- Depth
- Diameter
- Dimension origin
- General dimensioning symbols
- MAX dimensions
- MIN dimensions
- Model coordinate system
- Number of places
- Radius
- Reference
- Right-hand rule
- Spherical diameter
- Spherical radius
- Spotface
- Square

## What This Chapter Is About

This chapter introduces fifteen general dimensioning symbols. General dimensioning symbols are the symbols that describe a part characteristic and/or tolerance condition. In the past, the equivalent of the general dimensioning symbols were often described in local notes. Today, we have the general dimensioning symbols to replace many local notes. Recognizing and understanding these symbols is an important part of reading an engineering drawing.

The purpose of this chapter is to explain the general dimensioning symbols, show how they are used on engineering drawings, and illustrate their interpretation.

This chapter is important because nearly all engineering drawings use some of these symbols. The symbols provide a universal means of communicating information that used to appear in notes written in English, Spanish, or other languages.

## TERMS AND CONCEPTS

### General Dimensioning Symbols

The general dimensioning symbols are shown in Figure 4-1. The interpretation of each symbol is shown throughout this chapter.

| Symbol | Term | Figure |
|:---:|:---|:---:|
| R | Radius | |
| CR | Controlled radius | 4-2 |
| SR | Spherical radius | |
| ⌀ | Diameter | |
| S⌀ | Spherical diameter | 4-3 |
| □ | Square | |
| ⊔ | Counterbore | |
| ⌊SF⌋ | Spotface | 4-4 |
| ⊽ | Depth | |
| ∨ | Countersink | |
| X | Number of places or "by" | 4-5 |
| MAX | Maximum dimension | |
| MIN | Minimum dimension | 4-6 |
| ( ) | Reference | |
| ⊕— | Dimension origin | 4-7 |

*FIGURE 4-1  General Dimensioning Symbols*

### Radius, Controlled Radius, and Spherical Radius

This section covers the symbols for radius, controlled radius, and spherical radius. A *radius* is a straight line extending from the center to the periphery of a circle or sphere. The symbol for radius is an "R." A radius symbol creates a tolerance zone defined as the space between the two arcs (minimum and maximum radii). The part surface must be located within this zone.

Where a radius is specified, the part surface may have flats and reversals, as long as the arc surface is located within the tolerance zone. Where the center of a radius is not located, its location is often defined by the surfaces of the arc being tangent to adjacent part surfaces. Where an arc is located by its tangent surfaces, its tolerance zone is crescent shaped.

The tolerance zone for a controlled radius is the same as the tolerance zone for a radius. A *controlled radius* is a radius where the part surface must have a smooth curve within the tolerance zone. In a controlled radius, no flats or reversals are permitted on the arc surface.

A *spherical radius* is a radius that is a segment of a sphere. The tolerance zone is the space between two spherical segments (minimum spherical radius and maximum spherical radius). Where a spherical radius is specified, the spherical segment surface may have flats and reversals as long as the spherical segment is located within the tolerance zone.

Figure 4-2 shows the symbols for radius, controlled radius, and spherical radius and their interpretations.

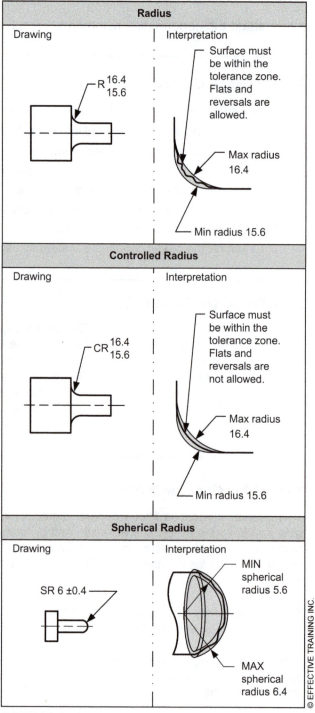

*FIGURE 4-2  Symbols and Interpretation for Radius, Controlled Radius, Spherical Radius*

## Diameter, Spherical Diameter, and Square

This section covers the symbols for diameter, spherical diameter, and square. A *diameter* is the length of a straight line that passes through the center of a circle or cylinder and touches two points on its edge. The diameter symbol (∅) is used two different ways on engineering drawings. One use is to indicate that a dimension applies to a circular or cylindrical shape. The second use is to indicate a cylindrical-shaped tolerance zone. We will discuss the second application later in this book.

Where the diameter symbol is placed in front of a dimension on an engineering drawing, it indicates that the dimension defines a circular or cylindrical shape. A *spherical diameter* is the length of a straight line that passes through the center of a sphere and touches two points on its edge.

The spherical diameter symbol (s∅) is used two different ways on engineering drawings. One use is to indicate that a dimension applies to a spherical shape. The second use is to indicate a spherical tolerance zone. We will discuss the second application later in this book.

Where the spherical diameter symbol is placed in front of a dimension on an engineering drawing, it indicates that the dimension defines a spherical shape.

The *square symbol* (□) indicates that a dimension applies to the length and width of a square shape. The symbols for diameter, spherical diameter and square, along with their applications, are shown in Figure 4-3.

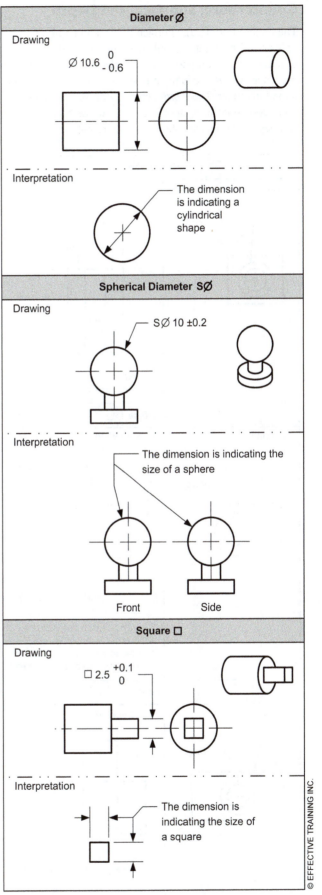

**FIGURE 4-3 Symbol Examples and Interpretations for Diameter, Spherical Diameter, and Square**

## Counterbore, Spotface, and Depth

The counterbore (⊔), spotface (⌷SF⌷), and depth (⊽) symbols are all drafting symbols that simply denote the shape of a geometry on a part. A *counterbore* is a flat-bottomed diameter. A *spotface* is the min size of the flat on a flat-bottomed diameter. The *depth symbol* indicates the depth of a hole, counterbore, or spotface. The counterbore, spotface, and depth symbols are shown in Figure 4-4.

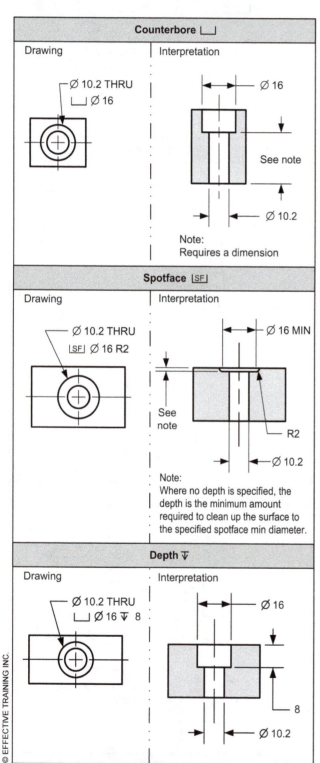

## Countersink, Number of Places, and "By"

The countersink (∨) symbol, the number of places (nX) indication, and the by indication (nXn) are drafting symbols.

The *countersink symbol* is a symbol that denotes the shape of geometry on a part. A *countersink* is an angular lead-in to enlarge and bevel the top of a hole so that a screw, nail, or bolt can be inserted flush with the surface.

The *number of places symbol* is a symbol that indicates how many places a dimension or shape apply on a drawing.

A *"by" symbol* is a symbol that indicates a relationship between two coordinate dimensions.

The countersink, number of places, and "by" symbols are shown in Figure 4-5.

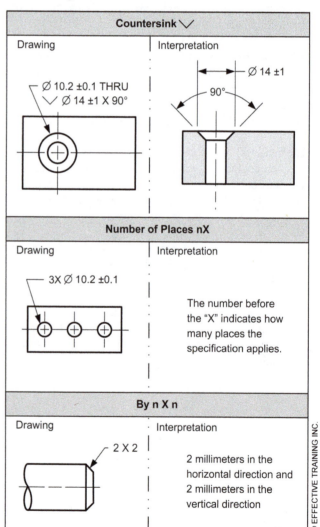

*FIGURE 4-5 Symbol Examples and Interpretations for Countersink, Number of Places, and "By"*

*FIGURE 4-4 Symbol Examples and Interpretations for Counterbore, Spotface, and Depth*

## Maximum, Minimum, and Reference Dimensions

The maximum, minimum, and reference dimension symbols are drafting symbols. A *maximum dimension* is the largest acceptable limit for a dimension. Its symbol is the letters "MAX" following the dimension value. Where a maximum dimension is indicated, the opposite end of the specification is usually zero.

A *minimum dimension* is the smallest acceptable limit for a dimension. Its symbol is the letters "MIN" following the dimension value. Where a minimum dimension is indicated, the opposite end of the specification is as large as the dimension could be and still fit on the part.

A *reference dimension* is a dimension used for information purposes only. Its symbol is a set of parentheses. The information enclosed in the parentheses is indicated as reference information. A reference dimension is not used to determine part acceptance and it is usually specified without a tolerance. The symbols for maximum and minimum dimensions and examples of their use are shown in Figure 4-6.

**Author's Comment**
Title block tolerances or general tolerances do not apply to MAX, MIN, or reference dimensions.

## Dimension Origin

The dimension origin symbol consists of a circle in place of an arrowhead on one end of a dimension line. The *dimension origin symbol* is used to denote that a dimension between two surfaces originates from a plane established from one surface and not the other. The high points of the surface establish a tangent plane designated as the origin (reference) plane for the measurement. The measurements are taken from the origin reference plane and define a zone within which the other surface must be located.

The symbol for dimension origin and an example of its use are shown in Figure 4-7.

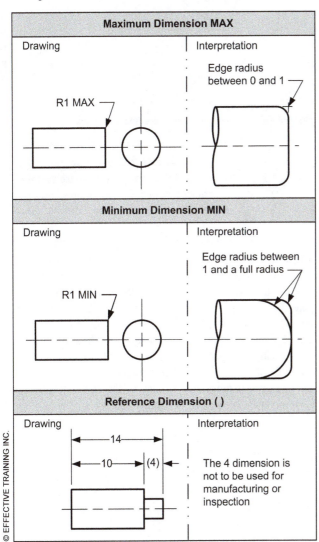

© EFFECTIVE TRAINING INC.

**FIGURE 4-6** *Symbol Examples and Interpretations for Maximum, Minimum, and Reference Dimensions*

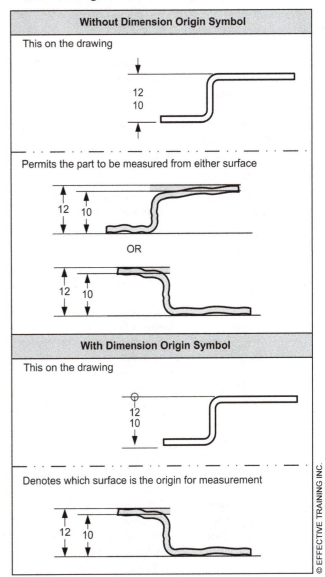

© EFFECTIVE TRAINING INC.

**FIGURE 4-7** *Dimension Origin Symbol Example and Interpretation*

## Model Coordinate System

A *model coordinate system* is a representation of a Cartesian coordinate system shown on a CAD model or on an engineering drawing. The symbol for a right-hand model coordinate system is shown in Figure 4-8. Each axis of the model coordinate system must be labeled with the positive direction shown. The labels for a model coordinate system are the upper case letters X, Y, and Z.

A rule of thumb to visualize the positive direction of the axes of a right-handed coordinate system is the "right-hand rule." The *right-hand rule* works as follows:

If you hold the back of your right hand against a drawing and hold the thumb and index finger at 90° to each other:

- The thumb pointing to the right represents the X-axis positive direction.
- The index finger pointing to the top of the drawing represents the Y-axis positive direction.
- The middle finger pointing at you represents the Z-axis positive direction.

An example of a right-hand model coordinate system shown in Figure 4-8.

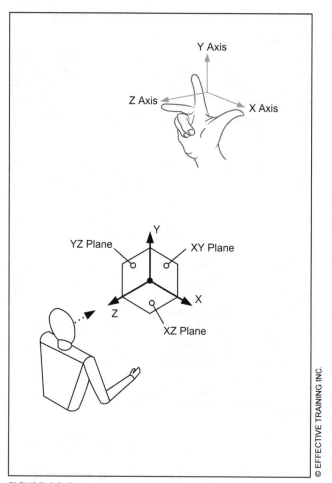

*FIGURE 4-8  Right-Hand Model Coordinate Systems*

© EFFECTIVE TRAINING INC.

The model coordinate system symbol is required on 3D CAD models and axonometric views on 2D drawings, but is optional on orthographic views. Showing a model coordinate system symbol on 2D orthographic views helps relate the 2D views to the 3D model Cartesian coordinate system. This helps the drawing users know which direction they are viewing the part when looking at each view. The labels are typically shown on the axes that are in the plane of the drawing. An example of the use of the model coordinate system symbol is shown in Figure 4-9.

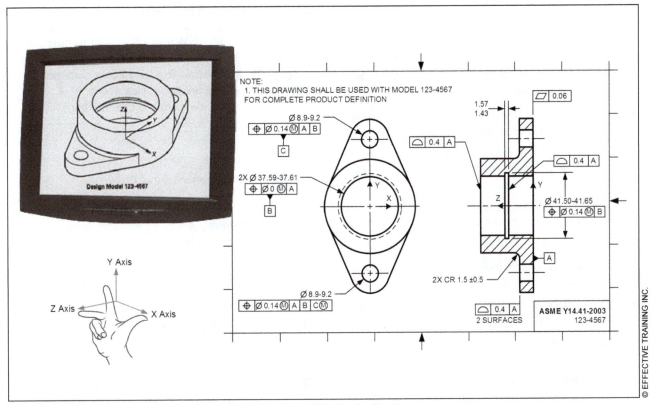

FIGURE 4-9  *Example of the Model Coordinate Symbol Used on a Model and a Drawing*

## SUMMARY

### Key Points

- Flats and reversals are permitted on the surface of a radius.

- Flats and reversals are not permitted on the surface of a controlled radius.

- A diameter symbol may be used to denote a circular or cylindrical shape or to denote a cylindrical tolerance zone.

- A spherical diameter symbol may be used to denote a spherical-shaped dimension or to denote a spherical-shaped tolerance zone.

- The square symbol is used to indicate that a dimension applies to both sides of a square shape.

- The counterbore symbol indicates a flat-bottomed diameter.

- The spotface symbol indicates the size of a flat area on a surface.

- The depth symbol indicates the depth of a counterbore, spotface, or hole.

- The countersink symbol indicates an angular lead in of a hole.

- The number of places symbol indicates that a dimension or part geometry applies multiple places on a drawing.

- The "by" symbol indicates a relationship between co-ordinate dimensions.

- Where a "MAX" dimension is indicated, the opposite end of the specification is usually zero.

- Where a "MIN" dimension is indicated, the opposite end of the specification is as large as the dimension could be and still fit on the part.

- Where a reference dimension is indicated, the information is not to be used to determine part acceptance.

- Where a dimension origin symbol is indicated on a dimension between two surfaces, the measurements are to be taken from the origin plane and define a zone within which the other surface must be located.

- A model coordinate system is a representation of a Cartesian coordinate system shown on a CAD model or on an engineering drawing.

### Additional Related Topics

*These topics are recommended for further study to improve your understanding of general dimensioning symbols.*

| Topic | Source |
|---|---|
| Model coordinate system symbol | *ASME Y14.41-2003* |
| Counterdrilled holes | *ASME Y14.5-2009, Para. 1.8.12* |
| Slope | *ASME Y14.5-2009, Para. 2.14* |
| Conical taper | *ASME Y14.5-2009, Para. 2.13* |
| Arc length | *ASME Y14.5-2009, Para. 1.3.8* |

## QUESTIONS AND PROBLEMS

**Website Bonus Materials**
Additional questions are available at our website. To access bonus materials for this textbook, please visit:
www.etinews.com/textbookbonus

### True and False

*Indicate if each statement is true or false.*

T / F   1. The symbol for critical radius is CR.

T / F   2. Non-mandatory information should be marked, "REF."

T / F   3. ⋁ is the symbol for countersink.

T / F   4. Where MIN is specified, the general tolerance applies.

T / F   5. "nX" means number of places, but "n X n" means "by."

T / F   6. ⌊SF⌋ stands for spherical feature.

T / F   7. Countersinks are specified using the diameter at the surface and the included angle.

## Multiple Choice

*Circle the best answer to each statement.*

1. Which of the following apply to a tangent radii with a specification of R15 ±0.5?
   A. The tolerance zone is the space between two tangent arcs of 14.5 and 15.5
   B. The resulting surface may have flats and reversals
   C. The specification of the -⌽ symbol
   D. Both A and B

2. The dim origin symbol aids in making a repeatable measurement by ____________ .
   A. Providing a single direction for the measurement and exact start point
   B. Eliminating variation due to convex forms
   C. Geometric tolerances
   D. Providing additional tolerance

3. Where the counterbore symbol (⌴) is specified, one additional consideration would be ________ .
   A. The specification of the ⌊SF⌋ symbol
   B. The specification of the SØ symbol
   C. The specification of the ↧ symbol
   D. The specification of the -⌽ symbol

4. Where the countersink symbol (⌵) is specified, one additional consideration would be ________ .
   A. To specify a depth
   B. To specify the angle
   C. To specify a datum reference
   D. To specify the origin symbol

5. Where the specification 12 MAX is applied on an engineering drawing:
   A. The limits for variations are between zero and 12
   B. The general tolerance applies
   C. The value of 12 is an absolute value
   D. The dimension is non-mandatory and won't be inspected

6. The model coordinate system symbol...
   A. Is a representation of a Cartesian coordinate system
   B. Is required for all 2D orthographic view drawings
   C. Is optional for 3D CAD models
   D. Uses the left hand rule to determine the positive direction of each axis

## Application Problems

*The application problems are designed to provide practice on applying the chapter concepts to situations that are similar to on-the-job conditions.*

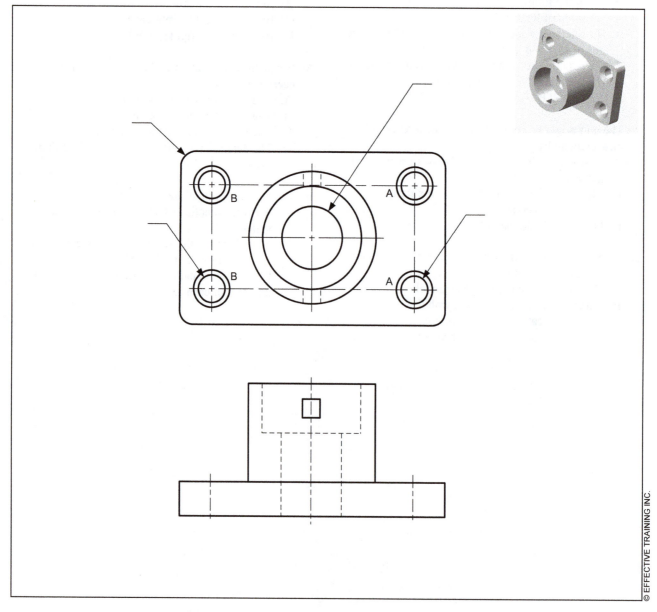

*Use the general dimensioning symbols to correctly specify the following requirements on the drawing above.*

1. The nominal height of the part is 40mm. The tolerance on the height of the part is 1 mm equally distributed. Provide instructions to the inspector that the height of the part must be verified by using a height gage when the bottom of the part is resting on a surface plate.

2. A 6 to 6.4mm square hole must go through both sides of the part with an equally distributed size tolerance of 0.4mm.

3. The nominal size of the two holes labeled "A" is 8.5mm. The top side of the hole must have a tapered lead in of 14 mm diameter with an included angle of 90°. The title block tolerance applies to the counter sink diameter.

4. The two holes labeled "B" have a nominal size of 8.5mm with an unequally distributed size tolerance of plus 0.2 and minus 0.1mm. The top surface around each hole must be cleaned to a 15 minimum diameter area.

5. The hole in the center of the part is a 16.2 diameter. The hole can get 0.2mm bigger, but not smaller. A second 30 mm flat bottomed coaxial hole must be machined to a min depth of 17.7mm and a max depth of 18.3mm.

6. The four corners of the part must have a radius with a minimum of 9mm and a maximum of 9.6mm.

# Key GD&T Terms

## Goal

Understand the key terms used in the GD&T language

## Performance Objectives

Upon completing this chapter, you should be able to:

1. Describe the terms "opposed," "partially opposed," and "non-opposed" (p.48)
2. Describe the terms "size dimension" and "actual local size" (p.48)
3. Explain why it is important to distinguish between a feature and a feature of size (p.49)
4. Describe the terms "feature" and "complex feature"  (p.50)
5. Describe the terms "feature of size," "regular feature of size," and "irregular feature of size" (p.50)
6. Classify regular and irregular features of size and non-features of size on a drawing (p.52)
7. Explain why it is important to distinguish between a regular and irregular feature of size (p.52)
8. Describe the terms "actual mating envelope," "related actual mating envelope," and "unrelated actual mating envelope" (p.55)
9. Describe the terms "axis" and "center plane" (p.57)
10. Describe the terms "maximum material condition," "least material condition," and "regardless of feature size" (p.57)
11. Calculate the maximum and least material condition of a feature of size (p.57)
12. Describe the term "pattern" (p.58)
13. List eight ways to indicate a pattern on a drawing (p.58)

## New Terms

- Actual local size
- Actual mating envelope
- Axis
- Center plane
- Complex feature
- Feature
- Feature of size
- Irregular feature of size
- Least material condition (LMC)
- Maximum material condition (MMC)
- Non-opposed
- Opposed
- Partially opposed
- Pattern
- Regardless of feature size (RFS)
- Regular feature of size
- Related actual mating envelope
- Simultaneous requirement
- Size dimension
- Unrelated actual mating envelope

## What This Chapter Is About

This chapter explains key GD&T terms, shows how they apply on engineering drawings, and illustrates their interpretations.

Learning GD&T is like learning a new language. First you learn some key words, then you build your vocabulary over time. The terms in this chapter are your start for learning the language of GD&T. In the following chapters, additional terms are introduced to increase your GD&T vocabulary.

Understanding the key terms used in the GD&T language is an important part of interpreting drawings. Understanding the vocabulary of any topic is the first step to becoming proficient in the topic.

# TERMS AND CONCEPTS

## Opposed, Partially Opposed, Non-Opposed

The concept of opposed surfaces or elements is important and deserves some discussion. In the Oxford English Dictionary, the word "opposite" is defined:

> Op.po.site Adj. 1, Situated on the other or further side; on either side of an intervening space or thing; facing: a crowd gathered on the opposite side of the street.

Figure 5-1 shows a simple example of this definition:

- In panel A, the planar surfaces are opposed.
- In panel B, the planar surfaces are partially opposed.
- In panel C, the planar surfaces are non-opposed (offset).

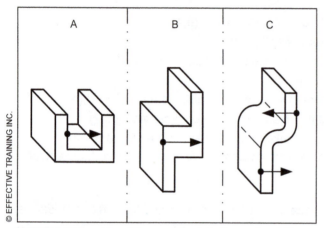

FIGURE 5-1  *Examples of Opposed, Partially Opposed, and Non-Opposed Surfaces*

In the language of GD&T, two planar surfaces are completely *opposed* if **all** rays normal from each planar surface intersects the other surface. Two surfaces are *partially opposed* if **some** of the rays projected normal from each planar surface intersect the other surface. Two surfaces are *non-opposed* if **none** of the rays projected from each surface intersect the other surface.

Figure 5-2 shows examples of opposed, partially opposed, and non-opposed surfaces or line elements on an engineering drawing.

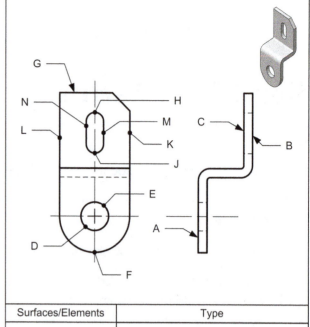

| Surfaces/Elements | Type |
|---|---|
| A & B | Non-opposed planar surfaces |
| B & C | Opposed planar surfaces |
| D & E | Opposed line elements |
| F & G | Non-opposed |
| H & J | Opposed line elements |
| K & L | Partially opposed planar surfaces |
| M & N | Opposed planar surfaces |

FIGURE 5-2  *Examples of Opposed, Partially Opposed, and Non-Opposed Surfaces or Line Elements on an Engineering Drawing*

## Size Dimension and Actual Local Size

A *size dimension* is a dimension across two opposed (or partially opposed) surfaces, line elements, or points. The drawing in Figure 5-3 shows two examples of size dimensions: the length of the pin (30-32) and the diameter of the pin (14.0-14.8).

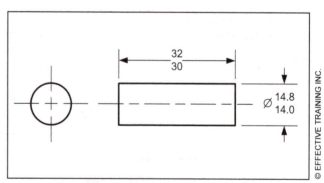

FIGURE 5-3 *Examples of Size Dimensions*

**Author's Comment**
This is only part of the information about size dimensions. In Y14.5, the default condition is that size dimensions are controlled by Rule #1. You will learn about Rule #1 in Chapter 7.

An *actual local size* is the measured value of any individual distance at any cross section of a regular feature of size. The measurement is often made with an instrument such as a micrometer or caliper. Figure 5-4 shows examples of actual local sizes.

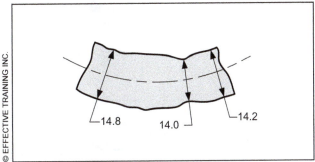

FIGURE 5-4 Examples of Actual Local Size

### Author's Comment

Actual mating envelope is covered later in this chapter. Be sure to compare the terms "actual local size" and "actual mating envelope" because it is important to understand their differences.

## Importance of Distinguishing Between a Feature and Feature of Size

The next several sections of this chapter explain features and features of size. In order to interpret a drawing, you need to be able to recognize the difference between features and features of size. Five important reasons are listed here and illustrated in Figure 5-5. The concepts are covered later in the book.

It is important to recognize the difference between a feature and a feature of size because:

1. The MMC and LMC modifiers may only be used where a feature control frame is applied to a feature of size.

2. Rule #1 only applies to regular features of size.

3. Many GD&T terms only apply to features of size, for example, maximum material condition, least material condition, regardless of feature size, virtual condition, and bonus tolerance.

4. Certain geometric tolerances may only be applied to a feature of size.

5. Certain geometric tolerances may only be applied to a feature.

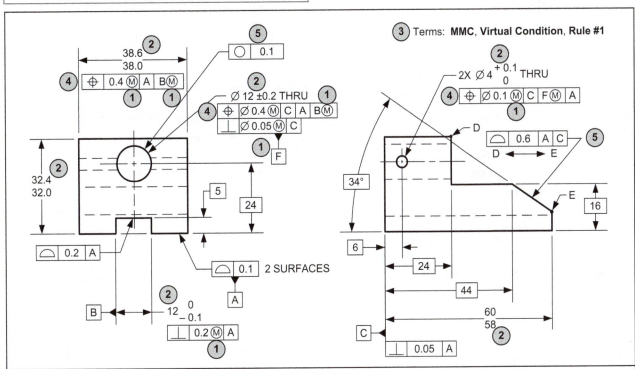

FIGURE 5-5 Examples of Why it is Important to Recognize the Difference Between a Feature and a Feature of Size

## Feature and Complex Feature

A *feature* is a physical portion of a part, such as a surface, pin, hole, or slot, or its representation on drawings, models, or digital data files. An easy way to remember this term is to think of a feature as a surface. The part in Figure 5-6 has seven features: the top and bottom, the left and right sides, the front and back, and the surface of the hole.

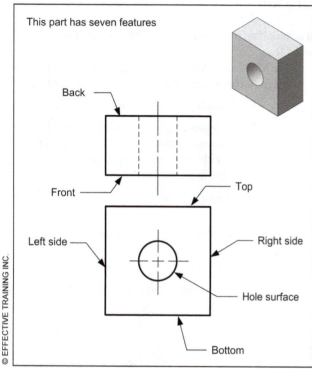

FIGURE 5-6 Examples of Features

---

**TECHNOTE 5-1**
**Feature**

A feature is any surface on a part.

---

**Author's Comment**
The Y14.5 standard overuses the term "feature" and defines it in a very general sense. As a result, the reader must try to determine the type of part geometry it refers to.

To make this book easier to read, I have added a chart in Appendix C that explains the relationship between the words used in this text and the part geometry being referred to.

A *complex feature* is a single surface of compound curvature or a collection of other features that constrains up to six degrees of freedom. In Figure 5-7, the entire outside surface of the part between points A and B is an example of a complex feature.

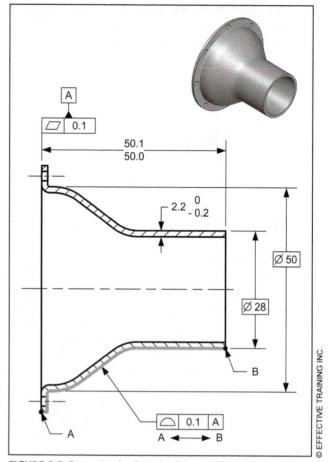

FIGURE 5-7 Example of a Complex Feature (Highlighted in Green)

## Feature of Size, Regular Feature of Size, and Irregular Feature of Size

This section covers the terms feature of size, regular feature of size, and irregular feature of size.

A *feature of size* is a general term that is used to refer to either a regular feature of size or an irregular feature of size.

A *regular feature of size* is one cylindrical or spherical surface, a circular element, a set of two opposed (or partially opposed) parallel elements, or opposed (or partially opposed) parallel surfaces, each of which is associated with a directly toleranced dimension (size dimension).

The definition of a regular feature of size involves four concepts. A regular feature of size must:

- Contain opposed (or at least partially opposed) surfaces or elements

- Contain sufficient opposed surfaces or elements to be able to establish an axis, center point, or center plane

- Be associated with a directly toleranced size dimension

- Be a shape that is cylindrical, spherical, circular, two opposed elements (lines), or two opposed parallel planes

Examples of regular features of size are shown in Figure 5-8.

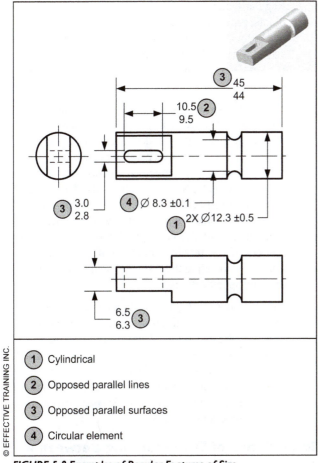

1 Cylindrical

2 Opposed parallel lines

3 Opposed parallel surfaces

4 Circular element

© EFFECTIVE TRAINING INC.

**FIGURE 5-8 Examples of Regular Features of Size**

An *irregular feature of size* has two types:

A. A directly toleranced feature or collection of features that may contain or be contained by an actual mating envelope: that is a sphere, cylinder, or pair of parallel planes. See Figure 5-9 for an example.

B. A directly toleranced feature or collection of features that may contain or be contained by an actual mating envelope: other than a sphere, cylinder, or pair of parallel planes. See Figure 5-10 for an example.

An irregular feature of size does not require opposed surfaces or elements, or association with a directly toleranced size dimension, but it must be able to establish an actual mating envelope.

An irregular feature of size may or may not have an actual local size.

**Author's Comment**
One important difference between regular and irregular features of size is that Rule #1 (explained in Chapter 7) only applies to regular features of size.

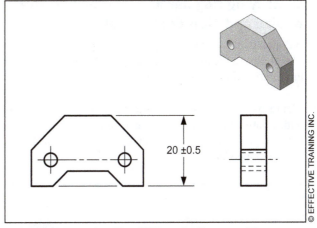

© EFFECTIVE TRAINING INC.

**FIGURE 5-9 Example of Type "A" Irregular Features of Size**

**Author's Comment**
In this text, the word "width" or the phrase "planar feature of size" is used as a shorter, more clear way than using the phrase "parallel plane regular feature of size."

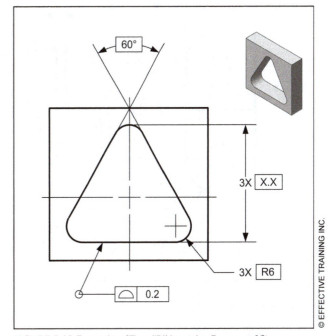

© EFFECTIVE TRAINING INC.

**FIGURE 5-10 Example of Type "B" Irregular Feature of Size**

**Author's Comment**
Irregular feature of size, like the one shown in Figure 5-9, must be defined with profile tolerances.

## Classifying Regular and Irregular Features of Size and Non-Features of Size

A drawing may contain many different types of dimensions. The ability to identify each type is helpful in applying and interpreting drawing specifications.

Figure 5-11 shows several different types of dimensions with a label classifying each dimension into one of three types: regular and irregular features of size and non-feature of size dimensions.

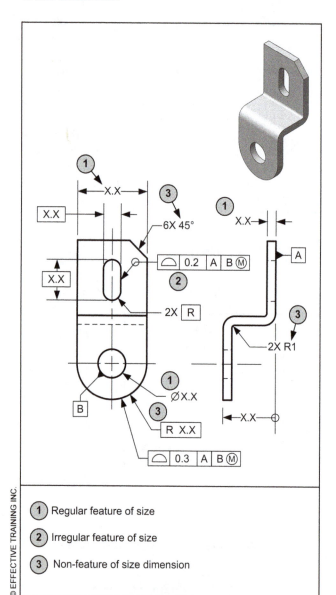

1. Regular feature of size
2. Irregular feature of size
3. Non-feature of size dimension

*FIGURE 5-11 Feature of Size Examples*

**Author's Comment**
Irregular features of size are introduced in this text. For a more complete coverage of this topic, see ETI's *Advanced Concepts of GD&T* textbook.

1. Regular features of size contain two or more opposed points, providing an actual local size. These may be defined with coordinate tolerances that control their size and form. Regular features of size were simply known as features of size in the Y14.5M-1994 standard.

2. Irregular features of size do not require opposed points, and may create an actual mating envelope that is not spherical, cylindrical, circular, opposed parallel lines, or opposed parallel planes. In some cases, these may be defined with coordinate tolerancing, but the "UNSAFE" conditions identified in Chapter 3 must be considered. The concept of irregular feature of size is a new, but powerful expansion of the feature of size concept. The use of irregular features of size requires some use of geometric tolerances to define the form and avoid incomplete drawing specifications.

3. Non-feature of size dimensions are highly prone to the "UNSAFE" conditions of coordinate tolerancing. With the exception of tangent radii and simple chamfers, these require geometric dimensioning and tolerancing. Non-feature of size dimensions of all types are frequently defined using geometric dimensioning and tolerancing to provide specifications with a single interpretation.

## Importance of Distinguishing Between a Regular and an Irregular Feature of Size

In order to correctly interpret a drawing, you need to be able to recognize the difference between a regular feature of size and an irregular feature of size. It is important to recognize this difference because Rule #1 (explained in Chapter 7) only applies to regular features of size. In the drawing in Figure 5-12, there are four regular features of size and one irregular feature of size.

> **TECHNOTE 5-2**
> **Feature of Size**
> Four important concepts to remember about a regular feature of size are:
> - It must have at least some opposed surfaces or elements.
> - It must be able to derive an axis, center plane, or center point.
> - It must be associated with a directly toleranced size dimension.
> - It must meet the requirements of Rule #1 (explained in Chapter 7).

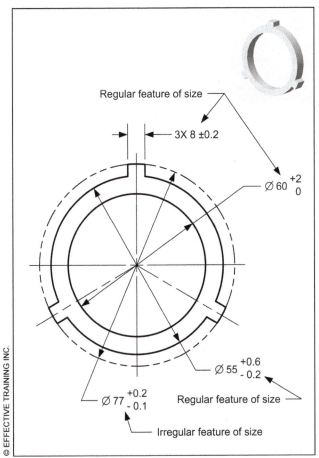

*FIGURE 5-12 Examples of Regular and Irregular Features of Size*

## Actual Mating Envelope, Unrelated Actual Mating Envelope, and Related Actual Mating Envelope

An *actual mating envelope* is a similar perfect feature counterpart of the smallest size that can be contracted about an external feature of size, or the largest size that can be expanded within an internal feature of size, so that it coincides with the surfaces at the highest points.

A similar perfect feature counterpart is a boundary that is the perfect inverse of the feature. Examples of similar perfect feature counterparts are a perfect cylinder or a set of perfect parallel planes.

Actual mating envelope refers to a similar perfect feature counterpart that would surround the high points of a feature of size. An actual mating envelope is a variable value and is always outside the material.

The actual mating envelope of an external feature of size is a similar perfect feature counterpart that can be circumscribed around the feature of size so that it just contacts the surface at the highest points.

Figure 5-13 shows an example of an actual mating envelope for an external feature of size.

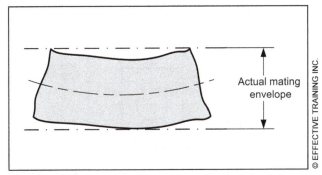

*FIGURE 5-13 Example of an Actual Mating Envelope for an External Feature of Size*

The actual mating envelope of an internal feature of size is a similar perfect feature counterpart that can be inscribed inside the feature of size so that it just contacts the surface at the highest points. Figure 5-14 shows an example of an actual mating envelope for an internal feature of size.

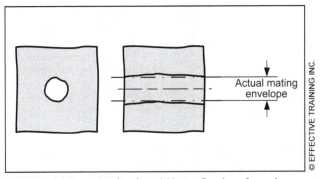

*FIGURE 5-14 Example of an Actual Mating Envelope for an Internal Feature of Size*

**Author's Comment**

The concept of actual mating envelope is used extensively in the language of GD&T. Where a geometric tolerance applies at RFS, the tolerance zone applies to the axis or center plane of the unrelated actual mating envelope.

There are two types of actual mating envelopes: related and unrelated. An *unrelated actual mating envelope* is a similar perfect feature counterpart expanded within an internal feature of size or contracted about an external feature of size and not constrained to any datums. A feature of size has only one unrelated actual mating envelope. An example is shown in Figure 5-15.

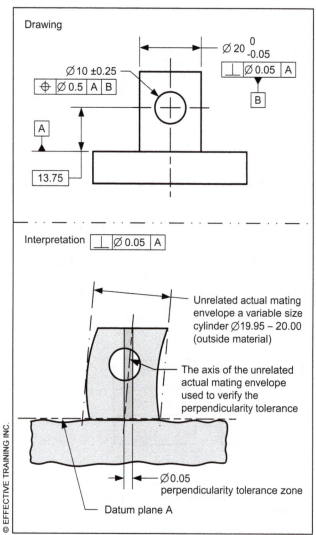

*FIGURE 5-15 Example of an Unrelated Actual Mating Envelope*

A *related actual mating envelope* is a similar perfect feature counterpart expanded within an internal feature of size, or contracted about an external feature of size, while constrained in either orientation or location (or both) to any applicable datums. An example is shown in Figure 5-16.

**Author's Comment**

The concept of unrelated mating envelope is generally used when interpreting the tolerance portion of a feature control frame and the concept of related actual mating envelope is generally used when interpreting datum features (secondary or tertiary) of a feature control frame.

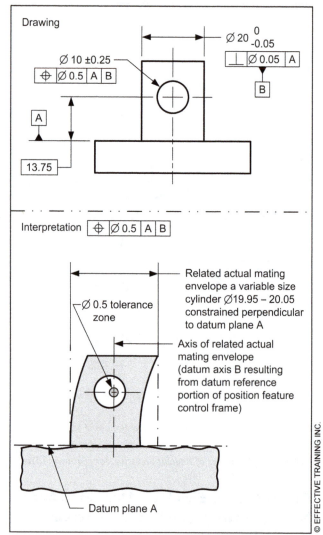

*FIGURE 5-16 Example of a Related Actual Mating Envelope*

If the location or orientation of a feature of size is defined with a geometric control at RFS, the axis of the unrelated mating envelope must be within the tolerance zone.

---

**TECHNOTE 5-3**
**Actual Mating Envelope**

- There are two types of actual mating envelopes: related and unrelated.

- A feature of size has only one unrelated actual mating envelope.

- A feature of size may have several related actual mating envelopes. Each geometric control with unique datum references can result in a different related actual mating envelope.

- Actual mating envelope is a variable value derived from the actual part.

- An actual mating envelope is always outside the material. It circumscribes or inscribes the applicable feature of size.

## Axis and Center Plane

An *axis* is a theoretical straight line about which a geometric object rotates or may be imagined to rotate. Where the term "axis" is used in this text, unless otherwise indicated, it refers to the axis of the unrelated actual mating envelope of a cylindrical feature of size.

A *center plane* is a theoretical plane about which a geometric object is equally disposed. Where the term "center plane" is used, unless otherwise indicated, it refers to the center plane of the unrelated actual mating envelope of a width feature of size.

See Chapter 15 for an explanation of the terms "datum axis" and "datum center plane."

## Maximum Material Condition, Least Material Condition, and Regardless of Feature Size

An important concept in GD&T is the ability to specify a geometric tolerance at various material conditions for a feature of size. A geometric tolerance can be specified to apply at the largest size, smallest size, or actual size of a feature of size.

*Maximum material condition* (MMC) is the condition in which a feature of size contains the maximum amount of material within the stated limits of size. The MMC for an external feature of size (e.g., shaft diameter) is the largest value. See Figure 5-17A.

The maximum material condition for an internal feature of size (e.g., hole diameter) is the smallest value. See Figure 5-17B. Figure 5-18 shows examples of maximum material condition.

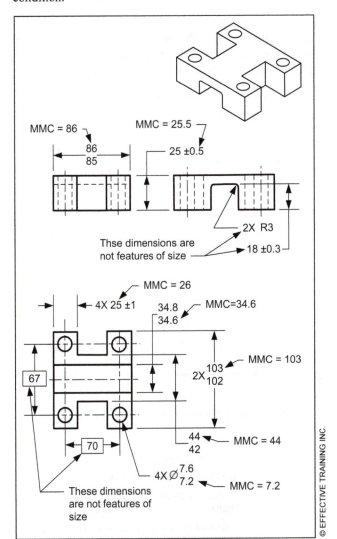

FIGURE 5-18 Examples of Identifying Maximum Material Condition (MMC) on a Drawing

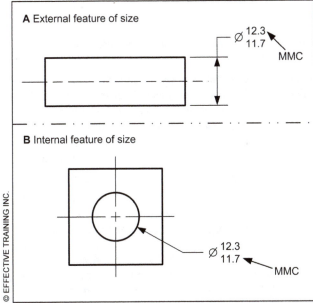

FIGURE 5-17 Maximum Material Condition (MMC)

---

**TECHNOTE 5-4**
**Maximum Material Condition**

- The maximum material condition of an external feature of size (e.g., shaft) is its largest size limit.

- The maximum material condition of an internal feature of size (e.g., hole) is its smallest size limit.

- In other words, the maximum material condition adds material to the part, increasing its mass or volume.

*Least material condition* (LMC) is the condition in which a feature of size contains the least amount of material within the stated limits of size. The least material condition for an external feature of size (i.e., shaft diameter) is the smallest value. See Figure 5-19A.

The least material condition for an internal feature of size (i.e., hole diameter) is the largest value. See Figure 5-19B. Figure 5-20 shows examples of least material condition.

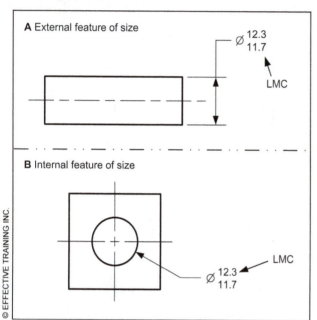

**A** External feature of size

Ø 12.3 / 11.7

LMC

**B** Internal feature of size

Ø 12.3 / 11.7    LMC

© EFFECTIVE TRAINING INC.

**FIGURE 5-19  Least Material Condition (LMC)**

---

## TECHNOTE 5-5
### Least Material Condition

- The least material condition of an external feature of size (e.g., shaft) is its smallest size limit.
- The least material condition of an internal feature of size (e.g., hole) is its largest size limit.
- In other words, the least material condition subtracts material from the part, decreasing its mass or volume.

**Author's Comment**
MMC and LMC are theoretical conditions. They do not exist on an actual part.

---

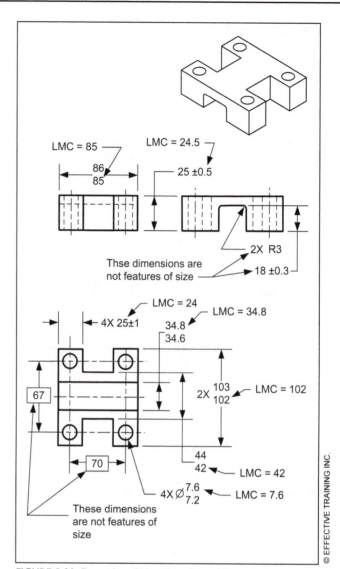

LMC = 85    LMC = 24.5

86 / 85    25 ±0.5

2X R3

Thse dimensions are not features of size    18 ±0.3

LMC = 24    LMC = 34.8

4X 25±1    34.8 / 34.6

67    2X 103 / 102    LMC = 102

70    44 / 42    LMC = 42

4X Ø 7.6 / 7.2    LMC = 7.6

These dimensions are not features of size

© EFFECTIVE TRAINING INC.

**FIGURE 5-20  Examples of Identifying Least Material Condition (LMC) on a drawing**

*Regardless of feature size* (RFS) indicates that a geometric tolerance applies at any increment of size of the actual mating envelope of the feature of size. Another way to visualize RFS is that the geometric tolerance applies at whatever size the toleranced feature of size is produced. There is no symbol for RFS, it is the default condition for all geometric tolerances.

An example of RFS is shown in Figure 5-21. In this example, the center plane of the part must be flat within 0.4, regardless of the feature size (the 10.3 ±0.4 dimension). The flatness tolerance is fully explained in Chapter 9.

**Author's Comment**
The terms maximum material condition, least material condition, and regardless of feature size only apply to toleranced features of size. Different terms are used for datum features of size.

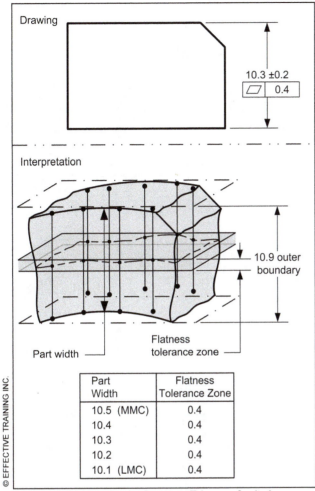

Drawing

10.3 ±0.2

▱ 0.4

Interpretation

10.9 outer boundary

Part width

Flatness tolerance zone

| Part Width | Flatness Tolerance Zone |
|------------|------------------------|
| 10.5 (MMC) | 0.4 |
| 10.4 | 0.4 |
| 10.3 | 0.4 |
| 10.2 | 0.4 |
| 10.1 (LMC) | 0.4 |

© EFFECTIVE TRAINING INC.

FIGURE 5-21  *Example of a Geometric Tolerance Applied Regardless of Feature Size*

## Determining Maximum and Least Material Condition

Every feature of size has a maximum and least material condition. Limit dimensions directly specify the maximum and least material condition of a feature of size. Where a drawing contains plus-minus tolerances, the maximum or least material conditions is calculated from the tolerances. Figure 5-22 shows examples of maximum and least material conditions.

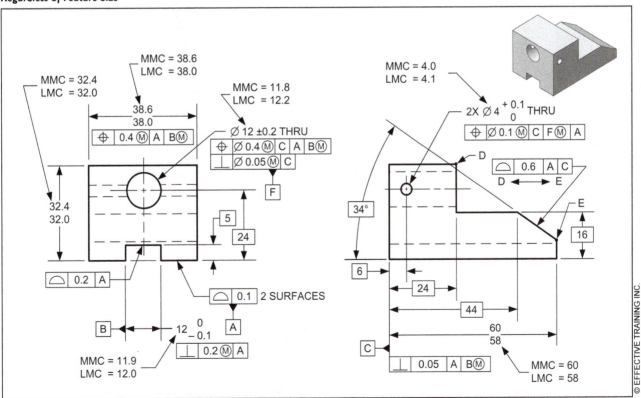

FIGURE 5-22  *Examples of Identifying Maximum and Least Material Conditions*

## Pattern and Pattern Indication

A *pattern* is two or more features or features of size to which a locational geometric tolerance is applied and is grouped by one of the following methods (n=number):

- nX

- INDICATED

- n COAXIAL HOLES

- ALL OVER

- ALL AROUND

- A ←→ B (between symbol)

- n Surfaces

- Simultaneous requirement

Where a pattern is indicated on a drawing, it groups the features or features of size. Where a geometric tolerance applies to a pattern, it applies to all of the features or features of size in the pattern simultaneously. Examples of pattern designations are shown in Figure 5-23. Additional examples of the pattern grouping methods are shown throughout this text.

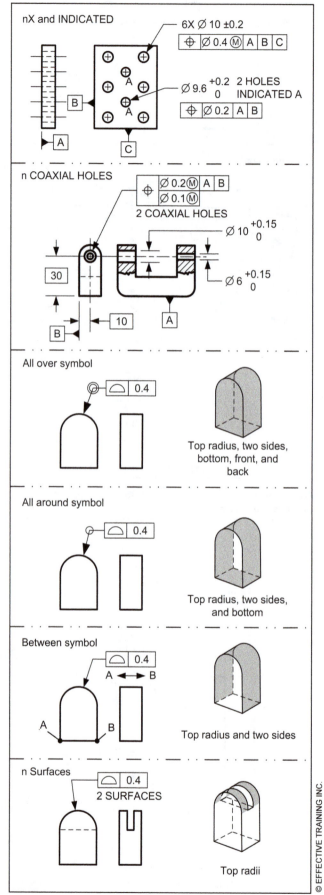

FIGURE 5-23  Examples of Pattern Designation

The last pattern creating method is simultaneous require-ment. A *simultaneous requirement* is where two or more geometric tolerances have identical datum references and are considered a pattern. Identical datum references require the same datum references, in the same order and with the same modifiers. In Figure 5-24, the position tolerances lo-cating the hole patterns create a single pattern of both hole patterns.

### Author's Comment
Simultaneous requirements are a complex topic with several variations. The topic is only introduced in this text. For additional information, see ETI's *Advanced Concepts of GD&T* (2009) textbook.

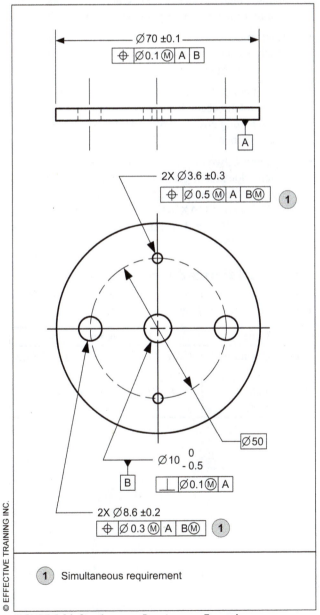

**FIGURE 5-24  Simultaneous Requirement Example**

# SUMMARY

## Key Points

- Two surfaces or elements are opposed where any ray normal to and extending from one of the surfaces or elements intersects the other surface or element.

- Two surfaces or elements are partially opposed where some rays normal to and extending from one of the surfaces or elements intersects the other surface or element.

- Two surfaces or elements are non-opposed where none of the rays normal to and extending from one of the surfaces or elements intersects the other surface or element.

- A size dimension is a dimension across two opposed parallel surfaces, line elements, or points.

- An actual local size is the measured value of any individual distance at any cross section of a feature of size.

- The MMC and LMC modifiers may only be used where a feature control frame is applied to a feature of size.

- Rule #1 only applies to regular features of size.

- Certain geometric tolerances may only be applied to a feature of size.

- Certain geometric tolerances may only be applied to a feature.

- A feature is a physical portion of a part such as a surface or hole, or its representation on drawings, models, or digital data files.

- A complex feature is a single surface of compound curvature or a collection of other features that constrains up to six degrees of freedom.

- A feature of size is a general term that is used to refer to either a regular feature of size or an irregular feature of size.

- A regular feature of size is one cylindrical or spherical surface, a circular element, a set of two opposed parallel elements or opposed parallel surfaces, each of which is associated with a size dimension.

- An irregular feature of size does not require opposed elements or surfaces, but it must be able to establish an actual mating envelope and be associated with a directly toleranced feature.

- An actual mating envelope is a similar perfect feature counterpart of smallest size that can be contracted about an external feature of size or the largest size that can be expanded within an internal feature of size so that it coincides with the surfaces at the highest points.

- There are two types of actual mating envelopes: related and unrelated.

- A feature of size may have several actual mating envelopes. Each geometric control with unique datum references can result in a different related actual mating envelope.

- Actual mating envelope is a variable value derived from the actual part.

- An actual mating envelope is always outside the material. It circumscribes or inscribes the applicable feature of size.

- An axis is a straight line about which a geometric object rotates or can be imagined to rotate.

- A center plane is the theoretical plane about which a geometric object is equally disposed.

- Maximum material condition (MMC) is the condition in which a feature of size contains that maximum amount of material within the stated limits of size.

- Least material condition (LMC) is the condition in which a feature of size contains that least amount of material within the stated limits of size.

- Regardless of feature size (RFS) indicates that a geometric tolerance applies at any increment of size of the actual mating envelope of the feature of size within its size tolerance.

- A pattern is two or more features or features of size to which a locational geometric tolerance is applied and is grouped by one of the following methods: nX, INDICATED, n COAXIAL HOLES, ALL OVER, ALL AROUND, A ↔ B, n Surfaces, AND simultaneous requirements.

## Additional Related Topics

*These topics are recommended for further study to improve your understanding of key GD&T terms.*

| Topic | Source |
|---|---|
| Irregular feature of size | *ASME Y14.5-2009, Paras. 1.3.32.2 & 4.17* |
| Simultaneous requirement | *ASME Y14.5-2009, Para. 4.19* |

# QUESTIONS AND PROBLEMS

**Website Bonus Materials**
Additional questions are available at our website. To access bonus materials for this textbook, please visit:
www.etinews.com/textbookbonus

## True and False

*Indicate if each statement is true or false*

T / F  1. A single surface of compound curvature is considered a complex feature.

T / F  2. An axis is considered a feature.

T / F  3. A pin ∅8 X 10mm long has hundreds of actual local sizes.

T / F  4. A pin ∅8 X 10mm long has only one unrelated actual mating envelope.

T / F  5. A triangular hole may be described as an irregular feature of size.

T / F  6. A sphere is an irregular feature of size.

T / F  7. A regular feature of size has both an MMC and an LMC.

## Multiple Choice

*Circle the best answer to each statement.*

1. Two types of actual mating envelopes are:
   A. Regular and irregular
   B. MMC and LMC
   C. Related and unrelated
   D. Opposed and non-opposed

2. Which of the following could be used to create a regular feature of size?
   A. A cylinder
   B. A set of parallel planes
   C. A circle
   D. All of the above

3. A center plane can be established from:
   A. Any regular feature of size
   B. A width feature of size
   C. Any feature of size
   D. All of the above

4. When working with a related actual mating envelope, the term related refers to the relationship...
   A. With adjacent features
   B. To specified datum features
   C. Between actual local sizes
   D. Between features in a pattern

5. One reason it's important to distinguish between regular and irregular features of size is:
   A. Rule #1 only applies to regular features of size
   B. Only regular features of size have a MMC condition
   C. Only regular features of size have an actual mating envelope
   D. Only regular features of size have an RFS condition

6. The maximum material condition of an external regular feature of size is the _____ size limit.
   A. Smallest
   B. Nominal
   C. Largest
   D. Basic

7. A regular feature of size must contain _____ surfaces or elements.
   A. Some opposed
   B. Non-opposed
   C. Perpendicular
   D. Adjacent

## Application Problems

*The application problems are designed to provide practice on applying the chapter concepts to situations that are similar to on-the-job conditions.*

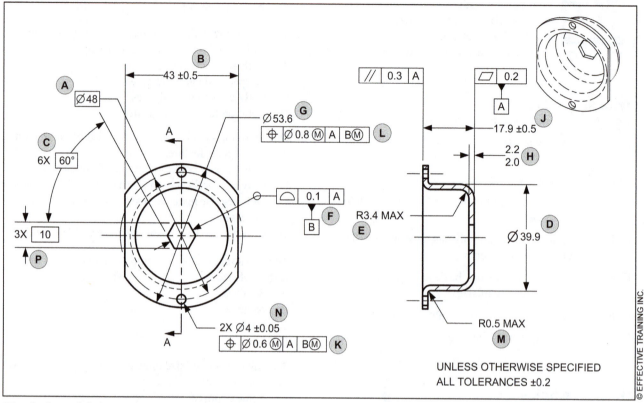

*Use the letter labels to answer questions 1–5.*

1. Identify all of the regular features of size dimensions. _____

2. Identify all of the irregular features of size.

3. Using the dimension labeled "J" describe:

    A. The actual mating envelope: _____

    B. The MMC and LMC limits: _____

    C. The type of opposition (fully opposed, partially opposed or non-opposed): _____

4. Which indications of a pattern are used on this drawing? _____

5. Fill in the chart below.

| Dimension | MMC | LMC |
|:---:|:---:|:---:|
| B | | |
| D | | |
| G | | |
| H | | |
| N | | |

# Symbols and Modifiers

## Goal

Recognize the symbols and modifiers used in GD&T

## Performance Objectives

Upon completing this chapter, you should be able to:

1. Identify the fourteen geometric characteristic symbols (p.64)
2. List the five types of geometric characteristic symbols (p.64)
3. Identify twenty-one geometric modifying symbols (p.64)
4. List the five types of geometry attributes (p.65)
5. Identify the parts of a feature control frame (p.66)
6. Explain how the location of a feature control frame affects its interpretation (p.67)
7. Explain how the continuous feature modifier affects a feature and a feature of size (p.68)

## New Terms

- Continuous feature symbol
- Feature control frame
- Geometric characteristic symbols
- Geometry attributes
- Modifiers

## What This Chapter Is About

This chapter introduces the geometric and modifying symbols, feature control frame, and continuous feature symbol used in GD&T. The symbols and terms in this chapter lay the foundation for understanding the concepts in the following chapters.

The purpose of this chapter is to help you recognize the names of the geometric and modifying symbols used on engineering drawings, and to explain the interpretation of the continuous feature symbol and feature control frames.

This chapter is important because understanding the GD&T symbols and modifiers is critical to being able to read engineering drawings that are based on Y14.5.

# TERMS AND CONCEPTS

## Geometric Characteristic Symbols

The *geometric characteristic symbols* are a set of 14 symbols used in the language of GD&T to describe the geometry attributes of a part. The symbols are shown in Figure 6-1.

| Type | Characteristic | | Symbol | Uses a Datum Reference |
|------|------|------|------|------|
| Form | Straightness | | — | Never |
| | Flatness | | ▱ | |
| | Circularity (roundness) | | ○ | |
| | Cylindricity | | ⌭ | |
| Profile | Profile of a line | | ⌒ | Sometimes |
| | Profile of a surface | | ⌓ | |
| Location | Position | | ⊕ | |
| | Concentricity | | ◎ | |
| | Symmetry | | ⌯ | |
| Orientation | Angularity | | ∠ | Always |
| | Perpendicularity | | ⊥ | |
| | Parallelism | | // | |
| Runout | Circular runout | | ↗ | |
| | Total runout | | ↗↗ | |

**FIGURE 6-1** *Geometric Characteristic Symbols*

## The Five Types of Geometric Characteristic Symbols

The geometric characteristics symbols are divided into five types: form, profile, orientation, location, and runout. Based on which category a symbol is in, it may never, sometimes, or always reference a datum. (See the fourth column in Figure 6-1.) The size and proportions for the symbols are given in Appendix D.

## Geometric Modifying Symbols

The language of GD&T also contains a set of symbols, called "modifiers." *Modifiers* are symbols or keywords that communicate additional information about the tolerancing of a part. The common modifiers used in geometric tolerancing are shown in Figure 6-2.

---

**TECHNOTE 6-1**
**General Dimensioning Symbols**

In addition to the modifiers, general dimensioning symbols (from Chapter 4) are also used for describing a part.

---

**Author's Comment**

In addition to the symbols shown in this chart, there are 15 general dimensioning symbols shown in Figure 4-1 that some people consider modifying symbols as well.

| Term | Symbol | Specified |
|------|------|------|
| Maximum material condition | Ⓜ | Inside a feature control frame |
| Maximum material boundary | Ⓜ | |
| Least material condition | Ⓛ | |
| Least material boundary | Ⓛ | |
| Translation * | ▷ | |
| Basic datum simulator location * | [BSC] or [BASIC] | |
| MMB datum simulator size * | DⓂ[Ø7.5] | |
| Tangent plane | Ⓣ | |
| Unequally disposed profile | Ⓤ | |
| Non-uniform profile | NON-UNIFORM | |
| Projected tolerance zone | Ⓟ | |
| Statistical tolerance * | ⟨ST⟩ | Inside or outside |
| Free state * | Ⓕ | |
| Independency | Ⓘ | Outside feature control frame |
| Continuous feature | ⟨CF⟩ | |
| Between | ↔ | |
| All around | ⊸ | |
| All over | ⊸ | |
| Boundary | BOUNDARY | |
| Each element * | EACH ELEMENT | |
| Each radial element * | EACH RADIAL ELEMENT | |

* Not covered in this textbook

**FIGURE 6-2** *Common GD&T Modifiers*

---

**Author's Comment**

The use/interpretation of symbols marked by an asterisk in the chart above are beyond the scope of this text. They are explained in ETI's *Advanced Concepts of GD&T (2009)* textbook..

© EFFECTIVE TRAINING INC.

## The Five Geometry Attributes of a Feature

A *geometry attribute* is a characteristic of a feature or feature of size. Each surface or feature of size has five geometry attributes: size, location, orientation, form, and surface texture. A simplified explanation of each geometry attribute is shown below.

**Size** – How large or small a feature of size is

**Location** – Where a feature of size (or surface) is relative to other features of size (or surfaces) or in a coordinate system

**Orientation** – The angular relationship between a surface or feature of size and a reference plane or axis

**Form** – The shape of a surface

**Surface texture** – The composite of roughness and waviness deviations that are typical of a real surface; surface texture only applies for a limited length (the sample length) over a surface where form may apply for the entire length of a surface

An example of deviation for each geometry attribute is shown in Figure 6-3.

The geometric tolerances specified on a drawing describe allowable deviations for four of the geometry attributes: size, location, orientation, and form. The fifth attribute, surface texture, has its own set of symbology.

A geometric tolerance may control the deviations of more than one of the five geometry attributes depending upon:

- The geometric characteristic symbol, and
- Whether the geometric tolerance is applied to a surface or a feature of size,
- The type of part geometry it is applied to

An example of where a geometric tolerance may control more than one geometry attribute is a position tolerance. A position tolerance is a location tolerance that may control deviations in both the location and orientation of a feature of size. See Chapters 20–23 for a complete explanation of position tolerances.

An example of a geometric tolerance that cannot vary in the types of geometry attributes it can control is a flatness tolerance. A flatness tolerance can only control form, it can never control orientation or location. See Chapter 9 for a complete explanation of the flatness tolerance.

**Author's Comment**
Surface texture is an important geometry attribute; however, it is a complex topic and not within the scope of this text (based on ASME Y14.5-2009). For additional information on surface texture, see ASME Y14.36 and ASME B46-1-2009.

Size – The amount the hole can deviate from its specified nominal size
Maximum size
Minimum size

Location – The amount the hole can deviate from the specified basic location

Orientation – The amount the hole can deviate from the specified basic angle

Form – The amount the hole can deviate from being perfectly round or straight

Surface Texture – The microscopic amount the surface of the hole can deviate from being perfectly smooth

© EFFECTIVE TRAINING INC.

*FIGURE 6-3 Examples of Deviations of the Five Geometry Attributes for a Hole*

The chart in Figure 6-4 shows the geometric tolerances and the geometry attributes that they can control.

- A gray cell indicates that the tolerance cannot be used to control that geometry attribute.

- A "P" in a cell indicates that the tolerance may be used to directly control that geometry attribute.

- An "I" in a cell indicates that the geometry attribute is indirectly controlled by the tolerance.

The geometry attributes that are directly and indirectly controlled by each geometric tolerance are covered in the following chapters.

| Tolerance | Geometry Attributes | | | | |
|---|---|---|---|---|---|
| | Size | Location | Orientation | Form | Surface Texture |
| Size dimension | P | | | | N/A |
| Rule #1 | | | | P | |
| Profile of a surface | P | P | P | P | N/A |
| Profile of a line | P | P | P | P | N/A |
| Position | | P | I | I | N/A |
| Concentricity | | P | I | I | N/A |
| Symmetry | | P | I | I | N/A |
| Circular runout | | P | I | I | N/A |
| Total runout | | P | P | I | N/A |
| Perpendicularity | | | P | I | N/A |
| Parallelism | | | P | I | N/A |
| Angularity | | | P | I | N/A |
| Flatness | | | | P | N/A |
| Straightness | | | | P | N/A |
| Circularity | | | | P | N/A |
| Cylindricity | | | | P | N/A |

| P | = Primary control of this geometry attribute |
|---|---|
| I | = Indirect control of this geometry attribute |
| N/A | = In most cases, these tolerances are specified in millimeters and surface texture is specified in micrometers. Therefore, these tolerances generally do not affect surface texture. |
| (gray) | = Cannot control this geometry attribute |

*FIGURE 6-4 Size and Geometric Tolerances That Can Be Used to Control Deviations on Geometry Attributes*

## Parts of a Feature Control Frame

Geometric tolerances are specified on a drawing through the use of a feature control frame. A *feature control frame* is a rectangular box that is divided into compartments within which the geometric characteristic symbol, tolerance value, modifiers, and datum references are placed. The compartments of a feature control frame are shown in Figure 6-5.

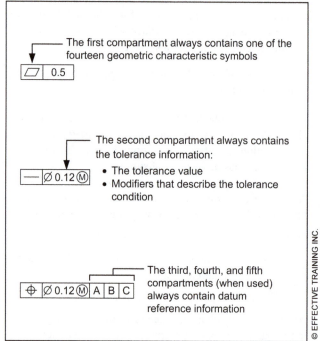

The first compartment always contains one of the fourteen geometric characteristic symbols

The second compartment always contains the tolerance information:
- The tolerance value
- Modifiers that describe the tolerance condition

The third, fourth, and fifth compartments (when used) always contain datum reference information

*FIGURE 6-5 Parts of a Feature Control Frame*

The first compartment of a feature control frame is called the geometric characteristic portion. It contains one of the fourteen geometric characteristic symbols.

The second compartment of a feature control frame is referred to as the tolerance portion. The tolerance portion of a feature control frame may contain several bits of information. For example, if the tolerance value is preceded by a diameter symbol ∅, the shape of the tolerance zone is a cylinder. If a diameter symbol is not shown in the front of the tolerance value, the shape of the tolerance zone is either parallel planes, parallel lines, or a uniform boundary in the case of profile. The tolerance value specified is always a total value.

When specifying a nondatum-related tolerance, a feature control frame will have two compartments. When specifying a datum-related tolerance, a feature control frame may have up to five compartments: the first contains the geometric characteristic symbol, the second contains the tolerance information, and up to three additional compartments for datum references. The third, fourth, and fifth compartments of a feature control frame are referred to as the datum reference portion of the feature control frame.

## Feature Control Frame Location

The location of a feature control frame on a drawing affects its interpretation. A feature control frame will have an entirely different interpretation if it is applied to a surface or to a feature of size. It is important to be able to understand whether a feature control frame applies to a surface or to a feature of size. There are several conventions on feature control frame location that indicate what the feature control frame applies to:

- A feature control frame applies to a surface when it is attached to an extension line of a surface, or has a leader line pointing to a surface or extension line of a surface.

- A feature control frame applies to a feature of size when it is associated with the size dimension by placing it next to the size dimension, or attaching it to the leader line or dimension line of the size dimension.

Figure 6-6 shows a drawing with several feature control frames specified. The feature control frames labeled #1 are applied to surfaces, and the feature control frames labeled #2 are applied to features of size.

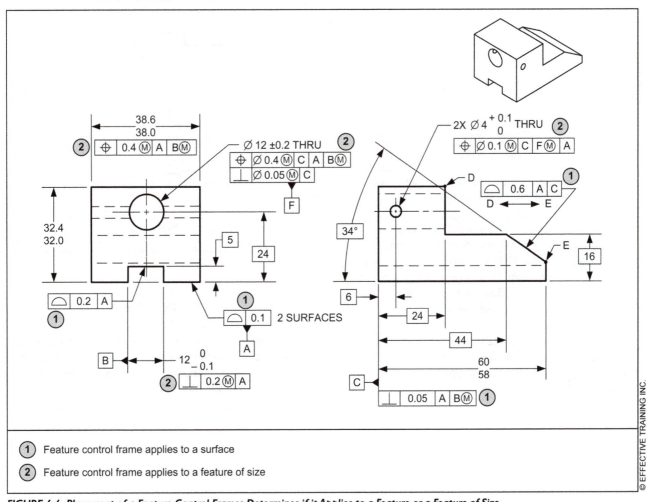

**FIGURE 6-6** *Placement of a Feature Control Frames Determines if it Applies to a Feature or a Feature of Size*

## The Continuous Feature Symbol

A *continuous feature symbol* indicates where a group of two or more features of size are to be treated geometrically as a single feature or feature of size.

The continuous feature symbol can be used to indicate that an interrupted feature or feature of size is considered as a single feature or feature of size. An interruption is where a surface contains a gap that divides the surface into multiple surfaces. Where used to indicate that multiple coplanar surfaces are to be treated as a single surface, the modifier may be placed on an extension line of the surfaces as shown in Figure 6-7. Where used to indicate that multiple features of size are to be treated as single feature of size, the continuous feature symbol is placed near the size dimension as shown in Figure 6-8A and B.

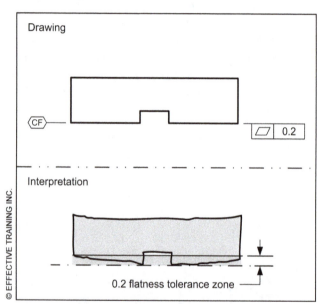

*Drawing*

*Interpretation*

0.2 flatness tolerance zone

© EFFECTIVE TRAINING INC.

*FIGURE 6-7 Continuous Feature Symbol Indicating Multiple Coplanar Surfaces Treated as a Single Surface*

### Author's Comment

The continuous feature modifier is new in the 2009 standard. Its use is not fully explained. The continuous feature modifier is defined in the standard as being applied to features of size.

The examples in the standard show the modifier being used on features and features of size. In this text, I use the continuous feature modifier to apply to both features and features of size.

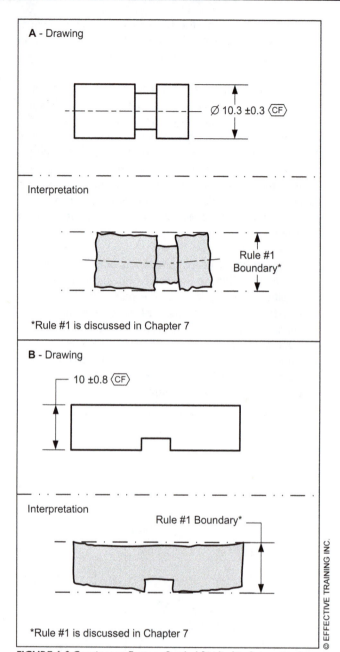

**A - Drawing**

Ø 10.3 ±0.3 (CF)

**Interpretation**

Rule #1 Boundary*

*Rule #1 is discussed in Chapter 7

**B - Drawing**

10 ±0.8 (CF)

**Interpretation**

Rule #1 Boundary*

*Rule #1 is discussed in Chapter 7

*FIGURE 6-8 Continuous Feature Symbol Applied to Interrupted Features of Size*

© EFFECTIVE TRAINING INC.

---

### TECHNOTE 6-2
### Extension Lines

The method of showing an extension line between two coplanar surfaces does not indicate that the surfaces are a single surface. It is used to indicate that the surfaces are nominally in-line.

# SUMMARY

## Key Points

- The fourteen geometric characteristic symbols are flatness, straightness, circularity, cylindricity, perpendicularity, angularity, parallelism, position, concentricity, symmetry, circular runout, total runout, profile of a line, and profile of a surface.

- A geometry attribute is a characteristic of a feature or feature of size. The five geometry attributes are size, location, orientation, form, and surface texture.

- The five categories of geometric controls are form, profile, orientation, location, and runout.

- The 21 geometric modifying symbols.

- The three parts of a feature control frame are the geometric characteristic portion, the tolerance portion, and the datum reference portion.

- The continuous feature modifier creates a single feature from multiple features or a single feature of size from multiple features of size of the same size.

## Additional Related Topics

*These topics are recommended for further study to improve your understanding of GD&T symbols and modifiers.*

| Topic | Source |
|---|---|
| Statistical tolerancing | *ASME Y14.5-2009, Para. 2.17* |
| Each element | *ASME Y14.5-2009, Para. 6.4.3* |
| Each radial element | *ASME Y14.5-2009, Para. 6.4.3* |
| Independency | *ASME Y14.5-2009, Para. 2.7.3* |
| Datum feature simulator size indicator | *ASME Y14.5-2009, Para. 4.11.6* |
| Continuous feature | *ASME Y14.5-2009, Para. 2.7.5* |

# QUESTIONS AND PROBLEMS

## True and False

*Indicate if each statement is true or false*

T / F    1. Symbols that communicate additional information about the tolerance of a feature are called modifiers.

T / F    2. The $\textcircled{1}$ symbol is specified in the tolerance compartment of a feature control frame.

T / F    3. A runout type tolerance must be specified with a datum reference.

T / F    4. A feature control frame is required to have at least five compartments.

T / F    5. The five categories of geometric tolerances are form, profile, orientation, location, and concentricity.

T / F    6. A position tolerance may be used with or without a datum reference.

T / F    7. An orientation tolerance may indirectly control form.

T / F    8. Surface texture is considered one of the five geometric attributes.

## Multiple Choice

*Circle the best answer to each statement.*

1. Which symbol is interpreted as "all over"?
   A. ◄──►          C. Ⓤ
   B. (symbol)      D. (symbol)

2. What is the name of the ◎ symbol?
   A. Coaxiality        C. All around
   B. Concentricity     D. Aligned

3. Which modifier may only be used inside a feature control frame?
   A. ⒸF         B. ▷          C. Ⓘ          D. ⓈT

4. What is the minimum number of compartments in a feature control frame?
   A. 1          C. 3
   B. 2          D. 4

5. What does the ⒸF modifier indicate?
   A. A critical feature
   B. A compound feature pattern exists
   C. Two features or features of size are to be treated as one feature of feature of size
   D. Two surfaces are produced with the same operation(s)

## Application Problems

*The application problems are designed to provide practice on applying the chapter concepts to situations that are similar to on-the-job conditions.*

1. Which of the geometric tolerances are applied to a feature of size?
   A. A, D, E        C. A, F, G
   B. A, D, F        D. E, F, G

2. Which of the geometric tolerances are applied to a feature?
   A. A & D          C. F & G
   B. D & G          D. D & E

3. What is the name of the modifier labeled "B"?
   A. Maximum material boundary
   B. Minimum material boundary
   C. Maximum material condition
   D. Minimum material condition

4. What is the name of the modifier labeled "C"?
   A. Diameter        C. Nominal
   B. Cylinder        D. Size

5. Which type of geometric characteristic symbol is the geometric tolerance labeled "G"?
   A. Form           C. Orientation
   B. Runout         D. Location

6. What is the name of the geometric tolerance labeled "E"?
   A. Profile of a surface    C. Perpendicularity
   B. Total runout            D. Profile of a line

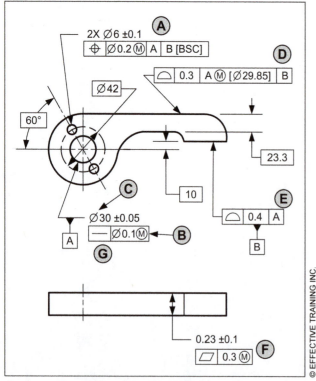

© EFFECTIVE TRAINING INC.

*Questions 1 – 6 refer to the drawing above.*

# GD&T Rules

## Goal

Understand the rules used in GD&T

## Performance Objectives

Upon completing this chapter, you should be able to:

1. Recognize the 16 fundamental dimensioning rules used in GD&T (p.72)
2. Explain the impact of each fundamental dimensioning rule on a drawing (p.73)
3. Explain Rule #1 (p.76)
4. Determine where Rule #1 applies to a feature of size (p.78)
5. Describe the Rule #1 boundary (p.79)
6. Explain how Rule #1 affects the interrelationship between features of size (p.79)
7. List two exceptions to Rule #1 (p.80)
8. List three ways Rule #1 can be overridden (p.80)
9. Describe the independency concept (p.81)
10. Explain Rule #2 (p.82)
11. Describe the verification principles for Rule #1 (p.82)
12. Explain what a fixed-limit gage, GO gage, and NOGO gage are (p.83)
13. Describe how to inspect a feature of size that is controlled by Rule #1 (p.84)

## New Terms

- Fixed-limit gage
- Fundamental dimensioning rules
- GO gage
- Independency concept
- NOGO gage
- Rule #1
- Rule #2

## What This Chapter Is About

The purpose of this chapter is to help you recognize how the GD&T rules affect engineering drawings, design intent, and inspection.

Definitions and rules comprise one of the six major components of the GD&T language. This chapter introduces the fundamental dimensioning rules, Rule #1, and Rule #2.

The fundamental rules in this chapter establish a set of default conditions that are essential for correctly interpreting engineering drawings.

# TERMS, RULES, AND CONCEPTS

## Fundamental Dimensioning Rules

The *fundamental dimensioning rules* are a set of general rules that apply to dimensioning and interpreting engineering drawings. These rules are automatically invoked where the ASME Y14.5-2009 Standard applies to a drawing.

The 16 fundamental dimensioning rules are paraphrased and illustrated in the charts and figures on the following pages. The engineering drawing in Figure 7-1 is used to illustrate how the rules apply. The letters on the drawing correspond to the letters used in the charts that relate the dimensioning rules to the parts of the drawing.

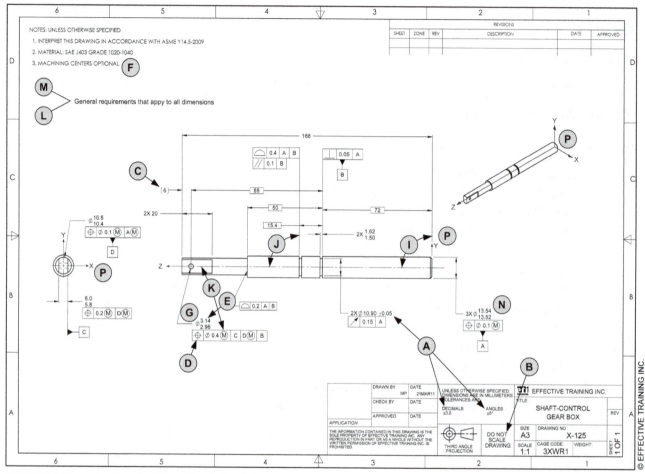

**FIGURE 7-1  Areas of a Drawing Affected by the Fundamental Dimensioning Rules**

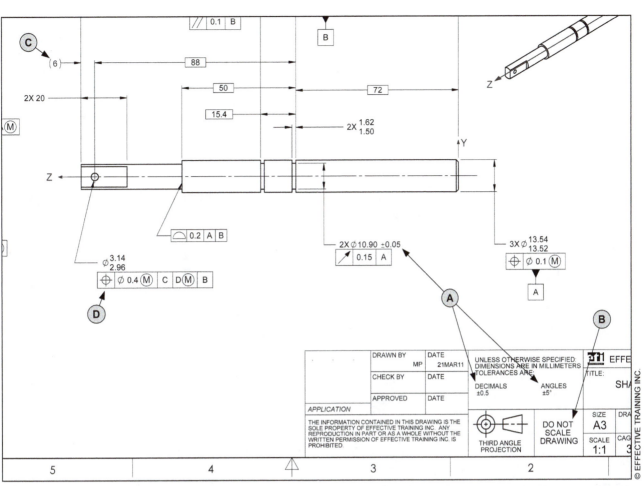

FIGURE 7-2  Examples of Fundamental Dimensioning Rules A–D

| Fundamental Dimensioning Rule | Impact on Drawings |
|---|---|
| a) Each dimension shall have a tolerance, except dimensions identified as reference, maximum, minimum, or stock (commercial stock) size. Tolerances may be applied directly to a dimension, through a general note, or through the title block. See Figure 7-2, areas labeled (A) for examples. | All dimensions require a tolerance to be assigned through one of the methods stated. |
| b) Dimensioning and tolerancing shall be complete so there is full definition of each part feature. Values may be expressed on a drawing, or in a CAD product definition data set. Scaling or assumption of a distance is not permitted. See Figure 7-2, area labeled (B) for an example. | • Dimensions and tolerances must be used to fully define (size, location, orientation, form, and surface texture) of each part feature and feature of size.<br>• Scaling of a drawing is not allowed.<br>• Assumption for a part distance is not allowed.<br>• Since drawings are legal documents, the dimensioning and tolerancing must be fully defined.<br>• Basic and nominal dimensions may be obtained from querying the model. |
| c) Each necessary dimension of an end product shall be shown. Use of reference dimensions should be minimized. See Figure 7-2, area labeled (C) for an example. | Reference dimensions cannot be used to establish part requirements. |
| d) Dimensions shall be selected and arranged to define the function and mating relationships of a part and shall not be subject to more than one interpretation. See Figure 7-2, area labeled (D) for an example. | • Dimensions and tolerances should define the part based on how it functions in the product.<br>• Dimensions and tolerances should have a single interpretation (i.e., coordinate tolerances should not be used to locate features). |

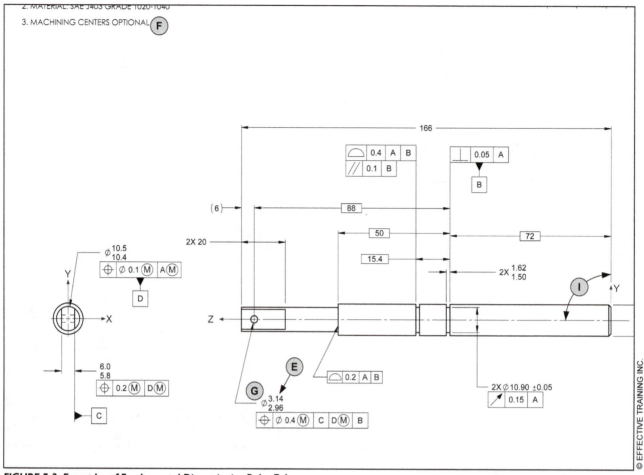

**FIGURE 7-3**  *Examples of Fundamental Dimensioning Rules E–I*

| Fundamental Dimensioning Rule | Impact on Drawings |
|---|---|
| e) The drawing should define a part without specifying manufacturing methods. See Figure 7-3, area labeled (E) for examples. However, where manufacturing, processing, quality assurance, or environmental information is essential to the definition of engineering requirements, it shall be specified on the drawing or in a document. | • A drawing should specify the diameter of a hole, and not specify manufacturing methods like "DRILL," "REAM," etc.<br>• Certain manufacturing or quality information may be included where it is necessary to fully define the engineering requirements of a part. |
| f) Nonmandatory processing information shall be identified such as "NONMANDATORY (MFG DATA)" or as reference information (e.g., finish allowance, shrink allowance, etc.) See Figure 7-3, area labeled (F) for an example. | Where processing information is shown on a drawing, it should be labeled nonmandatory or as reference information. |
| g) Dimensions should be arranged for optimum readability. Dimensions should be shown in true profile views and refer to visible outlines. See Figure 7-3, area labeled (G) for an example. | • Dimensions should be clear and readable.<br>• Dimensions should not be applied to hidden lines.<br>• Dimensions should only be applied to true profile views. |
| h) Wire, cables, sheets, rods, and other materials manufactured to gage or code numbers shall be specified by linear dimensions indicating the diameter or thickness. Gage or code numbers may be shown in parentheses following the dimension. | When used, commercial stock sizes should be shown with a linear dimension and tolerance followed by the note, "STOCK SIZE" or "STK THK." The gage or code number may be specified as reference. |
| i) A 90° angle applies where center lines or lines depicting features are shown at right angles and no angular dimension is specified on a 2D orthographic drawing. See Figure 7-3, area labeled (I) for an example. | • On 2D orthographic drawings, 90° angles do not have to be specified.<br>• This rule does not apply to 3D digital data sets.<br>• When used, the implied 90° angle is subject to a tolerance (usually a general tolerance or a title block tolerance). |

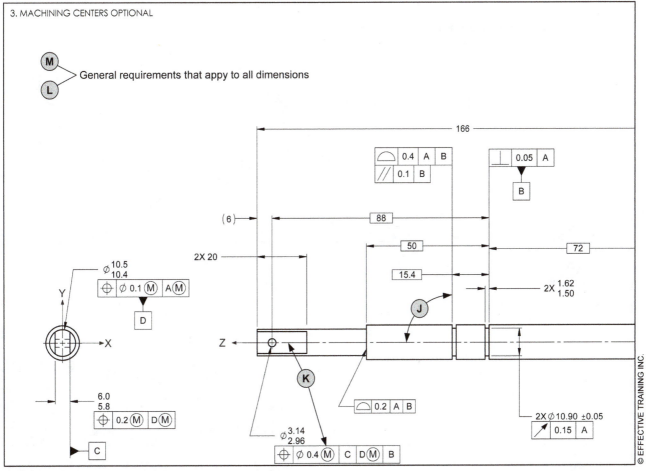

3. MACHINING CENTERS OPTIONAL

General requirements that appy to all dimensions

FIGURE 7-4  Examples of Fundamental Dimensioning Rules J–M

| Fundamental Dimensioning Rule | Impact on Drawings |
|---|---|
| j) A 90° basic angle applies where center lines of features in a pattern, or lines shown at right angles on a 2D orthographic drawing, are located and defined with basic dimensions, and no angular dimension is specified. See Figure 7-4, area labeled (J) for an example. | • On 2D orthographic drawings, 90° basic angles do not have to be specified.<br>• Geometric tolerances must be used to define the allowable deviation of implied 90° basic angles. |
| k) A zero basic dimension applies where axes, center planes, or surfaces are shown coincident on a drawing, and geometric tolerances establish the relationship among the features or features of size. See Figure 7-4, area labeled (K) for an example. | • Zero basic dimensions do not have to be specified.<br>• Geometric tolerances must be used to define the allowable deviation of zero basic dimensions. |
| l) Unless otherwise specified, all dimensions and tolerances are applicable at 20° C (68° F). Compensation may be made for measurements made at other temperatures. See Figure 7-4, area labeled (L) for an example. | • Establishes the temperature for analysis and inspection of parts.<br>• Allows compensation for measurements made at other temperatures. |
| m) Unless otherwise specified, all tolerances apply in a free-state condition, except for nonrigid parts defined with a restraint note. See Figure 7-4, area labeled (M) for an example. | • Clamping is not allowed during inspection unless a note on the drawing describes the clamp conditions.<br>• Nonrigid parts are typically defined with a restraint note that overrides this rule. |

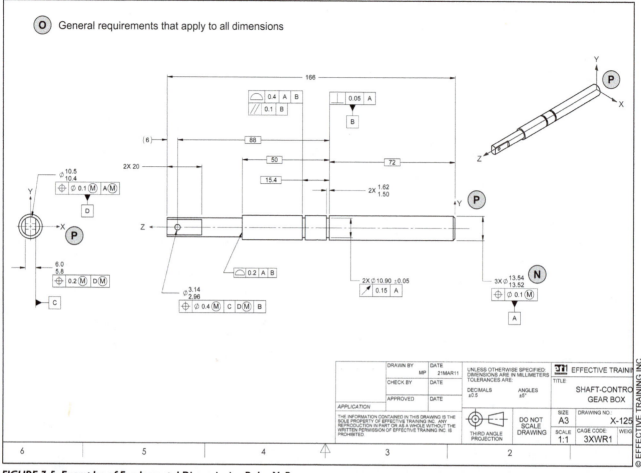

**FIGURE 7-5** *Examples of Fundamental Dimensioning Rules N–P*

| Fundamental Dimensioning Rule | Impact on Drawings |
|---|---|
| n) Unless otherwise specified, all tolerances apply for the full length, width, and depth of a feature or feature of size. See Figure 7-5, area labeled **N** for an example. | Defines the extent of coverage for a tolerance. |
| o) Dimensions and tolerances only apply at the drawing level where they are specified. A dimension specified on a detail drawing is not required at the assembly level drawing. See Figure 7-5, area labeled **O** for an example. | • A dimension only applies for the drawing level where it is specified. • Bolt torque, welding, brazing, press fits, or other assembly operations may distort the features at assembly; therefore, detail dimensions do not apply at the assembly level. • Assembly drawings must repeat detail drawing dimensions when they are to apply in the assembly. |
| p) Unless otherwise specified, where a coordinate system is shown on a drawing, it shall be right-handed. Each axis shall be labeled, and the positive direction shall be shown. Model coordinate systems shall be in compliance with ASME Y14.41. See Figure 7-5, areas labeled **P** for examples. | • Establishes a convention for reading coordinate systems on engineering drawings. • Unless otherwise specified, coordinate systems are right-handed. • Model coordinate systems must be in compliance with ASME Y14.41. |

# Rule #1

Rule #1 is referred to as the "Envelope Principle." It is part of the rules and definitions section of the six major components of the GD&T language shown in Figure 3-11.

Rule #1 is a dimensioning rule that ensures that individual regular features of size will assemble with one another.

Where Rule #1 applies to a regular feature of size, it establishes an overall boundary (or envelope) for the regular feature of size that is equal to its MMC. Where Rule #1 applies to an external regular feature of size, the maximum boundary (or envelope) of the regular feature of size is equal to its largest size limit (MMC).

Where Rule #1 applies to an internal regular feature of size, the minimum boundary (or envelope) of the regular feature of size is equal to its smallest size limit (MMC). To design two regular features of size that will assemble, the designer can use their MMC's.

*Rule #1* – The form of an individual regular feature of size is controlled by its limits of size.

Rule #1 requirements:

- The surface or surfaces of a regular feature of size shall not extend beyond a boundary of perfect form at MMC.

- Where a regular feature of size is produced at MMC, it must have perfect form.

- Where the actual local sizes of a regular feature of size are different than the MMC, a local variation in form equal to the amount of departure from MMC is allowed.

- Rule #1 only applies to individual regular features of size.

In industry, Rule #1 is often paraphrased as "perfect form at MMC" or "the envelope principle." In Rule #1, the words "perfect form" means perfect flatness, straightness, circularity, and cylindricity. If a planar regular feature of size is not at MMC, then form (flatness) variation is allowed.

There are two components to Rule #1: the envelope principle and the allowable variations of form as the size departs from MMC.

**Envelope Principle**

Let's look at how Rule #1 affects the diameter of a pin. When the pin diameter is at MMC, the pin must have perfect form. In this case, perfect form means perfect straightness, perfect roundness (i.e., perfect cylindricity). In theory, this would allow the pin to fit through a boundary equal to its MMC size. See Figure 7-6A.

Where Rule #1 applies to a hole, the hole must have perfect form at MMC. For a hole, MMC is the smallest value, and perfect form means perfect roundness and perfect straightness. In theory, this would allow the hole to assemble with a pin equal to its MMC size. See Figure 7-6B.

**Variations of Form**

If the size of the pin is less than MMC, the pin can have variation of form (straightness and roundness variation) equal to the amount the pin departs from MMC. An example of the effects of Rule #1 on an external and internal regular feature of size is shown in Figure 7-6A.

If the size of the hole is greater than MMC, the hole can have deviations in form (straightness and roundness) equal to the amount the hole departs from MMC. An example of the effects of Rule #1 on an internal regular feature of size is shown in Figure 7-6B.

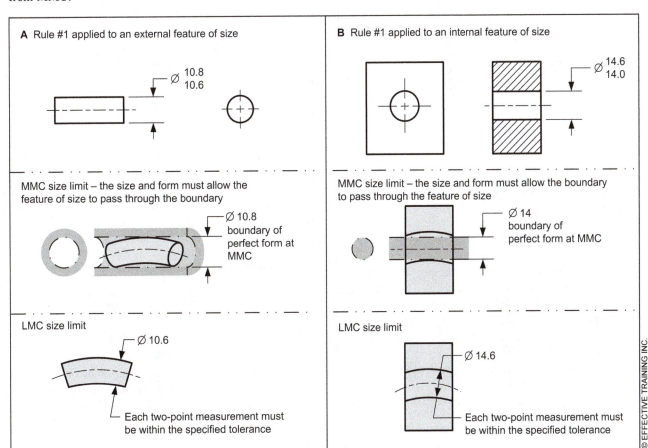

*FIGURE 7-6 Rule #1 Examples*

On a planar regular feature of size (as in Figure 7-7) perfect form refers to perfect flatness. In this figure, if the 10.2-10.8 dimension was produced at 10.7, then form variation equal to the amount of the departure from MMC (10.8-10.7=0.1) is allowed. If this dimension was produced at LMC (10.2), form variation equal to the amount of departure from MMC (0.6) would be permitted.

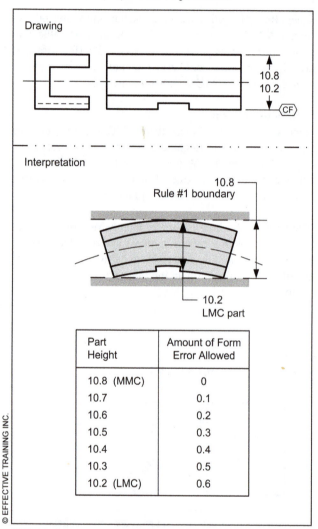

FIGURE 7-7  *Interpretation of Rule #1 Applied to a Planar Regular Feature of Size*

---

**TECHNOTE 7-1**
**Rule #1**

Rule #1 – The form of an individual regular feature of size is controlled by its limits of size. The following requirements are established by Rule #1:

- The surface or surfaces of a regular feature of size shall not extend beyond a boundary of perfect form at MMC.

- Where a regular feature of size is produced at MMC, it must have perfect form.

- Where the actual local size of a regular feature of size is not at MMC, a local variation in form is allowed equal to the amount of departure from MMC.

- Rule #1 only applies to regular features of size.

---

## Where Rule #1 Applies

Where the ASME Y14.5 standard applies to a drawing, Rule #1 automatically applies to all individual regular features of size.

Figure 7-8 contains several regular features of size. The 70 dia dimension is an irregular feature of size (type A), so Rule #1 does not apply. The 150° ±1° dimension is an angular dimension, not a feature of size, so Rule #1 does not apply.

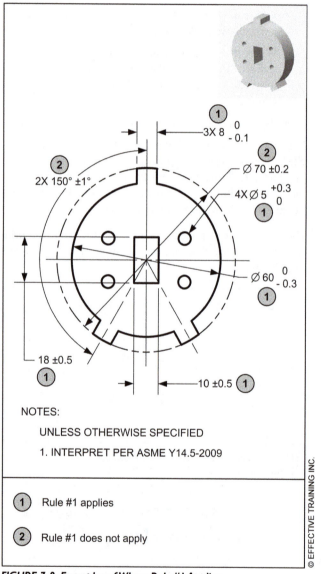

FIGURE 7-8  *Examples of Where Rule #1 Applies*

---

**TECHNOTE 7-2**
**Rule #1**

- Rule #1 applies to individual *regular* features of size.
- Rule #1 does not apply to *irregular* features of size.

## Rule #1 Boundary

The Rule #1 boundary applied to a regular feature of size is the same shape of the true geometric form shown on the drawing. Based on the true geometric form of a regular feature of size, the boundary will be one of the following five shapes:

- a cylinder
- a sphere
- a circle
- two parallel planes
- two parallel lines

The boundary applies to the entire length, width, and depth of a regular feature of size. Rule #1 protects the basic but essential function of assembly by ensuring that local variations do not violate the boundary of perfect form at MMC.

## Rule #1 and the Interrelationship Between Features of Size

A part often contains many features of size. Rule #1 controls the form of each regular feature of size but does not affect the location, orientation, or relationship between features of size. Features of size shown perpendicular, symmetrical, or coaxial must be controlled for location and orientation to avoid incomplete drawing specifications. Often implied 90° angles are toleranced by a general angular tolerance note or by an angular tolerance in the drawing title block.

In Figure 7-9, the dimensions shown are all regular features of size. The orientations between these features of size are labeled A, B, and C. Rule #1 does not affect the variation of these angles.

There is a coaxial relationship shown between diameters labeled "D" and "E." Rule #1 does not affect the coaxiality between these diameters. A geometric tolerance such as a position tolerance would need to be specified to control this relationship.

---

### TECHNOTE 7-3
### Rule #1 Limitation

Rule #1 does not control the location, orientation, or alignment between features of size.

---

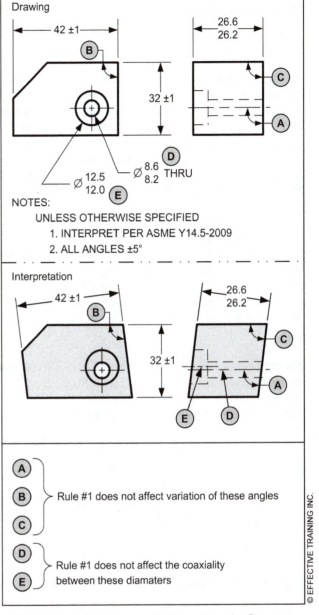

**FIGURE 7-9  Rule #1's Effects on the Interrelationship Between Features of Size**

## Exceptions to Rule #1

There are two exceptions to Rule #1:

- Nonrigid parts
- Stock sizes

Nonrigid parts are not subject to Rule #1 when in the unrestrained condition, because they cannot hold their form without restraint. Examples include o-rings, belts, gaskets and other nonrigid parts that are flexible and do not maintain their nominal shape in the free state. A part is typically designated as nonrigid by the specification of a restraint note. Requiring such parts to meet the requirement of perfect form at MMC would be impractical.

Rule #1 does not apply to stock sizes, such as bar stock, tubing, sheet metal, or structural shapes. Stock sizes are specifications that refer to a material size established before the part is in its produced state. Paragraph 2.7.2 of Y14.5 states that, unless geometric tolerances are specified on these surfaces, industry standards for these items govern the surfaces that remain in the as-furnished (purchased) condition on the part.

For example, if a dimension on a stamping is followed by a designation such as "GAGE" or "STOCK SIZE," it refers to the size of the sheet metal as produced from the mill before the part is formed. Therefore, Rule #1 does not apply.

---

**TECHNOTE 7-4**
**Rule #1 Exceptions**

There are two exceptions to Rule #1:
1. A feature of size on a nonrigid part in the unrestrained condition
2. Stock sizes

---

## Three Ways Rule #1 Can Be Overridden

Unless otherwise specified, Rule #1 applies to each regular feature of size on a drawing. There are three ways Rule #1 can be overridden:

- Applying a straightness tolerance to the size or dimension of a cylindrical regular feature of size

- Applying a flatness tolerance to the size dimension of a planar regular feature of size

- Specifying the independency symbol to a regular feature of size dimension

### Overriding Rule #1 With a Straightness or Flatness Tolerance

Where a straightness or flatness tolerance apply to the size dimension of a regular feature of size, the axis or center plane of the regular feature of size is permitted to have form variation when the size dimension is at MMC. Therefore, Rule #1 does not apply.

### Overriding Rule #1 With the Independency Symbol

Where the independency symbol is applied to the size dimension of a regular feature of size, the characteristics of size and form are independent, so Rule #1 does not apply. Examples of how to override Rule #1 with a straightness tolerance, a flatness tolerance, and the independency symbol are shown in Figure 7-10.

---

**TECHNOTE 7-5**
**How to Override Rule #1**

There are three ways to override Rule #1:

1. Applying a straightness tolerance to the size dimension of a cylindrical regular feature of size.

2. Applying a flatness tolerance to the size dimension of a planar regular feature of size.

3. Specifying the independency symbol to a regular feature of size dimension.

---

**Author's Comment**

*Theory vs. reality of inspection:* When inspecting a part dimension, the measurements taken (reality) verify that the part is as close to the theoretical definition as practical. In most cases, the method and number of measurements taken involve judgment by the inspector.

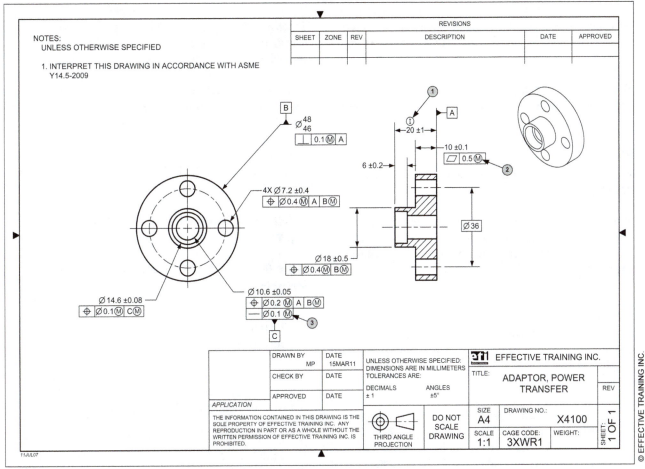

FIGURE 7-10  Three Methods to Override Rule #1 (Labeled 1, 2, and 3)

## The Independency Concept

For features of size that do not have the function of assembly and Rule #1 is not required, the independency concept may be invoked.

> *Independency concept* – size and form are independent; Rule #1 does not apply

The independency concept is invoked by applying the independency symbol to a feature of size dimension. Where the independency symbol is indicated, perfect form at MMC (Rule #1) is not required. An example of the independency concept is shown in Figure 7-11.

**Author's Comment**

The independency concept is new to the Y14.5 standard. The use and interpretation of the independency concept, along with geometric requirements, is not shown in the standard. Therefore, I recommend it be used with caution.

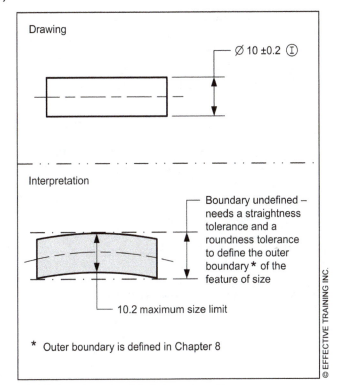

FIGURE 7-11  Example of the Independency Concept

When used as shown in Figure 7-11, all size measurements are taken at cross sections along the feature of size ( a series of actual local size measurements. There is no requirement for an overall measurement.

When the independency principle is used, a form tolerance, such as a straightness or flatness tolerance, should be specified to a surface of the feature of size to define its boundary.

## Rule #2

Rule #2 is a rule about the default conditions in a feature control frame.

*Rule #2* – RFS applies with respect to the individual tolerance, and RMB applies, with respect to the individual datum feature reference, where no modifying symbol is specified.

An example of Rule #2 is shown in Figure 7-12.

RMB, MMB, and LMB are all modifiers that define size conditions for datum features. They are explained in Chapters 15 and 16.

**Author's Comment**
Certain geometric tolerances always apply RFS. The chart in Figure 8-11 shows geometric tolerances that cannot be modified to MMC or LMC.

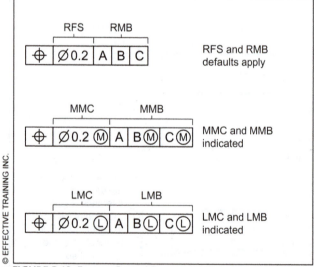

*FIGURE 7-12 Feature Control Frames with Rule #2 and Modifiers Indicated*

Where a feature control frame is applied to a feature of size dimension, MMC or LMC may be applied to the tolerance portion of the feature control frame. RMB or LMB may be applied to the datum references of a feature control frame.

Where a geometric tolerance is applied on an RFS basis, the geometric tolerance is limited to the specified value, regardless, of the size, of the feature of size.

Rule #2 is overridden for any segment of a feature control frame where an MMC, LMC, MMB, or LMB modifier is specified.

## VERIFICATION PRINCIPLES AND METHODS

### Rule #1 and Size Dimension Verification Principles

This section contains a simplified explanation of the verification principles and methods that can be used to inspect a size dimension with Rule #1 applied.

When a size dimension is subject to Rule #1, there are two parts to the verification. The first part is the MMC size and Rule #1 boundary. The second part is the LMC actual local size.

The MMC size and Rule #1 boundary are both verified by ensuring that the feature of size does not violate its MMC perfect form boundary for its full length.

The LMC actual local size is verified by ensuring that each actual local size is not less than its LMC size limit (for an external feature of size), or greater than its LMC size limit (for an internal feature of size).

There are several methods that can be used to verify Rule #1 and a size dimension. In the following sections we will look at using gages. Let's start by introducing several types of gages.

## Fixed-Limit Gage, GO Gage, and NOGO Gage

A *fixed-limit gage* is a device of defined geometric form and size used to assess the conformance of a feature of size of a workpiece to a dimensional specification. A fixed-limit gage does not return a measured value, but instead returns an "accept or reject" indication for a dimension.

A *GO gage* is a fixed-limit gage that checks a feature of size for acceptance within MMC perfect form boundary. A GO gage must be at least as long as the feature of size it is checking to ensure it is verifying the entire MMC boundary. The check involves the full engagement of the part and the gage. Simply put, the gage must go into or around the feature of size to pass the verification. Figure 7-13 shows an example of a GO gage.

**Author's Comment**
All gages have tolerances. However, because gage tolerances are not part of the scope of this text, gage dimensions are shown with no tolerances. For information on gage tolerances, see ASME Y14.43-2011.

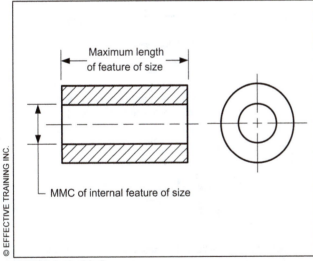

© EFFECTIVE TRAINING INC.

Maximum length of feature of size

MMC of internal feature of size

*FIGURE 7-13  GO Gage*

A *NOGO gage* is a fixed-limit gage that checks a feature of size for violation of its LMC actual local size. The check involves the gage not fitting into or around the feature of size. Since a NOGO gage is checking an actual local size, it is designed to check one cross section of the feature of size at a time and not the entire length at the same time. Figure 7-14 shows an example of a NOGO gage.

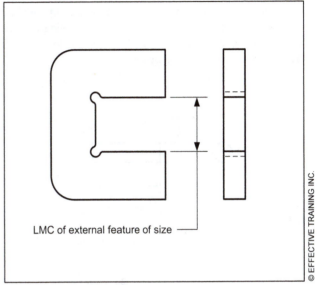

LMC of external feature of size

© EFFECTIVE TRAINING INC.

*FIGURE 7-14  NOGO Gage*

**Author's Comment**
In theory, a NOGO gage should result in a two point contact. In practice, a NOGO gage is often made thick enough for handling and durability.

## Inspecting a Regular Feature of Size Controlled by Rule #1

When inspecting a regular feature of size, both the effects of Rule #1 and the size must be verified. The maximum size and the Rule #1 envelope can be verified with a Go gage.

Figure 7-15 shows examples of a GO gage and a NOGO gage used to verify the LMC actual local size and the MMC envelope of a external regular feature of size. In panel A, the maximum size and Rule #1 boundary are being verified with a GO gage. In panel B, the LMC actual local size is being verified with a NOGO gage.

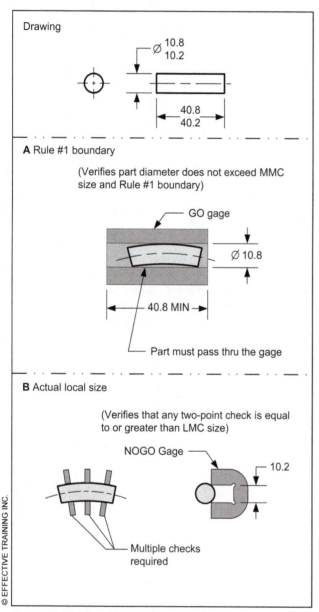

FIGURE 7-15  Inspecting a Feature of Size Using a GO Gage

Figure 7-16 shows examples of a GO gage and a NOGO gage used to verify the LMC actual local size and the MMC envelope of a internal regular feature of size. In panel A, the minimum size and Rule #1 boundary are being verified with a GO gage. In panel B, the LMC actual local size is being verified with a NOGO gage.

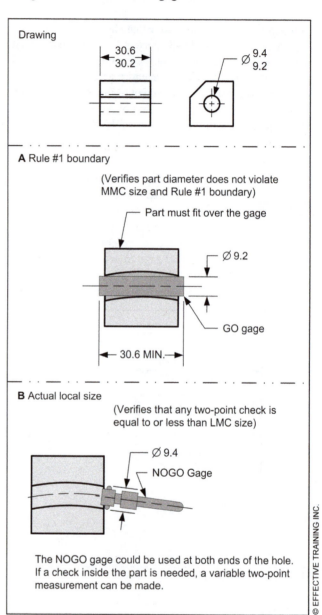

FIGURE 7-16  Inspecting a Feature of Size Using a NOGO Gage

# SUMMARY

## Key Points

- When ASME Y14.5 applies to a drawing, the sixteen fundamental dimensioning rules apply.

- Rule #1 is referred to as the "envelope principle."

- Rule #1 states, "The form of an individual regular feature of size is controlled by its limits of size."

- As a result of Rule #1, where a regular feature of size is produced at MMC, it must have perfect form.

- Rule #1 only applies to regular features of size.

- Rule #1 does not control the location, orientation, or relationship between features of size.

- There are two exceptions to Rule #1, nonrigid parts (in the unrestrained state) and stock sizes.

- There are three ways to override Rule #1:
  o  Applying a straightness tolerance to the size dimension of a cylindrical regular feature of size.
  o  Applying a flatness tolerance to the size dimension of a planar regular feature of size.
  o  Specifying the independency symbol to a regular feature of size dimension.

- When inspecting a regular feature of size both the effects of Rule #1 and the LMC actual local size must be verified.

- A fixed-limit gage is a device of defined geometric form and size used to assess the conformance of a feature of size of a workpiece to a dimensional specification.

- A GO gage is a fixed-limit gage that checks a feature of size for acceptance within MMC perfect form boundary.

- A NOGO gage is a fixed-limit gage that checks a feature of size for violation of LMC actual local size.

- Rule #2 – RFS applies with respect to the individual tolerance, and RMB applies, with respect to the individual datum feature reference, where no modifying symbol is specified.

- Independency concept – size and form are independent; Rule #1 does not apply.

## Additional Related Topics

*These topics are recommended for further study to improve your understanding of GD&T rules.*

| Topic | Source |
|---|---|
| Temperature and humidity environment for dimensional measurement | *ASME B89.6.2-1973 (R2003)* |
| Nonrigid parts | ETI's *Advanced Concepts of GD&T* textbook |

# QUESTIONS AND PROBLEMS

## True and False

*Indicate if each statement is true or false.*

T / F   1. Rule #1 protects the function of assembly.

T / F   2. One of the fundamental dimensioning rules requires all dimensions apply in the free-state condition for rigid parts.

T / F   3. The fundamental dimensioning rules that apply on a drawing must be listed in the general notes.

T / F   4. Where Rule #1 applies to a drawing, it limits the form of every feature of size on the drawing.

T / F   5. Rule #1 limits the variation between features of size on a part.

T / F   6. The designer must specify on the drawing which features of size use Rule #1.

T / F   7. Rule #1 applies to nonrigid parts (in the unrestrained state).

T / F

T / F   8. A GO gage is a fixed-limit gage.

9. Rule #1 requires that the form of an individual regular feature of size is controlled by its limits of size

## Multiple Choice

*Circle the best answer to each statement.*

1.  Why do fundamental dimensioning rules require applicable dimensions and tolerances to be specified on a detail drawing and repeated on an assembly drawing?
    A.  Torque may change the form of a feature
    B.  Heat from welding or brazing operations may change part geometry
    C.  Force (press) fit operations may alter the size of a feature
    D.  All of the above

2.  According to the fundamental dimensioning rules, the temperature for dimensional measurements is _____ .
    A.  20° F
    B.  20° C
    C.  20° K
    D.  Temperature must be specified on the drawing

3.  Rule #1 can be overridden by specifying:
    A.  An independency symbol on a feature of size dimension
    B.  A flatness tolerance on a planar feature of size dimension
    C.  A straightness tolerance on a cylindrical feature of size dimension
    D.  All of the above

4.  When inspecting a feature of size where Rule #1 applies, two items must be inspected, the MMC boundary and the:
    A.  MMC size
    B.  LMC local size
    C.  Minimum size
    D.  Related actual mating envelope

5.  Rule #2 can be summarized as...
    A.  The default material conditions for feature control frames are RFS and RMB
    B.  Every dimension requires a tolerance
    C.  Perfect form at MMC
    D.  The default material conditions for feature control frames are MMC and MMB

6.  Where the independency concept is applied to a feature of size...
    A.  Rule #1 cannot be applied to the same feature of size
    B.  Rule #1 can be applied to the same feature of size
    C.  A fixed-limit gage may be used to verify the independency concept
    D.  The feature of size does not have an MMC

## Application Problems

*The application problems are designed to provide practice on applying the chapter concepts to situations that are similar to on-the-job conditions.*

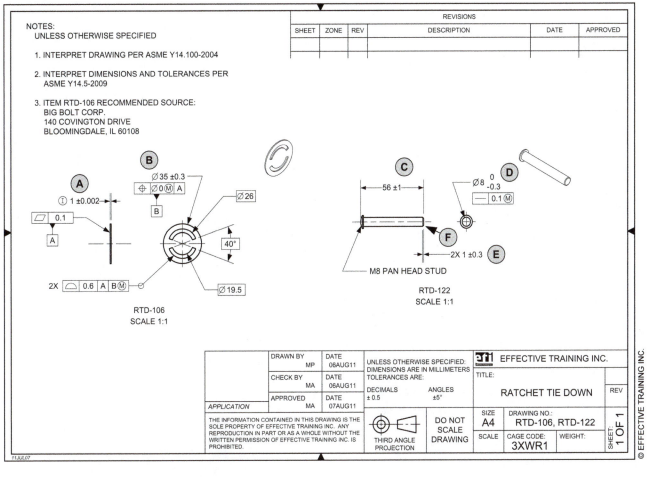

*Application questions 1–5 refer to the drawing above.*

1. Do the fundamental dimensioning rules apply to this drawing?_____ How do you know?_____

2. Which dimensions does Rule #1 apply to on this drawing? _____

3. Describe the Rule #1 boundary for the applicable dimensions. _____

_____

4. What controls the angular relationship between the surface labeled F and the diameter labeled D?_____

5. List the dimensions that have Rule #1 overridden. _____

# GD&T Concepts

## Goal

Understand the concepts of worst-case boundary, virtual condition, and bonus tolerance

## Performance Objectives

Upon completing this chapter, you should be able to:

1. Describe the terms, "worst-case boundary," "inner boundary," and "outer boundary" (p.90)
2. Determine if a geometric tolerance affects the worst-case boundary of a feature of size (p.91)
3. Explain the virtual condition concept and its uses (p.91)
4. Describe the worst-case boundary and virtual condition resulting from form tolerances (p.92)
5. Describe the worst-case boundary and virtual condition resulting from geometric tolerances applied to a feature of size (p.92)
6. State the worst-case boundary formulas (p.94)
7. Describe the effects of multiple geometric tolerances applied to a feature of size (p.94)
8. Explain the concept of bonus tolerance and when it applies (p.95)
9. Calculate the maximum permissible bonus tolerance for a geometric tolerance (p.96)
10. Identify which geometric tolerances get a bonus tolerance (p.96)
11. Describe the verification principles and methods for virtual condition (p.97)

## New Terms

- Bonus tolerance
- Functional gage
- Inner boundary
- Outer boundary
- Virtual condition (VC)
- Worst-case boundary

## What This Chapter Is About

When designing a part, it is important to be able to calculate the worst-case or extreme boundary for surfaces of features of size. Understanding the worst-case boundary enables the designer to design mating parts and gages, and also to make tolerance stacks and other engineering calculations. Depending upon the dimensions indicated, there are several names for worst-case boundaries.

This chapter introduces several worst-case boundary concepts. The terms, "inner boundary" and "outer boundary" are introduced. The concepts of virtual condition and bonus tolerance are covered, and the verification principles for virtual condition acceptance boundaries are discussed.

When designing parts, it is important to be able to calculate the boundary conditions of surfaces of features of size and any additional tolerance (bonus) permissible. Virtual condition and bonus tolerance are important concepts because, where properly applied, they protect the function of assembly and provide additional tolerances for manufacturing.

# TERMS AND CONCEPTS

## Worst-Case, Inner, and Outer Boundaries

It is important to be able to calculate the boundary condition for a feature of size for several reasons: ensure that mating parts will assemble, design gages, make engineering calculations, etc. The boundary concepts in this section are explained using examples of how boundaries affect assembly.

A *worst-case boundary* is a general term used to refer to an inner boundary, outer boundary, or virtual condition. The worst-case boundary affects how parts assemble. Figure 8-1 shows an example.

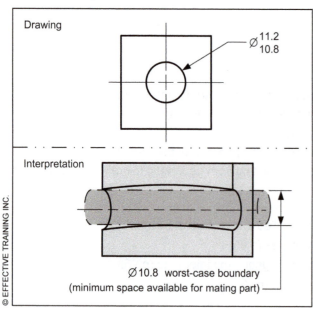

*FIGURE 8-1 Worst-Case Boundary*

There are several specific terms used to describe boundary conditions based on the tolerancing applied to a feature of size: inner boundary, outer boundary, and virtual condition.

*Inner boundary* (of an internal feature of size) is a worst-case boundary generated by the smallest feature of size (MMC) minus the effects of the applicable geometric tolerance and any additional tolerance (bonus) that may apply.

*Inner boundary* (of an external feature of size) is a worst-case boundary generated by the smallest feature of size (LMC) minus the effects of the applicable geometric tolerance and additional tolerances (bonus) that may apply.

*Outer boundary* (of an external feature of size) is a worst-case boundary generated by the largest feature of size (MMC) plus the effects of the applicable geometric tolerance.

*Outer boundary* (of an internal feature of size) is a worst-case boundary generated by the largest feature of size (LMC) plus effects of the applicable geometric tolerance and additional tolerances (bonus) that may apply.

**Author's Comment**

In this text, inner boundary will be used with internal features of size, and outer boundary will be used with external features of size.

Depending upon the geometric tolerances applied to a feature of size, its worst-case boundary may be referred to as an outer (or inner) boundary or a virtual condition. Where no geometric tolerances are applied to an external feature of size, its worst-case outer boundary is its Rule #1 MMC boundary. See Figure 8-2.

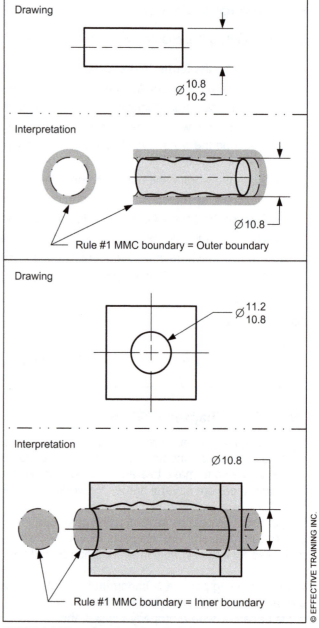

*FIGURE 8-2 Worst-Case Boundaries Without Geometric Tolerances*

Where no geometric tolerances are applied to an internal feature of size, its worst-case inner boundary is equal to its Rule #1 MMC boundary.

## Effects of a Geometric Tolerance on Worst-Case Boundary

It is important to recognize where a geometric tolerance does or does not affect a worst-case boundary. The main criterion for determining if a worst-case boundary is affected is whether a geometric tolerance is applied to a feature or a feature of size.

In most cases, where a feature control frame is associated with a feature (surface), it does not affect the worst-case boundary. However, where it is associated with a feature of size (placed next to or below a size dimension), it does affect the worst-case boundary. Figure 8-3 shows examples of a feature control frame applied to a feature and to a feature of size dimension.

## Virtual Condition and Its Uses

*Virtual condition (VC)* is a fixed-size boundary generated by the collective effects of a considered feature of size's specified MMC or LMC and the geometric tolerance for that material condition. VC boundaries are often used as the acceptance boundaries for verifying geometric tolerances.

### Author's Comment
In this text, we will learn about virtual condition using the MMC modifier. Virtual condition using the LMC modifier is beyond the scope of this text.

The virtual condition at MMC of an external feature of size is a fixed-size (worst-case) boundary generated by the collective effects of the largest feature of size (MMC) plus the applicable geometric tolerance.

The virtual condition at MMC of an internal feature of size is a fixed-size (worst-case) boundary generated by the collective effects of the smallest feature of size (MMC) minus the applicable geometric tolerance.

The concept of virtual condition is useful for:

1. Establishing boundaries to ensure assembly
2. Allowing bonus tolerances
3. Establishing sizes for functional gages
4. Establishing acceptance boundaries for inspection

### Design Tip
In most cases, a virtual condition is the acceptance boundary for verifying the geometric tolerance.

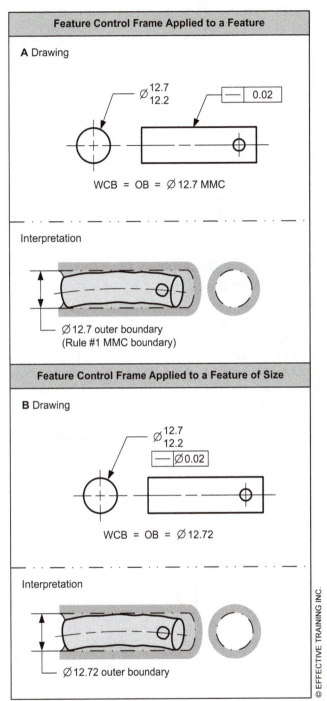

**FIGURE 8-3** *Feature Control Frame Placement Effect on Worst-Case Boundary*

### TECHNOTE 8-1
### Worst-Case Boundary

In most cases, where a feature control frame is applied to a feature (a surface), it does not affect the worst-case boundary of the associated feature of size. If it applies to the size dimension of a feature of size, it does affect the worst-case boundary.

*Author's Comment*
Remember, in addition to verifying the boundary of a feature of size, its size must also be within the size dimension limits.

The modifier specified in the tolerance portion of the feature control frame determines which boundary term should be used. If the MMC modifier is specified, the worst-case boundary is referred to as the virtual condition. If the LMC modifier is specified, the worst-case boundary is referred to as the LMC virtual condition (acceptance boundary). If no modifier is specified, the worst-case boundary is referred to as an inner or outer boundary.

## Worst-Case Boundaries That Result From Form Tolerances

Let's look at the case where a straightness tolerance is applied to an external feature of size dimension. A straightness tolerance overrides Rule #1; therefore, only one worst-case boundary exists based on the effects of the straightness tolerance. An example is shown in Figure 8-4. In this example, the virtual condition (acceptance boundary) is established by adding the straightness tolerance value to the MMC of the size dimension. This boundary could be used to calculate the dimensions of the mating part for assembly.

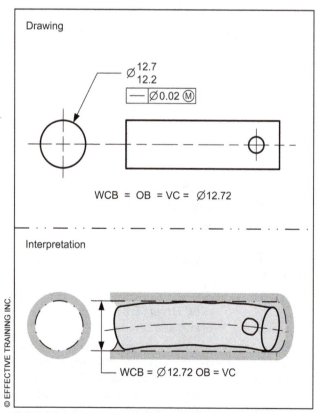

FIGURE 8-4 Worst-case Boundary Resulting From a Form Tolerance Applied to a Size Dimension

## Worst-Case Boundaries That Result From Geometric Tolerances Applied to a Feature of Size

Where a geometric tolerance is applied to a feature of size, it does not override Rule #1. Therefore, two worst-case boundaries exist: the Rule #1 MMC boundary, and the worst-case boundary resulting from the effects of the geometric tolerance. The worst-case boundary generated by a geometric tolerance is constrained relative to the datums referenced in the feature control frame.

Where a geometric tolerance at RFS (default condition per Rule #2) is applied to an external feature of size, it generates a worst-case outer boundary. The outer boundary is calculated by adding the geometric tolerance value to the MMC size limit. An example is shown in Figure 8-5A.

Where a geometric tolerance at RFS is applied to an internal feature of size, it generates a worst-case inner boundary. The inner boundary is calculated by subtracting the geometric tolerance value from the MMC size limit. An example is shown in Figure 8-5B.

Where a geometric tolerance at MMC is applied to an external feature of size, it generates a virtual condition outer boundary. The virtual condition (acceptance boundary) is calculated by adding the geometric tolerance value to the MMC size limit. An example is shown in Figure 8-6A.

Where a geometric tolerance at MMC is applied to an internal feature of size, it generates a virtual condition (inner boundary). The virtual condition (acceptance boundary) is calculated by subtracting the geometric tolerance value from the MMC size limit. An example is shown in Figure 8-6B.

*Author's Comment*
The worst-case boundaries for datum features are referred to as regardless of material boundary (RMB), maximum material boundary (MMB), and least material boundary (LMB). These boundaries are covered in Chapter 15 on size datum features.

Virtual condition boundaries are acceptance boundaries that guarantee mating parts will assemble. Mating features of size are often toleranced so that their virtual condition boundaries are equal for a worst-case zero clearance condition.

The worst-case boundaries in Figure 8-5 and the virtual condition (acceptance boundaries) in Figure 8-6 are used to determine if two features of size, at their worst-case will assemble.

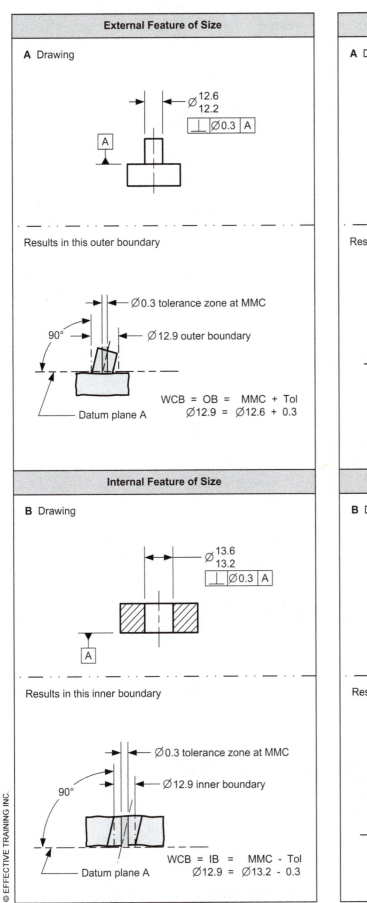

FIGURE 8-5  *Inner and Outer Boundary Examples*

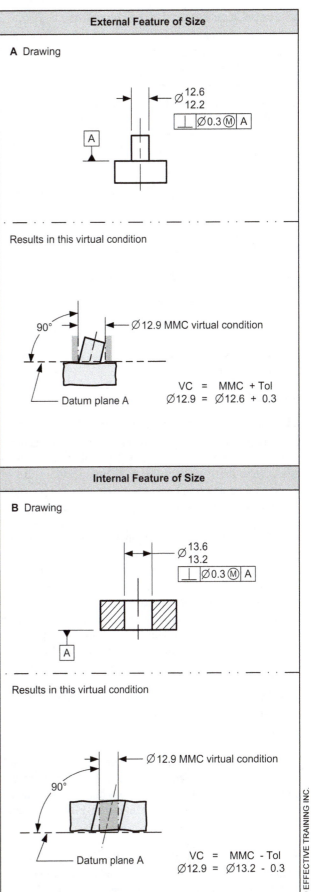

FIGURE 8-6  *Virtual Condition Examples*

## Worst-Case Boundary Formulas

There are several formulas for calculating worst-case boundaries. The chart in Figure 8-7 shows formulas that apply for various feature of size specifications.

| Feature of Size Specification | Internal or External | Formula for Worst-Case Boundary |
|---|---|---|
| Size dimension only | Internal | IB = MMC* |
| | External | OB = MMC* |
| Size dimension with GD&T @ RFS | Internal | IB = MMC - TOL |
| | External | OB = MMC + TOL |
| Size dimension with GD&T @ MMC | Internal | VC = IB = MMC - TOL |
| | External | VC = OB = MMC + TOL |
| Terms | IB = Inner boundary<br>OB = Outer boundary<br>VC = Virtual condition boundary<br>TOL = Geometric tolerance value | |
| * Rule #1 boundary | | |

FIGURE 8-7 Worst-Case Boundary Formulas

## Effects of Multiple Geometric Tolerances

On complex drawings, it is common to have several geometric tolerances applied to a feature of size dimension. Where this occurs, the feature of size will have several different worst-case boundaries. Each worst-case boundary is related to the datums referenced by the geometric tolerance.

Figure 8-8 shows a part with a pin diameter that has a position and perpendicularity tolerance applied to it. In this case, the pin diameter has three worst-case boundaries. First, the Rule #1 MMC boundary which controls the form of the pin. Second, an orientation virtual condition boundary that is constrained in rotational degrees of freedom only relative to the datums referenced. Third, a location virtual condition boundary that is constrained in translation and rotational degrees of freedom relative to the datums referenced.

Panel A shows the 10.2 diameter MMC boundary based on Rule #1. This boundary is not related to any datums and limits the form deviation of the pin diameter.

Panel B shows a 10.3 diameter virtual condition resulting from the perpendicularity tolerance. This boundary is perpendicular to datum plane A, but not related to datum planes B and C. This boundary limits the orientation deviation of the pin.

Panel C shows a 10.4 diameter virtual condition resulting from the position tolerance. This boundary is perpendicular to datum plane A and fixed at the basic location relative to datum planes B and C. This boundary limits the location deviation of the pin.

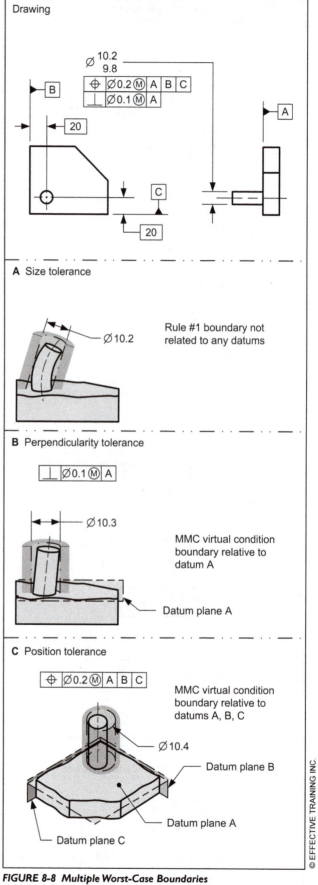

FIGURE 8-8 Multiple Worst-Case Boundaries

- A virtual condition boundary is considered a worst-case boundary

- The size requirement for the feature of size also applies

- A feature of size may have several virtual condition boundaries

- A virtual condition boundary is an acceptance boundary used to verify a geometric tolerance

## Bonus Tolerance

Bonus tolerance is an important concept in dimensioning parts, so it should be understood by every drawing user. Bonus tolerances can significantly reduce costs in manufacturing and inspection.

A *bonus tolerance* is a potential additional tolerance for a geometric tolerance. Where a geometric tolerance at MMC (or LMC) is applied to a feature of size dimension, a bonus tolerance is permissible. Where the MMC modifier is specified, it indicates that the stated geometric tolerance applies when the feature of size is at its MMC size limit. When the actual mating envelope of the feature of size departs from MMC towards LMC, an increase in the stated tolerance (equal to the amount of departure) is permitted. This increase or additional tolerance is called the "bonus tolerance." This bonus tolerance may occur as additional form, location, or orientation tolerance. A bonus tolerance is never additive to the size tolerance.

The term "bonus tolerance" is not officially defined in the Y14.5 standard, but the concept is explained in paragraph 2.8.2. Both the term and concept have been commonly used in industry for more than 40 years. This introduction to bonus tolerance is based on using the MMC modifier.

Here is how the bonus tolerance concept works. As the actual mating envelope of the feature of size departs from MMC towards LMC, the difference (clearance) between the actual mating envelope and the virtual condition boundary increases. As the clearance increases, the feature of size may have additional location, orientation, or form deviations and still not violate its virtual condition.

Because the virtual condition is a fixed size, it can be verified using a functional gage. The term "functional gage" is defined and its use is further explained in the last section of this chapter.

Although a flatness tolerance is used in the example in Figure 8-9, the bonus tolerance concept applies to any geometric tolerance that may use the MMC (or LMC) modifier.

Figure 8-9 shows a 0.2 flatness tolerance at MMC applied to the washer thickness dimension. This results in a 2.7 mm virtual condition (acceptance boundary) that may be verified with a functional gage. As the actual mating envelope of the washer thickness departs from MMC towards LMC, it could have more flatness deviation and still pass through the gage.

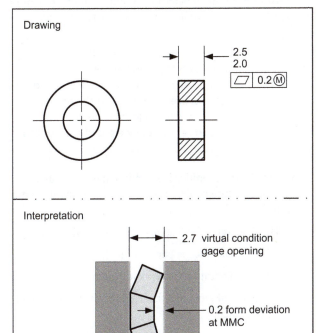

| Flatness Tolerance | Virtual Condition Acceptance Boundary | Actual Local Size | Total Tolerance | Bonus Tolerance |
|---|---|---|---|---|
| 0.2 | 2.7 | 2.5 MMC | 0.2 | 0.0 |
| | | 2.4 | 0.3 | 0.1 |
| | | 2.3 | 0.4 | 0.2 |
| | | 2.2 | 0.5 | 0.3 |
| | | 2.1 | 0.6 | 0.4 |
| | | 2.0 LMC | 0.7 | 0.5 |

NOTE:
Maximum bonus tolerance occurs when the actual mating envelope of the width is at LMC.

**FIGURE 8-9 Bonus Tolerance Example**

**Author's Comment**
A bonus tolerance is never additive to the size tolerance.

## Calculating Maximum Permissible Bonus Tolerance

The maximum amount of bonus tolerance permissible is equal to the difference between MMC and LMC of the actual mating envelope of the toleranced feature of size. Figure 8-10 shows how to determine the maximum amount of bonus tolerance permissible for a geometric tolerance.

**Author's Comment**
In addition to assembly, the MMC modifier may also be used is with features of size that are not critical to the function of the part, (e.g., clearance holes, holes used to reduce the weight of a part, etc.).

Where the MMC modifier is specified in the tolerance portion of a feature control frame, four items should come to mind:

- The function of the feature of size is usually assembly.
- A bonus tolerance is permissible.
- A virtual condition (acceptance boundary) exists.
- A functional gage may be used to verify the geometric tolerance.

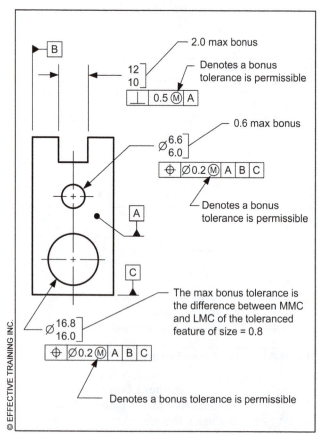

*FIGURE 8-10  Maximum Permissible Bonus Tolerance*

## Geometric Tolerances That Can Use Bonus Tolerance Concept

Not all geometric tolerances may use the bonus tolerance concept; only those that can be applied to a feature of size and specified at MMC (or LMC) may do so. Figure 8-11 shows which geometric tolerances may or may not be specified at MMC (or LMC) when applied to a feature of size.

| Geometric Tolerance Applied to a Feature of Size | May Be Applied to a Feature of Size | May Use MMC or LMC Resulting in Bonus Tolerance |
|---|---|---|
| Flatness | Yes | Yes |
| Straightness | | |
| Perpendicularity | | |
| Angularity | | |
| Parallelism | | |
| Position | | |
| Profile of surface* | | No |
| Profile of line* | | |
| Circularity | No | |
| Cylindricity | | |
| Symmetry | | |
| Concentricity | | |
| Circular runout** | | |
| Total runout** | | |

\* Profile tolerances can be used to define an irregular feature of size, but they cannot be used to apply to an axis, derived median line, center plane, or derived median plane of a feature of size.

\*\* Runout tolerances are an exception. They are shown in the standard, and on many drawings as applied to the size dimension of a feature of size, but they are actually a surface tolerance.

*FIGURE 8-11  Geometric Tolerances That May Result in a Bonus Tolerance*

## TECHNOTE 8-3
### Bonus Tolerance

- Bonus tolerance is an additional tolerance for a geometric tolerance

- Bonus tolerance is only permissible when an MMC (or LMC) modifier is indicated in a geometric tolerance.

- The actual amount of bonus tolerance is the amount the actual mating envelope departs from MMC (or LMC).

- The maximum amount of bonus tolerance allowed is equal to the size tolerance of the feature of size (MMC-LMC).

# VERIFICATION PRINCIPLES AND METHODS

## Virtual Condition Verification Principles

This section provides a simplified description of the verification principles and methods used to verify a virtual condition (acceptance boundary).

Verifying a VC (acceptance boundary) consists of two parts:

1. Establishing the relationship specified in the feature control frame between the part and the datum reference frame (if one is specified).

2. Verifying that the surface(s) of the feature of size does not violate the VC (acceptance boundary).

Where a geometric tolerance applies at MMC, a virtual condition acceptance boundary exists, and the functional parameter is usually assembly. If the surface of the toleranced feature of size does not violate its acceptance boundary, the feature of size will assemble.

If the surface of the toleranced feature of size does not violate the VC (acceptance boundary), the feature of size passes. If the surface of the toleranced feature of size violates the acceptance boundary, the feature of size does not pass.

The surface(s) of an external feature of size can be verified to its VC (acceptance boundary) by passing the feature of size into the acceptance boundary. The surface(s) of an internal feature of size can be verified to its VC (acceptance boundary) by passing the feature of size over (around) its acceptance boundary.

In the following sections, we will look at using a functional gage to verify that features of size do not violate their virtual condition acceptance boundaries. Let's start with an explanation of a functional gage.

Because the VC (acceptance boundary) is a fixed-size boundary, it can be verified using a functional gage. A *functional gage* is a fixed-limit gage used to verify virtual condition acceptance boundaries.

An external feature of size must fit into an internal fixed-size gage element. An internal feature of size must fit around an external fixed-size gage element. The gage elements are made to the virtual condition size of the feature of size. A functional gage is sometimes referred to as an attribute gage because it only provides attribute data (i.e., pass/fail, go/nogo, or fit/does not fit), not variable (measured value) data. An example of a functional gage is shown in Figure 8-12.

**Author's Comment**
Where a geometric tolerance applies at MMC (or LMC), it may be visualized with an axis/center plane or surface (VC boundary) interpretation. Both interpretations are valid; however, the standard indicates that where the two interpretations do not agree, the surface (VC boundary) interpretation takes precedence. See Appendix F.

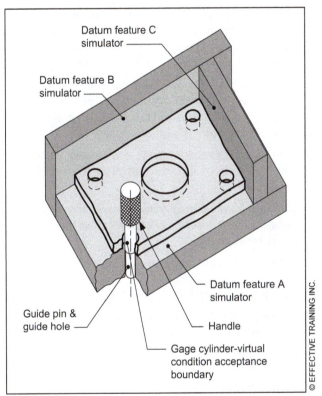

Datum feature C simulator

Datum feature B simulator

Datum feature A simulator

Handle

Guide pin & guide hole

Gage cylinder-virtual condition acceptance boundary

© EFFECTIVE TRAINING INC.

*FIGURE 8-12 Functional Gage for Verifying the Virtual Condition for the Location of a Hole*

**Author's Comment**
The drawings of functional gages in this text are simplified drawings. They show the boundaries and datum feature simulators (where applicable), but since they are simplified, they do not represent the actual gage construction or gage tolerances.

# SUMMARY

## Key Points

- The worst-case boundary of a feature of size is a general term used to refer to an inner boundary, outer boundary, or virtual condition.

- The inner boundary (of an internal feature of size) is a worst-case boundary generated by the smallest feature of size (MMC) minus the effects of the applicable geometric tolerance.

- Outer boundary (of an external feature of size) is a worst-case boundary generated by the largest feature of size (MMC) plus the effects of the applicable geometric tolerance.

- Virtual condition (of an external feature of size) is a fixed-size (worst-case) boundary generated by the collective effects of the largest feature of size (MMC) plus the applicable geometric tolerance.

- Virtual condition (of an internal feature of size) is a fixed-size (worst-case) boundary generated by the collective effects of the smallest feature of size (MMC) minus the applicable geometric tolerance.

- It is important to be able to calculate the boundary condition for a feature of size for several reasons: ensuring mating parts will assemble, designing gages, making engineering calculations, etc.

- When a feature control frame is associated with a feature (surface), it does not affect the worst-case boundary.

- When a feature control frame is associated with a size dimension (feature of size), it does affect the worst-case boundary.

- If a feature of size dimension does not have any geometric tolerances applied to it, the worst-case boundary is the Rule #1 MMC boundary.

- Where a straightness or flatness tolerance is applied to a feature of size dimension, Rule #1 is overridden and the worst-case boundary includes the effects of the geometric tolerance.

- A feature of size may have multiple worst-case boundaries relative to different datum reference frames.

- A bonus tolerance is an additional tolerance for a geometric tolerance.

- Bonus tolerance is only permissible when an MMC (or LMC) modifier is indicated in a geometric tolerance.

- The actual amount of bonus tolerance is the amount the actual mating envelope departs from MMC (or LMC).

- A functional gage is a fixed limit gage used to verify virtual condition (acceptance boundaries).

- Where the MMC modifier is specified in a feature control frame, four items should come to mind:
  o The function of the feature of size is usually assembly.
  o A bonus tolerance is permissible.
  o A virtual condition (acceptance boundary) exists
  o A functional gage may be used to verify the geometric tolerance.

- In addition to assembly, the MMC modifier may also be used with features of size that are not critical to the function of the part (e.g., clearance holes, holes used to reduce the weight of the part).

## Additional Related Topics

*These topics are recommended for further study to improve your understanding of GD&T concepts.*

| Topic | Source |
|---|---|
| LMC virtual condition | *ASME Y14.5-2009, Para. 2.7.5* |
| Resulting condition boundary | *ASME Y14.5-2009, Para. 2.11* |
| Non-rigid parts | ETI's *Advanced Concepts of GD&T* textbook |

## QUESTIONS AND PROBLEMS

### Website Bonus Materials
Additional questions are available at our website. To access bonus materials for this textbook, please visit:

www.etinews.com/textbookbonus

www.etinews.com

### True and False

*Indicate if each statement is true or false.*

T / F    1. A virtual condition is a fixed-size boundary.

T / F    2. Where a geometric tolerance applies to a feature of size, it affects the worst-case boundary.

T / F    3. Where a geometric tolerance is specified with a tolerance at MMC, a virtual condition boundary exists.

T / F    4. A virtual condition (acceptance boundary) is often verified with a variable gage.

T / F    5. A bonus tolerance is permissible for all geometric tolerances.

T / F    6. The amount of bonus tolerance comes from the feature control frame tolerance value.

### Multiple Choice

*Circle the best answer to each statement.*

1. Which geometric tolerance can permit a bonus tolerance?
   A. Profile of a line
   B. Total runout
   C. Angularity
   D. Symmetry

2. To establish a virtual condition (acceptance boundary) _____ is required.
   A. An Ⓜ symbol in the tolerance portion of a feature control frame
   B. An Ⓜ symbol in the datum portion of a feature control frame
   C. A ⌀ symbol in the tolerance portion of a feature control frame
   D. All of the above

3. The concept of virtual condition (acceptance boundary) is useful for:
   A. Establishing boundaries to ensure assembly
   B. Allowing bonus tolerances
   C. Establishing sizes for functional gages
   D. All of the above

4. Where no geometric tolerances are specified, the worst-case boundary of a hole is referred to as:
   A. An inner boundary
   B. A nominal boundary
   C. A virtual condition
   D. A related actual mating envelope

5. The virtual condition of _____ feature of size is a constant boundary generated by the collective effects of the largest feature of size plus the applicable geometric tolerance.
   A. An internal
   B. An external
   C. A nominal
   D. All of the above

6. A worst-case boundary may be equal to a_____ boundary.
   A. Virtual condition
   B. Outer
   C. Inner
   D. All of the above

## Application Problems

*The application problems are designed to provide practice on applying the chapter concepts to situations that are similar to on-the-job conditions.*

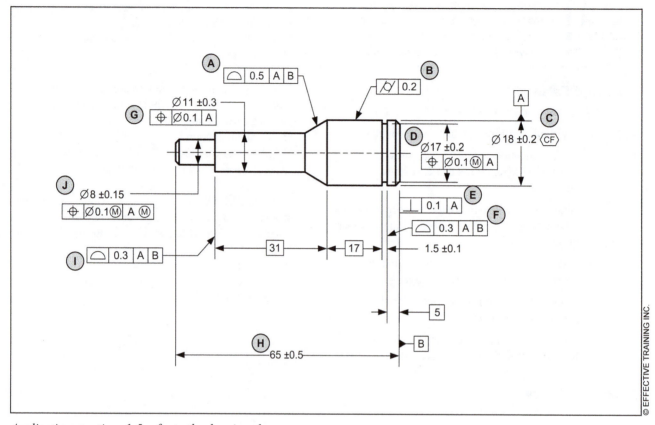

*Application questions 1-5 refer to the drawing above.*

1.  What is the worst-case boundary of the dimension labeled "H?" _____

    What type of boundary is this? _____

2.  What is the worst-case boundary of the dimension labeled "C?" _____

    What type of boundary is this? _____

3.  What is the worst-case boundary of the dimension labeled "D?" _____

    What type of boundary is this? _____

4.  How many worst-case boundaries are there for the dimension labeled "J?" _____

    Name the type(s). _____

5.  Calculate the amount of bonus tolerance permissible for the position tolerance applied to the dimension labeled "J."

    _____

# Flatness Tolerance

## Goal

Interpret the flatness tolerance

## Performance Objectives

Upon completing this chapter, you should be able to:

1. Describe the terms "flatness," "derived median plane," and "flatness tolerance" (p 102)
2. Describe the tolerance zones for a flatness tolerance (p 103)
3. Describe Rule #1's effects on flatness deviation (p 104)
4. Describe the modifiers that may be used in a flatness tolerance (p 105)
5. Recognize when a flatness tolerance is applied to a planar surface or a feature of size (p 105)
6. Evaluate if a flatness tolerance specification is standard-compliant (p 106)
7. Describe real-world applications of a flatness tolerance (p 107)
8. Interpret a flatness tolerance applied to a planar surface (p 107)
9. Interpret a flatness tolerance at RFS applied to a feature of size dimension (p 108)
10. Interpret a flatness tolerance at MMC applied to a feature of size dimension (p 108)
11. Interpret a flatness tolerance by using the "Significant Seven Questions" (p 109)
12. Understand flatness verification principles (p 110)
13. Describe how to verify a flatness tolerance applied to a planar surface (p 110)
14. Describe how to verify a flatness tolerance (at MMC) applied to a feature of size (p 111)

## New Terms

- Derived median plane
- Dial indicator
- Flatness
- Flatness tolerance
- Surface plate

## What This Chapter Is About

This chapter explains how to interpret the flatness tolerance, where to use it, and how to inspect it.

The flatness tolerance is one of the fourteen geometric characteristic symbols and one of the four direct form tolerances. A flatness tolerance can limit the flatness of a planar surface or ensure a virtual condition boundary for a planar feature of size.

Flatness tolerances are often used to ensure good joint design by defining the allowable flatness deviation of a planar surface. Flatness tolerances can also be used to define the allowable flatness deviation of a derived median plane of a feature size to ensure assembly.

## TERMS AND CONCEPTS

### Flatness

Y14.5 defines flatness as a perfect plane. *Flatness* is the condition of a surface or a derived median plane having all of its elements in one plane. An example of flatness of a surface is shown in Figure 9-1.

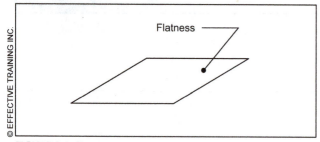

FIGURE 9-1  Flatness

### Derived Median Plane

A *derived median plane* is an imperfect plane formed by the center points of all line segments bounded by the feature of size. These line segments are perpendicular to the center plane of the unrelated actual mating envelope of the feature of size. An example of a derived median plane is shown in Figure 9-2

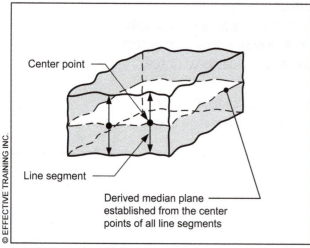

FIGURE 9-2 Derived Median Plane

The condition of flatness applies to a planar surface (feature) or to a feature of size dimension (derived median plane). For a part surface to meet the condition of flatness, it must have all of its elements in one plane. For a feature of size dimension to meet the condition of flatness, it must have all of the points of its derived median plane in one plane. However, because flatness (the perfect condition) cannot be produced, it is necessary to specify some allowance for flatness deviation.

### Flatness Tolerance

A *flatness tolerance* is a geometric tolerance that defines the allowable flatness deviation permitted on a planar surface or derived median plane.

A flatness tolerance is categorized as a form control. A flatness tolerance:

- Is one of the four direct form tolerances
- Never uses a datum reference
- Can only be applied to a single planar feature or a single planar regular feature of size
- Zone is the space between two parallel planes

An example of a flatness tolerance is shown in Figure 9-3.

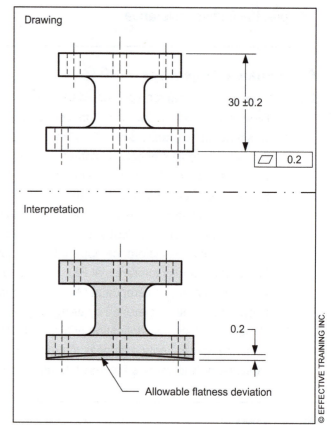

FIGURE 9-3  Flatness Tolerance Example

**Author's Comment**
In the previous versions of the Y14.5 standard, straightness at MMC was applied to planar features of size to create a boundary larger than Rule #1. This practice no longer applies. Now a flatness tolerance at MMC is applied to planar features of size to create a boundary larger than Rule #1.

## Flatness Tolerance Zones

The tolerance zone for a flatness tolerance is the space between two parallel planes. The distance between the planes is equal to the flatness tolerance value. If a flatness tolerance is applied to a surface, the high and low points of the surface must be within the tolerance zone. If a flatness tolerance is applied to a feature of size at RFS, median plane must be within the tolerance zone.

If a flatness tolerance is applied to a feature of size at MMC, the tolerance zone can be viewed two ways:

- Two parallel planes within which the derived median plane must be located

- As a VC boundary that the feature of size may not violate

In this text, we will use the surface interpretation (VC boundary). Appendix F explains why the surface interpretation for tolerances applied at MMC is preferred.

Figure 9-4 shows examples of flatness tolerance zones.

A part surface may have any type of form deviation within a flatness tolerance zone. Two typical form deviations are waviness and warping.

## Rule #1's Effects on Flatness Deviation of a Surface

Wherever Rule #1 applies to a planar regular feature of size, it invokes an indirect flatness deviation requirement on both surfaces. This requirement is a result of the interrelationship between Rule #1 (perfect form at MMC) and the size dimension.

Where the feature of size is at MMC, both surfaces must be perfectly flat. As the size dimension departs from MMC, a flatness deviation equal to the amount of departure is permitted.

Since Rule #1 provides an indirect flatness tolerance, a flatness tolerance should not be specified unless it is a refinement of the size tolerance. Figure 9-5 shows an example of how Rule #1 affects the flatness on a surface.

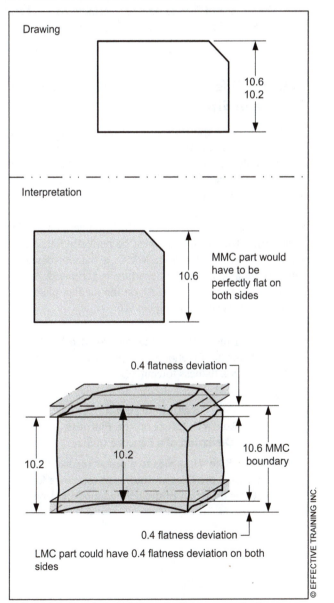

*FIGURE 9-5 How Rule #1 Affects the Flatness on a Surface*

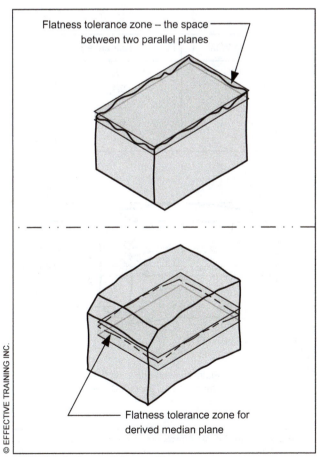

*FIGURE 9-4 Flatness Tolerance Zones*

© EFFECTIVE TRAINING INC.

---

## TECHNOTE 9-1
### How Rule #1 Affects Flatness of a Surface

- Wherever Rule #1 applies to a planar regular feature of size, it invokes an indirect flatness deviation requirement on both surfaces.
- The amount of flatness deviation permitted is equal to the limits of the size dimension (size tolerance).

## Rule #1's Effects on Flatness Deviation of a Feature of Size

Wherever Rule #1 applies to a planar regular feature of size, an automatic flatness deviation requirement exists for the derived median plane of the feature of size. This automatic requirement is a result of the interrelationship between Rule #1 and the size dimension.

Where the feature of size is at MMC, the derived median plane of the feature of size, must be perfectly flat. As the feature of size departs from MMC, a flatness deviation equal to the amount of the departure is permitted. An example of the effects of Rule #1 on the median plane of a feature of size is shown in Figure 9-6.

If the limit of the flatness deviation provided by Rule #1 is sufficient for the function of the application, there is no need to specify a flatness tolerance.

## TECHNOTE 9-2
### Rule #1's Effect on the Flatness Deviation of a Feature of Size

Wherever Rule #1 applies to a planar regular feature of size, it provides an automatic tolerance for the flatness deviation of the derived median plane.

Drawing

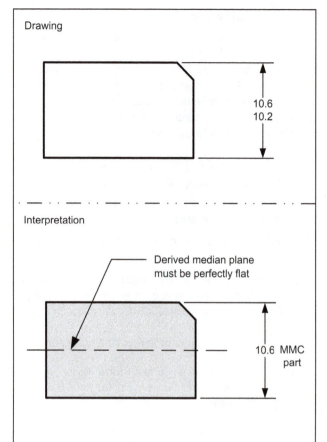

10.6
10.2

Interpretation

Derived median plane must be perfectly flat

10.6 MMC part

| Part Width | Flatness Tolerance Zone |
|---|---|
| 10.6 MMC | 0 |
| 10.5 | 0.1 |
| 10.4 | 0.2 |
| 10.3 | 0.3 |
| 10.2 LMC | 0.4 |

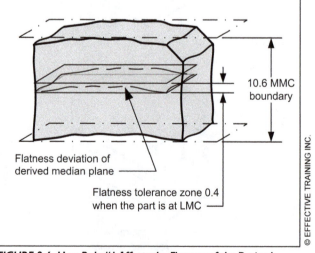

10.6 MMC boundary

Flatness deviation of derived median plane

Flatness tolerance zone 0.4 when the part is at LMC

© EFFECTIVE TRAINING INC.

**FIGURE 9-6  How Rule #1 Affects the Flatness of the Derived Median Plane**

## Modifiers Used With Flatness Tolerances

A flatness tolerance has certain requirements that must be followed. The chart in Figure 9-7 shows the requirements of a flatness tolerance application.

| Tolerance Modifier | Can Be Applied to | Effect | Functional Application |
|---|---|---|---|
| * (F) | Feature or feature of size | Release the restraint requirement | Nonrigid parts with restraint notes |
| * (ST) | Feature or feature of size | Requires statistical process controls | Statistically derived tolerances or tolerances used in statistical tolerance analyses |
| (M) | Feature of size | Permits a bonus tolerance<br><br>Permits use of a functional gage | Assembly |
| * (L) | Feature of size | Permits a bonus tolerance<br><br>Requires variable gaging | Minimum distance |

* These modifiers are only introduced in this textbook.

FIGURE 9-7  Modifiers Used With Flatness Tolerances

## Determining Whether a Flatness Tolerance Is Applied to a Planar Surface or a Feature of Size

It is important to be able to determine whether a flatness tolerance applies to a surface or to a feature of size dimension, because the interpretation for each is different. Figure 9-8 shows an example of a flatness tolerance applied to a surface and a feature of size.

The location of the flatness tolerance indicates what it is applied to. Where the leader line of a flatness tolerance is directed to a surface, or attached to an extension line from the surface, it indicates that it apples to the feature (surface). Where a flatness tolerance is located beneath or beside a size dimension, it indicates that it applies to the feature of size.

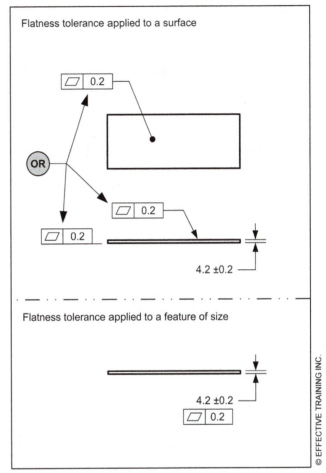

FIGURE 9-8  Flatness Tolerance Applied to a Surface and a Feature of Size

## Evaluate if a Flatness Tolerance Specification Is Standard-Compliant

For a flatness tolerance to be a standard-compliant specification (legal), it must satisfy certain requirements. In order to make it easier to remember what to look for when evaluating the correctness of a geometric tolerance on a drawing, I created a mnemonic. The requirements are divided into four areas represented by the initials "CARE."

CARE stands for:

C - Check for datums

A - Assess the application

R - Review the modifiers

E - Evaluate the tolerance value

A flowchart that identifies the requirements for a standard-compliant flatness tolerance is shown in Figure 9-9.

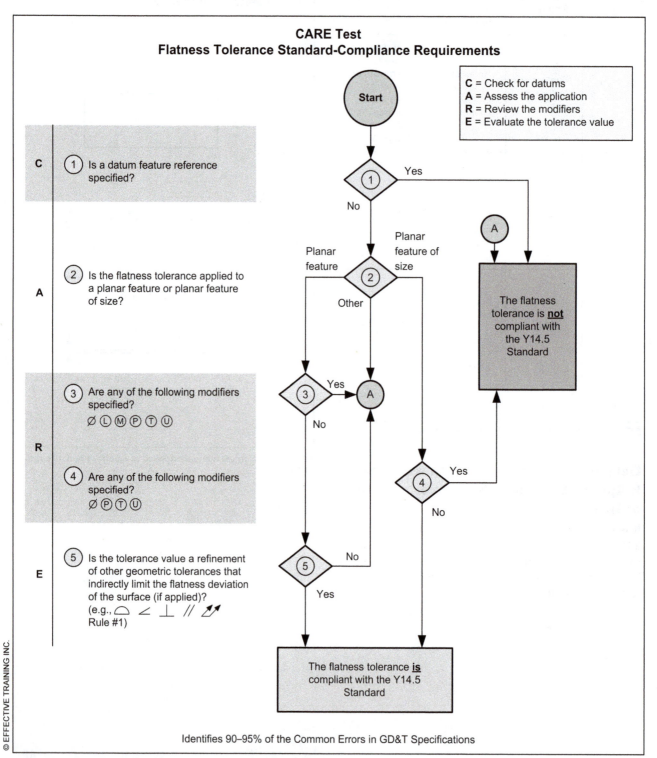

**FIGURE 9-9 Test For a Standard-Compliant Flatness Specification**

# FLATNESS APPLICATIONS

## Real-World Applications of a Flatness Tolerance

Common real-world applications of a flatness tolerances are:

- Sealing (gasket joint)
- Support (distribute loads or forces)
- Appearance (customer appeal)
- Measurement repeatability (qualify a surface for making more repeatable measurements related to it)
- Assembly (mating planar features of size)
- Reducing manufacturing costs on size dimensions by overriding Rule #1 and allowing the feature of size to have more flatness deviation than its size limits allow

## Flatness Tolerance Applied to a Planar Surface

Figure 9-10 shows an application of flatness applied to a planar surface. Where a flatness tolerance is applied to a surface, the following conditions apply:

- The tolerance zone is the space between two parallel planes.
- The tolerance zone may be oriented to best fit the high and low points of the surface.
- All surface elements of the controlled surface must be within this zone.
- The distance between the parallel planes is equal to the flatness tolerance value.
- The tolerance zone extends across the length and width of the surface.

Where a flatness tolerance is applied to a surface that is associated with a size dimension, and Rule #1 applies to the size dimension, the flatness tolerance value must be less than the size tolerance. Where a flatness tolerance is applied to a surface that is associated with a size dimension, and Rule #1 does not apply to the size dimension (i.e., overridden, independency, stock size, etc.), the flatness tolerance value may be larger than the size tolerance.

In Figure 9-10, there are three requirements that must be met. The size dimension requires that the actual local sizes and the Rule #1 boundary be verified. The flatness tolerance requires that the flatness deviation on the bottom surface be verified.

Drawing

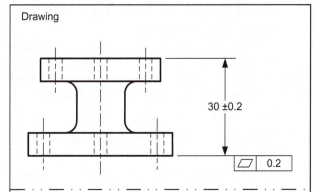

30 ±0.2

⌭ 0.2

**THE SIZE REQUIREMENT**
Each two-point measurement must be within the dimensional limits

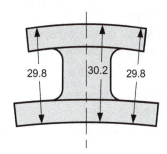

29.8    30.2    29.8

**THE RULE #1 BOUNDARY REQUIREMENT**
The part must be within a 30.2 boundary

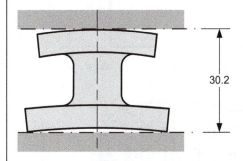

30.2

**THE FLATNESS REQUIREMENT**
All the surface elements must be within two parallel planes 0.2 apart

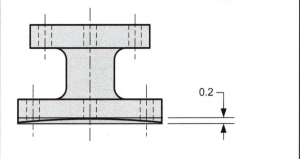

0.2

*FIGURE 9-10  Flatness Applied to a Surface*

## Flatness Tolerance (at RFS) Applied to a Feature of Size Dimension

The interpretation of a flatness tolerance applied to a feature of size is significantly different from a flatness tolerance applied to a surface. Where a flatness tolerance at RFS is applied to a feature of size, the following conditions apply:

- Rule #1 is overridden.

- The flatness tolerance value may be larger than the feature of size dimension tolerance.

- The tolerance zone is the space between two parallel planes.

- The distance between the planes is equal to the flatness tolerance value.

- The tolerance zone extends across the length and width of the feature of size.

- The derived median plane of the toleranced feature of size must be located within the tolerance zone.

- The feature of size must also be within its size limits.

Figure 9-11 shows an example of a flatness tolerance (at RFS) applied to a feature of size.

## Flatness Tolerance (at MMC) Applied to a Feature of Size Dimension

The interpretation of a flatness tolerance applied to a planar feature of size is significantly different from a flatness tolerance applied to a surface. Where a geometric control is applied on an MMC basis a surface interpretation (VC boundary) is often used. In this text, the surface interpretation is illustrated for MMC applications although a center plane interpretation may also be valid.

Where a flatness tolerance at MMC is applied to a feature of size, the following conditions apply:

- Rule #1 is overridden.

- The flatness tolerance value may be larger than the feature of size dimension tolerance.

- A bonus tolerance is permissible.

- Using the surface interpretation, the tolerance zone is the space within the virtual condition.

- The virtual condition size equals:
  - o  External feature of size = MMC + flatness tol value
  - o  Internal feature of size = MMC - flatness tol value

- The flatness of the derived median plane is verified by passing the feature of size through its virtual condition.

- The feature of size must also be within its size limits.

---

**TECHNOTE 9-3**
**Flatness of a Feature of Size**

Wherever a flatness tolerance with the MMC modifier is associated with a feature of size dimension, four things should come to mind:

- Rule #1 is overridden

- A bonus tolerance is permissible

- A functional gage may be used

- The function is likely assembly

---

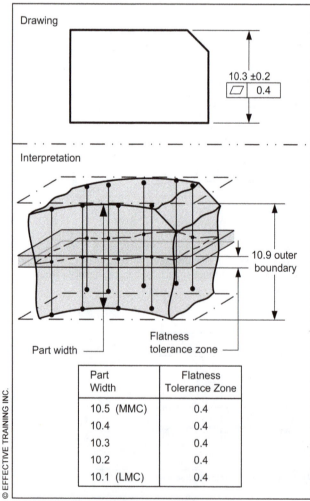

Drawing

10.3 ±0.2

▱ 0.4

Interpretation

10.9 outer boundary

Part width — Flatness tolerance zone

| Part Width | Flatness Tolerance Zone |
|---|---|
| 10.5  (MMC) | 0.4 |
| 10.4 | 0.4 |
| 10.3 | 0.4 |
| 10.2 | 0.4 |
| 10.1  (LMC) | 0.4 |

**FIGURE 9-11  Flatness Tolerance (at RFS) Applied to a Feature of Size**

**Author's Comment**
Although it is permissible to apply a flatness tolerance at LMC or RFS to a feature of size, the use of this condition is rare. This text does not cover this rare use of a flatness tolerance.

Figure 9-12 shows an example of a flatness tolerance (at MMC) applied to a feature of size. Where a flatness tolerance is applied to a feature of size, it can be specified at RFS (by default), at MMC, or at LMC. Remember, RFS is automatic where no modifier is shown.

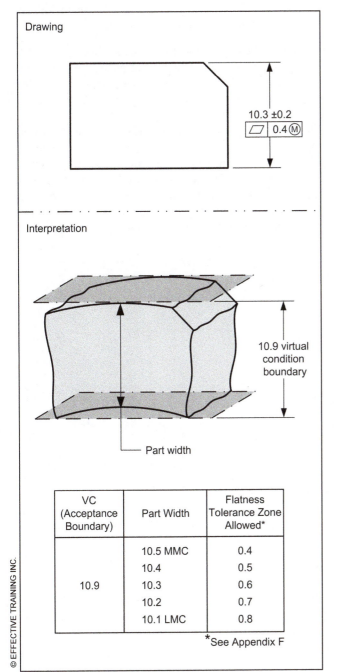

| VC (Acceptance Boundary) | Part Width | Flatness Tolerance Zone Allowed* |
|---|---|---|
| 10.9 | 10.5 MMC | 0.4 |
| | 10.4 | 0.5 |
| | 10.3 | 0.6 |
| | 10.2 | 0.7 |
| | 10.1 LMC | 0.8 |

*See Appendix F

**FIGURE 9-12  Flatness Tolerance (at MMC) Applied to a Feature of Size**

## The Significant Seven Questions

This section uses the Significant Seven Questions to interpret a flatness tolerance application. The questions are explained in Appendix B.

Figure 9-13 contains a flatness tolerance application. The Significant Seven Questions are answered in the interpretation section of the figure. Notice how each answer increases your understanding of the geometric tolerance.

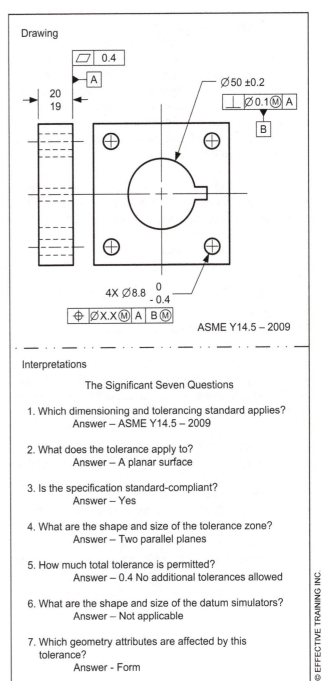

Interpretations

The Significant Seven Questions

1. Which dimensioning and tolerancing standard applies?
   Answer – ASME Y14.5 – 2009

2. What does the tolerance apply to?
   Answer – A planar surface

3. Is the specification standard-compliant?
   Answer – Yes

4. What are the shape and size of the tolerance zone?
   Answer – Two parallel planes

5. How much total tolerance is permitted?
   Answer – 0.4 No additional tolerances allowed

6. What are the shape and size of the datum simulators?
   Answer – Not applicable

7. Which geometry attributes are affected by this tolerance?
   Answer - Form

**FIGURE 9-13 The Significant Seven Questions Applied to a Flatness Tolerance**

# VERIFICATION PRINCIPLES AND METHODS

## Verification Principles for Flatness Tolerances

This section contains a simplified explanation of the verification principles and methods that can be used to inspect a flatness tolerance applied to a planar surface or a planar (width) feature of size (at MMC).

When a flatness tolerance is applied to a planar surface, the flatness deviation is verified by finding the distance between two parallel planes that just contain all of the points of the surface. The inspection method must be able to collect a set of points and determine the minimum distance between two planes that contain the set of points. Since a flatness tolerance is not related to any datums, care must be taken so that the location or orientation of the toleranced feature does not have an effect on the measurement.

When a flatness tolerance (MMC) is applied to a feature of size, the flatness deviation is often verified by passing the feature of size through a virtual condition boundary. There are many ways flatness tolerances can be verified. In this section, we will look at two of them.

## Verifying a Flatness Tolerance Applied to a Planar Surface

Figure 9-14A shows a flatness tolerance applied to the bottom of a part. When verifying this part, three separate requirements must be checked: the size dimension, the Rule #1 boundary, and the flatness requirement. There are many different verification methods that could be used. In this example, we will use a dial indicator and a surface plate.

A *surface plate* is a solid, flat plate used as the main horizontal reference plane (datum feature simulator; see Chapter 13) for precision inspection and tooling setup. A surface plate is often used as the baseline for all measurements to the workpiece; therefore, the top surface is finished extremely flat with accuracy up to 0.00025mm (for an inspection quality grade A granite plate, based on U.S. Fed spec GGG-P-463c). Most surface plates are made from granite or cast iron because of their dimensional stability and ability to be lapped to very tight flatness tolerances.

Surface plates are a very common tool in the manufacturing industry and are often permanently attached inspection devices, such as the table of a coordinate measuring machine. See Figure 9-14B.

A *dial indicator,* also known as a dial gage and probe indicator, is an instrument used to accurately measure small linear distances, and is frequently used in inspection of parts. It is named so because the measurement results are displayed in a magnified way by means of a dial. Dial indicators may be used to check the variation during the inspection process of a part, as well as many other situations where a small measurement needs to be registered or indicated. Dial indicators typically measure ranges from 0.25mm to 300mm, with graduations of 0.001mm to 0.01mm.

The dial indicator is mounted to the surface plate and is set to zero at the level of the surface plate by using a gage block. This establishes the reference plane (i.e., the first plane of the tolerance zone). See Figure 9-14C.

The part is then placed on the surface plate contacting at its high points (Figure 9-14D). As the part is moved randomly around on the surface plate, the dial indicator traces a path across the part surface and measures the distance between the surface plate and the low points of the part surface. If the indicator reading is larger than the flatness tolerance value, the surface is not within its flatness specification.

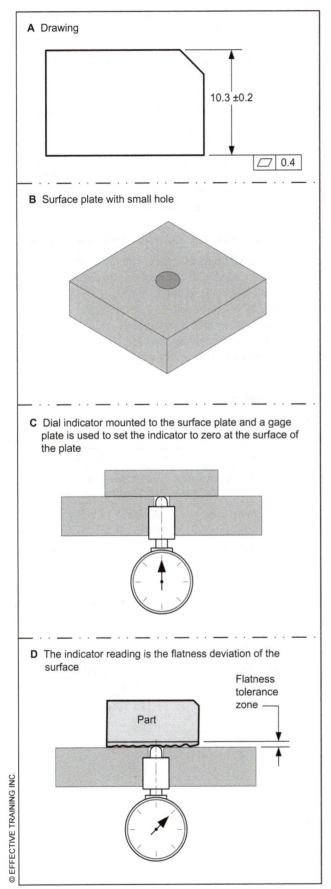

**A** Drawing

10.3 ±0.2

⬭ 0.4

**B** Surface plate with small hole

**C** Dial indicator mounted to the surface plate and a gage plate is used to set the indicator to zero at the surface of the plate

**D** The indicator reading is the flatness deviation of the surface

Flatness tolerance zone

Part

FIGURE 9-14 *Verifying a Flatness Tolerance Applied to a Planar Surface*

© EFFECTIVE TRAINING INC.

## Verifying a Flatness Tolerance (at MMC) Applied to a Feature of Size

Figure 9-12 shows a part with a flatness tolerance (at MMC) applied to a feature of size dimension. When verifying this part, two separate requirements must be checked: the size dimension, and the virtual condition resulting from the flatness tolerance. Chapter 5 discussed how to verify the size dimension. Figure 9-15 shows how the flatness tolerance requirements could be verified.

The most common verification method is to use a functional gage. The functional gage verifies that the feature of size fits into its virtual condition. The length and width of the gage would be at least as long and as wide as the feature of size it is verifying. The feature of size would have to fit through the gage for the flatness tolerance requirement to be accepted. The size tolerance must be verified separately.

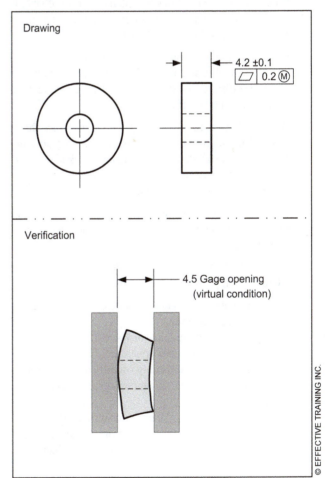

Drawing

4.2 ±0.1

⬭ 0.2 Ⓜ

Verification

4.5 Gage opening (virtual condition)

© EFFECTIVE TRAINING INC.

FIGURE 9-15 *Verifying a Flatness Tolerance (at MMC) Applied to a Feature of Size*

## SUMMARY
## Key Points

- Flatness is the condition of a surface or derived median plane having all of its elements in one plane.

- A derived median plane is an imperfect plane formed by the center points of all line segments bounded by the feature of size.

- A flatness tolerance is a geometric tolerance that defines the allowable flatness deviation permitted on a surface or derived median plane.

- A flatness tolerance may be applied to an individual surface or to an individual feature of size.

- A flatness tolerance is applied to a surface by directing its leader line or attaching it to a planar surface or to an extension line of a planar surface.

- Wherever Rule #1 applies to a planar feature of size, it invokes an automatic flatness deviation requirement on both surfaces and the median plane.

- Where a flatness tolerance is applied to a surface that is associated with a size dimension, and Rule #1 applies to the size dimension, the flatness tolerance value must be less than the size tolerance.

- Where a flatness tolerance is applied to a surface that is associated with a size dimension, and Rule #1 does not apply (i.e., overridden, independency, stock size, etc.) to the size dimension, the flatness tolerance value may be larger than the size tolerance.

- Where a flatness tolerance is located beneath or beside a size dimension, it indicates that it applies to the feature of size.

- The "CARE" test can be used to determine if a flatness tolerance complies with the standard.

- Real-world applications for a flatness tolerance are:
  o Sealing (gasket joint)
  o Support (distribute loads or forces)
  o Appearance (customer appeal)
  o Measurement repeatability (qualify a surface for making more repeatable measurements related to it).
  o Assembly (mating planar features of size)
  o Reducing manufacturing costs on size dimensions by allowing a planar width to vary more than its size limits

- Where a flatness tolerance is applied to a surface, the tolerance zone is the space between two parallel planes.

- Where a flatness tolerance applies to a planar regular feature of size (at MMC):
  o Rule #1 is overridden.
  o The flatness tolerance value may be larger than the feature of size dimension tolerance.
  o A bonus tolerance is permissible.
  o Using the surface interpretation, the part surfaces may not violate their virtual condition.
  o The virtual condition boundary size equals:
     External feature of size = MMC + flatness tol value
     Internal feature of size = MMC - flatness tol value
  o The flatness of the derived median plane is verified by passing the feature of size through its virtual condition.
  o The feature of size must also be within its size limits.

- Wherever a flatness tolerance with the MMC modifier is associated with a feature of size dimension, four things should come to mind:
  o Rule #1 is overridden.
  o A bonus tolerance is permissible.
  o A functional gage may be used.
  o The function is probably assembly.

## Additional Related Topics

*These topics are recommended for further study to improve your understanding of flatness.*

| Topic | Source |
|---|---|
| Flatness and independency | ETI's *Advanced Concepts of GD&T* textbook |
| Flatness per unit area | ASME Y14.5-2009, Para. 5.4.2.2 |
| Flatness and the CF modifier | ETI's *Advanced Concepts of GD&T* textbook |
| Modifying notes: SHALL NOT BE CONVEX SHALL NOT BE CONCAVE | ETI's *Advanced Concepts of GD&T* textbook |
| Calculating flatness tolerance value | ETI's *Advanced Concepts of GD&T* textbook |
| Using flatness in a stack up | ETI's *Tolerance Stacks Using GD&T* textbook |

# QUESTIONS AND PROBLEMS

**Website Bonus Materials**
Additional questions are available at our website. To access bonus materials for this textbook, please visit:
www.etinews.com www.etinews.com/textbookbonus

## True and False

*Indicate if each statement is true or false.*

T / F  1. A flatness tolerance must be applied RFS.

T / F  2. A flatness tolerance must be applied to a planar feature or width feature of size.

T / F  3. Flatness is the condition of being perfectly planar.

T / F  4. A flatness tolerance may override Rule #1.

T / F  5. A flatness tolerance may have a bonus tolerance.

T / F  6. A flatness tolerance may have a datum reference if the datum feature is planar.

## Multiple Choice

*Circle the best answer to each statement.*

1. Which of the following modifiers cannot be used with a flatness tolerance?
   A. Ⓜ              C. ∅
   B. ⟨ST⟩           D. Ⓕ

2. Which flatness tolerance could pass the CARE test?
   A. ⌓ ∅0.1          C. ⌓ 0.1 A
   B. ⌓ 0.1Ⓜ          D. None of the above

3. One method of inspecting a flatness tolerance (at MMC) applied to a feature of size is:
   A. A surface plate and dial indicator
   B. A functional gage made to the virtual condition size
   C. A surface plate and a feeler gage
   D. A caliper

4. The tolerance zone for flatness applied to a surface is:
   A. The space between two parallel lines
   B. The space between two parallel lines parallel to a datum plane
   C. The space between two parallel planes
   D. The space between a reference plane and a surface point

5. How does Rule #1 affect the application of a flatness tolerance applied to a surface?
   A. Where Rule #1 exists, flatness cannot be applied
   B. The flatness tolerance value must be equal to the size tolerance
   C. The flatness tolerance value must be less than the size tolerance
   D. The flatness tolerance value must be greater than the size tolerance

6. When verifying a flatness tolerance applied to a surface, the inspection method must measure the _____ of the surface.
   A. High and low points
   B. Median plane
   C. High points
   D. Virtual condition

## Application Problems

*The application problems are designed to provide practice applying the chapter concepts to situations that are similar to on-the-job conditions.*

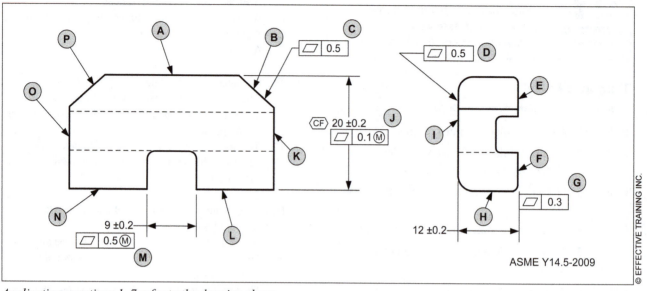

*Application questions 1–7 refer to the drawing above.*

1. What does the flatness tolerance labeled "G" apply to?
   A. Surface F
   B. Surfaces E and F
   C. Surfaces D, E, H, and I
   D. The derived median plane of 12 ±0.2

2. What is the maximum flatness deviation allowed by the flatness tolerance labeled "J"?
   A. 0.1
   B. 0.3
   C. 0.5
   D. None of the above

3. Is the flatness tolerance labeled "D" standard-compliant? _____ If no, describe why. (Hint: use the CARE test.)_____
   _____
   _____

4. Is the flatness tolerance labeled "G" standard-compliant? _____ If no, describe why. (Hint: use the CARE test.)_____
   _____
   _____

5. Is the flatness tolerance labeled "M" standard-compliant? _____ If no, describe why. (Hint: use the CARE test.)_____
   _____
   _____

6. What is the maximum flatness deviation allowed on the surface labeled "I"? _____

7. Use the Significant Seven Questions to interpret the flatness tolerance labeled "M."

   1) Which dimensioning and tolerancing standard applies?
   _____

   2) What does the tolerance apply to?_____
   _____

   3) Is the specification standard-compliant?_____

   4) What are the shape and size of the tolerance zone?
   _____

   5) How much total tolerance is permitted?_____

   6) What are the shape and size of the datum simulators?
   _____

   7) Which geometry attributes are affected by this tolerance?
   _____

# Straightness Tolerance

## Goal

Interpret the straightness tolerance

## Performance Objectives

Upon completing this chapter, you should be able to:

1. Describe the terms, "straightness," "derived median line," and "straightness tolerance" (p.116)
2. Describe the tolerance zones for a straightness tolerance (p.117)
3. Describe Rule #1's effects on straightness deviation (p.118)
4. Describe the modifiers that may be used with a straightness tolerance (p.119)
5. Determine if a straightness tolerance is applied to line elements or to a feature of size (p.119)
6. Evaluate if a straightness tolerance is standard-compliant (p.120)
7. Describe real-world applications for a straightness tolerance (p.121)
8. Interpret the straightness tolerance applied to line elements of a cylindrical surface (p.121)
9. Interpret a straightness tolerance at MMC applied to a feature of size dimension (p.122)
10. Interpret a straightness tolerance by using the "Significant Seven Questions" (p.123)
11. Understand straightness verification principles (p.124)
12. Describe how to verify a straightness tolerance applied to the elements of a cylindrical surface (p.124)
13. Draw a functional gage for verifying a straightness at MMC application (p.125)

## New Terms

- Derived median line
- Optical comparator
- Straightness
- Straightness tolerance

## What This Chapter Is About

This chapter explains how to interpret the straightness tolerance, where to use it, and how to inspect it.

The straightness tolerance is one of the fourteen geometric characteristic symbols and one of the four direct form tolerances.

Straightness tolerances are important to ensuring good joint design by defining the allowable straightness deviation of a surface line element. Straightness tolerances can also be used to define the allowable straightness deviation of a derived median line of a feature of size to ensure assembly.

## TERMS AND CONCEPTS

### Straightness

Y14.5 defines straightness as a perfect line element. *Straightness* is the condition where an element of a surface or a derived median line is a straight line. An example of straightness of a surface line element is shown in Figure 10-1.

Straight line

**FIGURE 10-1  Straightness**

### Derived Median Line

A *derived median line* is an imperfect line formed by the center points of all cross sections of a feature of size. These cross sections are perpendicular to the axis of the unrelated actual mating envelope of the feature of size. An example of a derived median line is shown in Figure 10-2.

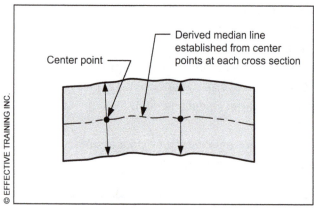

Center point

Derived median line established from center points at each cross section

**FIGURE 10-2  Derived Median Line**

The condition of straightness applies to line elements on a surface (feature) or to a feature of size dimension (derived median line). Therefore, in order for a part surface to meet the condition of straightness, each of its line elements must be perfectly straight. In order for a feature of size to meet the condition of straightness, its derived median line must be perfectly straight. Because straightness (the perfect condition) cannot be produced, it is necessary to specify some allowance for straightness deviation.

### Straightness Tolerance

A *straightness tolerance* is a geometric tolerance that defines the straightness deviation permitted on individual surface elements or on a derived median line. A straightness tolerance may be applied to line elements of a planar surface or to line elements of a cylindrical surface or the derived median line of a cylinder. A straightness tolerance may not be applied to a width feature of size dimension.

A straightness tolerance:

- Is one of the four direct form tolerances

- Never uses a datum reference

- Can only be applied to line elements of a single planar feature or a single cylindrical feature of size

- Where applied to a planar feature, it limits straightness deviations of the line elements

- Where applied to a cylindrical feature of size at RFS, it limits straightness deviations of the derived median line

- Where applied to a cylindrical feature of size at MMC, it limits the straightness deviations of the derived median line.

Examples of straightness tolerances are shown in Figure 10-3.

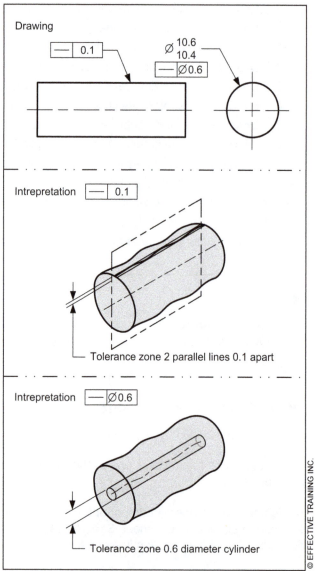

Drawing

Intrepretation

Tolerance zone 2 parallel lines 0.1 apart

Intrepretation

Tolerance zone 0.6 diameter cylinder

**FIGURE 10-3 Straightness Tolerance Examples**

## Straightness Tolerance Zones

A straightness tolerance can have two tolerance zone shapes: the space between two parallel lines and the space within a cylinder.

Where a straightness tolerance is applied to a surface, it applies independently to each line element of the surface. The tolerance zone is the space between two parallel lines equal to the tolerance value.

The line elements may have any type of form deviation within the tolerance zone. Typical form deviations are barreling, waisting, bending, and random. A straightness tolerance does not limit taper. Surface texture deviations are not included in its measurement of straightness deviations. Figure 10-3 shows a straightness tolerance zone applied to a surface line element.

Where a straightness tolerance is applied RFS to a cylindrical feature of size, the tolerance zone is the space within a cylinder that the derived median line must fit within. The derived median lined may have any type of form deviation within the tolerance zone. Typical form deviations are random and bending. Figure 10-3 shows a straightness tolerance RFS applied to a cylindrical feature of size.

Where applied at MMC, the tolerance zone can be interpreted two ways:

- As a cylinder that the derived median line must fit within
- As a VC boundary that the surface may not violate

In this text, we will use the surface interpretation (VC boundary). Appendix F explains why the surface interpretation is preferred.

Figure 10-4 shows typical straightness deviations.

***Author's Comment***
I have not included any examples of straightness tolerances applied to line elements of planar surfaces in this text for two reasons:

1. Using a straightness tolerance on line elements of a planar surface is rare. I have not seen it used in industry. The only place I have seen straightness applied to line elements of a planar surface used is in the Y14 Standard.

2. The interpretation of a straightness tolerance applied to line elements of a planar surface as defined in the Y14.5 Standard is not mathematically robust. The direction of the line element tolerance zone is not uniquely defined.

***Author's Comment***
In the previous versions of the Y14.5 Standard, the application of a straightness tolerance was allowed on both cylindrical and planar features of size. However, now the straightness tolerance can only be applied to cylindrical features of size.

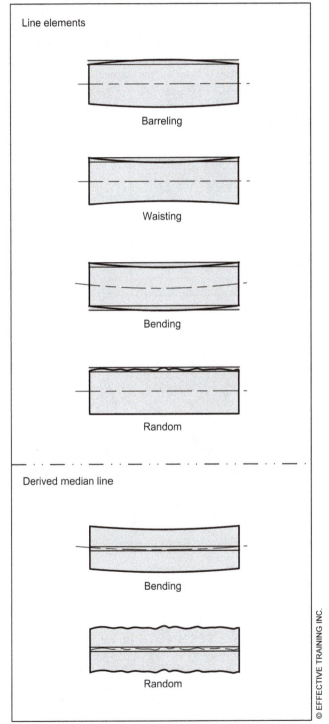

FIGURE 10-4 Types of Straightness Deviations

© EFFECTIVE TRAINING INC.

## Rule #1's Effects on the Straightness Deviation of Surface Elements

Wherever Rule #1 applies to a cylindrical feature of size, it invokes an indirect straightness deviation requirement on the surface elements. This requirement is a result of the interrelationship between Rule #1 (perfect form at MMC) and the size dimension. When the feature of size is at MMC, all line elements of the surface(s) must be perfectly straight. As the size dimension departs from MMC, a straightness deviation equal to the amount of departure is permitted.

Since Rule #1 provides an indirect straightness tolerance, a straightness tolerance should not be specified on surface elements, unless it is less than the dimensional limits of the size dimension (i.e., size tolerance). Figure 10-5 shows an example of how Rule #1 affects the straightness of surface elements.

## Rule #1's Effects on Straightness Deviation of a Regular Feature of Size

Wherever Rule #1 applies to a cylindrical regular feature of size, an indirect straightness deviation requirement exists for the derived median line of the feature of size. This indirect requirement is a result of the interrelationship between Rule #1 and the size dimension. Where the feature of size is at MMC, the derived median line of the feature of size must be perfectly straight. As the feature of size departs from MMC, a straightness deviation equal to the amount of the departure is permitted. An example of the effects of Rule #1 on the median line of a feature of size is shown in Figure 10-6.

Where the limit of the allowable straightness deviation provided by Rule #1 is sufficient for the function of the application, there is no need to specify a straightness tolerance.

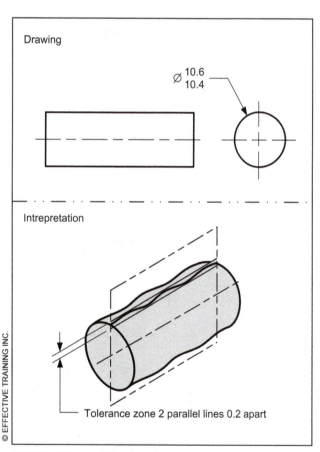

FIGURE 10-5 *Rule #1's Effects on the Straightness Deviations of Surface Elements*

---

### TECHNOTE 10-1
**How Rule #1 Affects Straightness Deviation of Surface Elements**
- Wherever Rule #1 applies to a regular feature of size, it invokes an indirect straightness deviation requirement on the surface elements.
- The amount of straightness deviation permitted is equal to the limits of the size dimension.

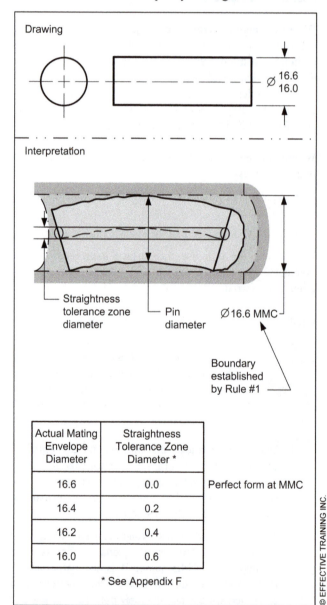

| Actual Mating Envelope Diameter | Straightness Tolerance Zone Diameter * | |
|---|---|---|
| 16.6 | 0.0 | Perfect form at MMC |
| 16.4 | 0.2 | |
| 16.2 | 0.4 | |
| 16.0 | 0.6 | |

\* See Appendix F

FIGURE 10-6 *Rule #1's Effect on Straightness Deviation of a Derived Median Line*

## Modifiers Used With Straightness Tolerances

A straightness tolerance has certain requirements that must be followed. The chart in Figure 10-7 shows the requirements of a straightness tolerance application.

| Tolerance Modifier | Can Be Applied To | Effect | Functional Application |
|---|---|---|---|
| ⌀ | Cylindrical feature of size | Changes the tolerance zone to a cylinder | Assembly |
| * Ⓕ | Feature or feature of size | Release the restraint requirement | Non-rigid parts with restraint notes |
| * ⟨ST⟩ | Feature or feature of size | Requires statistical process controls | Statistically derived tolerances or tolerances used in statistical tolerance analysis |
| Ⓜ | Cylindrical feature of size | Permits a bonus tolerance<br><br>Permits use of a functional gage | Assembly |
| * Ⓛ | Cylindrical feature of size | Permits a bonus tolerance<br><br>Requires variable gaging | Minimum distance (e.g., wall thickness, minimum machine stock, etc.) |

\* These modifiers are only introduced in this text.

FIGURE 10-7 Modifiers Used With a Straightness Tolerance

## How to Determine Whether a Straightness Tolerance is Applied to a Surface Element or a Feature of Size

It is important to be able to determine whether a straightness tolerance applies to surface elements or to a feature of size dimension because the interpretation of each is different.

Figure 10-8 shows an example of a straightness tolerance

applied to surface elements and a feature of size. The location of a straightness tolerance indicates what it is applied to. When the leader line of a straightness tolerance is directed to a surface — or attached to an extension line from the surface — it indicates that it applies to the surface elements. When a straightness tolerance is located beneath or beside a size dimension, it indicates that it applies to the derived median line of the feature of size.

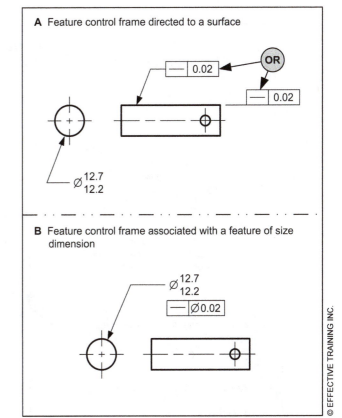

FIGURE 10-8 Application of the Straightness Tolerance

Two points to remember about specifying a straightness tolerance are:

- A straightness tolerance applied to surface line elements is view dependent. It should be shown in the view where the surface elements are shown as a line.

- When a straightness tolerance is applied to a feature of size, it can only apply to cylindrical features of size.

## Evaluate if a Straightness Tolerance Specification is Standard-Compliant

For a straightness tolerance to be a standard-compliant specification (legal), it must satisfy certain requirements.

In order to make it easier to remember what to look for when evaluating the correctness of a geometric tolerance on a drawing, I created a mnemonic.

The requirements are divided into four areas represented by the initials "CARE":

    C – Check for datums

    A – Assess the application

    R – Review the modifiers

    E – Evaluate the tolerance value

A flowchart that identifies the requirements for a standard-compliant straightness tolerance is shown in Figure 10-9.

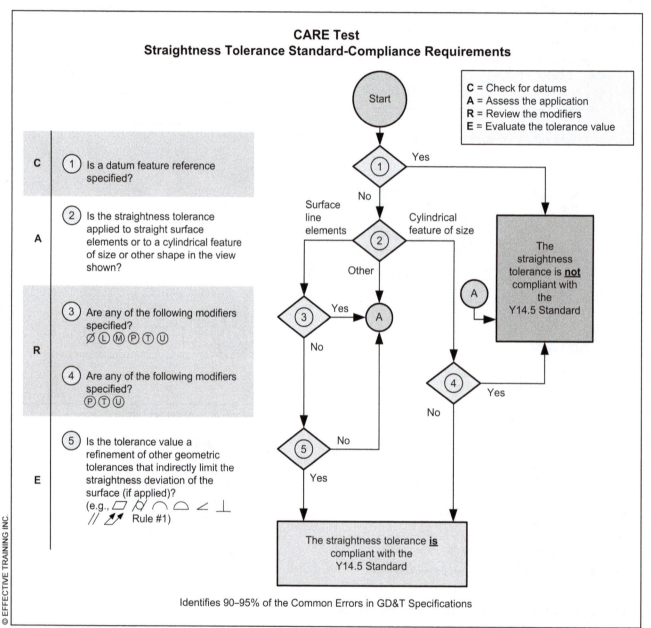

*FIGURE 10-9 Test for a Standard-Compliant Straightness Specification*

# STRAIGHTNESS APPLICATIONS

## Real-World Applications for a Straightness Tolerance

Four common real-world applications of a straightness tolerance are:

- Support (equal load along the line elements)
- Path of motion (wheel and track assemblies)
- Assembly (mating cylindrical features of size)
- Reducing manufacturing costs on size dimensions and allowing the feature of size to have more straightness deviation than its size limits allow.

## Straightness Tolerance Applied to Surface Elements

A straightness tolerance is applied to surface elements by directing its leader line to a surface or an extension line of a surface in a view where the line elements are shown as a line. The feature control frame contains a straightness symbol and a tolerance value.

Where a straightness tolerance is applied to surface elements:

- The tolerance zone is the space between two parallel lines.
- A straightness tolerance is view dependent. It must be applied to the view where the surface elements are shown as a line.
- The tolerance zone may be oriented to best fit the high and low points of the surface line element.
- Each surface element of the controlled surface must be within this zone.
- The distance between the parallel lines is equal to the straightness tolerance value.
- The tolerance zone extends across the length of the surface.

Figure 10-10 shows an example of a straightness tolerance applied to a surface.

Where a straightness tolerance is applied to surface elements that are associated with a size dimension, and Rule #1 applies to the size dimension, the straightness tolerance value must be less than the size tolerance.

***Author's Comment***
A straightness tolerance does not affect taper.

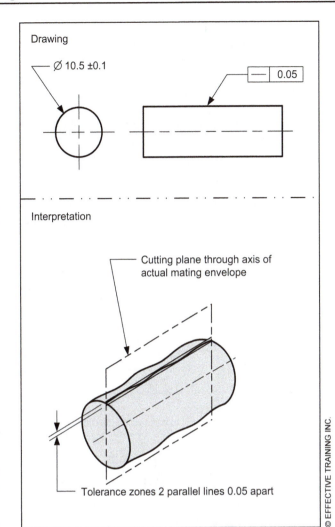

Drawing

Ø 10.5 ±0.1

— 0.05

Interpretation

Cutting plane through axis of actual mating envelope

Tolerance zones 2 parallel lines 0.05 apart

**FIGURE 10-10  Straightness Applied to a Surface**

© EFFECTIVE TRAINING INC.

***Design Tip***
Rule #1 is an indirect form control. There is a control of straightness deviation from the MMC boundary size and the size deviations. The straightness control provided by Rule #1 are often omitted during inspection. If it desired to have the straightness of the surface elements a surface inspected, a straightness tolerance should be specified in the drawing.

Where a straightness tolerance is applied to a surface element that is associated with a size dimension, and Rule #1 does not apply to the size dimension (i.e., overridden, independency, stock size, etc.), the straightness tolerance value may be larger than the size tolerance.

Let's look at another straightness tolerance application. In Figure 10-11, there are three requirements that must be met. The size dimension requires that the actual local sizes be verified. Rule #1 requires that the MMC boundary be verified. The straightness tolerance requires that the straightness deviation of the surface elements be verified.

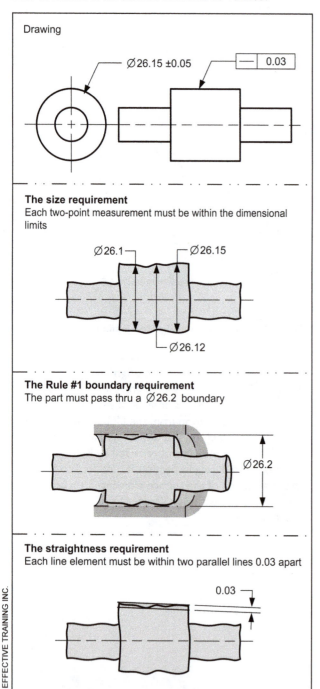

FIGURE 10-11 *Straightness Tolerance Application*

## Straightness Tolerance (at MMC) Applied to a Feature of Size

The interpretation of a straightness tolerance applied to a cylindrical feature of size is significantly different from a straightness tolerance applied to a surface. Where a straightness tolerance (at MMC) is applied to a feature of size, the following conditions apply:

- Rule #1 is overridden.

- The straightness tolerance value may be larger than the feature of size dimension tolerance.

- The diameter symbol modifier must be specified.

- The tolerance zone is a virtual condition boundary.

- The virtual condition size equals:
  o External feature of size = MMC + straightness tolerance value
  o Internal feature of size = MMC - straightness tolerance value

- A bonus tolerance is permissible.

- The straightness of the derived median line is verified by passing the feature of size through or over its virtual condition.

- The feature of size must also be within its size limits.

Figure 10-12 shows an example of a straightness tolerance applied to a feature of size.

***Author's Comment***
Although it is permissible to apply a straightness tolerance at LMC or RFS to a feature of size, the use of these conditions is rare. Therefore, this text does not cover these rare uses of a straightness tolerance.

Drawing

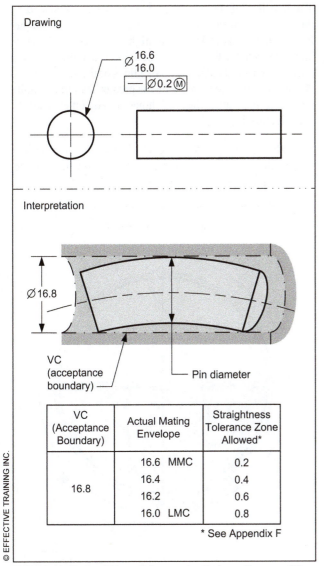

FIGURE 10-12 *Straightness Tolerance Applied to a Feature of Size*

When a straightness tolerance is applied to a feature of size, it can be specified at RFS (by default), at MMC, or at LMC. Remember, RFS applies where no modifier is shown.

---

**TECHNOTE 10-3**
**Straightness of a Feature of Size**

Wherever a straightness tolerance with the MMC modifier is associated with a feature of size dimension, four things should come to mind:

- Rule #1 is overridden
- A bonus tolerance is permissible
- A functional gage may be used
- The function is frequently assembly

---

## The Significant Seven Questions

This section uses the Significant Seven Questions to interpret a straightness tolerance application. The questions are explained in Appendix B.

Figure 10-13 contains a straightness tolerance application. The Significant Seven Questions are answered in the interpretation section of the figure. Notice how each answer increases your understanding of the geometric tolerance.

Drawing

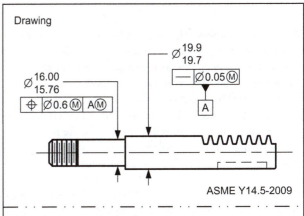

ASME Y14.5-2009

Interpretation

The Significant Seven Questions

1. Which dimensioning and tolerancing standard applies?
   Answer – ASME Y14.5-2009

2. What does the tolerance apply to?
   Answer – The derived median line of the diameter

3. Is the specification standard-compliant?
   Answer – Yes

4. What are the shape and size of the tolerance zone?
   Answer – A 19.95 virtual condition boundary

5. How much total tolerance is permitted?
   Answer – 0.25 at LMC

6. What are the shape and size of the datum simulators?
   Answer – Not Applicable

7. Which geometry attributes are affected by this tolerance?
   Answer - Form

FIGURE 10-13 *The Significant Seven Questions Applied to a Straightness Tolerance*

# VERIFICATION PRINCIPLES AND METHODS

## Verification Principles for Straightness Tolerances

This section contains a simplified explanation of the verification principles and methods that can be used to inspect a straightness tolerance applied to line elements of a planar or cylindrical surface or to a cylindrical feature of size (at MMC).

When a straightness tolerance is applied to a planar or cylindrical surface, the straightness deviation is verified by finding the distance between two parallel lines that all of the points of the surface element are located between. The inspection method must be able to collect a set of points and determine the distance between two lines that contains the set of points. Since location and orientation are not requirements of a straightness tolerance, the inspection method should not be influenced by these characteristics.

When a straightness tolerance (MMC) is applied to an external cylindrical feature of size, the straightness deviation is often verified by passing the feature of size through a virtual condition (the acceptance boundary).

When a straightness tolerance (MMC) is applied to an internal cylindrical feature of size, the straightness deviation is often verified by passing the feature of size over a virtual condition (the acceptance boundary).

There are many ways straightness tolerances can be verified. In this section, we will look at two of them.

## Verifying a Straightness Tolerance Applied to Surface Elements

Figure 10-10 shows a part with a straightness tolerance applied to the surface elements of the diameter. When verifying this part, three separate requirements must be checked, the size dimension, the Rule #1 boundary, and the straightness requirement. Figure 10-14 shows how the straightness requirements could be verified.

There are many different verification methods that could be used. In this example, we will use an optical comparator. An *optical comparator* (often called a comparator or profile projector) is a device that applies the use of optics to measure part geometry.

This is a simplified description of how an optical comparator works: A fixture is used to locate and orient the part for optimal viewing. A light beam is projected across the part surface and a set of mirrors and optics are used to cast a magnified image of the part profile on the viewing screen.

The first line of the tolerance zone may be established by the cross hair of the viewing screen. The image of the part line element is adjusted so its high points touch the cross hair. The distance between the high points and the low point may be measured using graduated lines of the screen or other means. If the measured value is greater than the straightness tolerance value, the straightness deviation exceeds the specification limit.

Since the straightness tolerance applies to each line element independently, this process is repeated for several line elements of the toleranced surface.

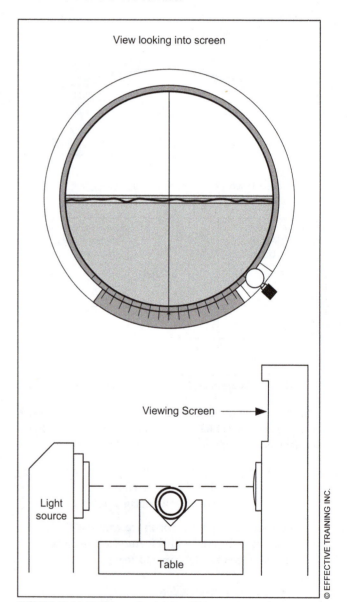

FIGURE 10-14 *Verifying a Straightness Tolerance*

© EFFECTIVE TRAINING INC.

## Verifying a Straightness Tolerance (at MMC) Applied to a Feature of Size

Figure 10-15 shows a part with a straightness tolerance (at MMC) applied to a feature of size dimension. When verifying this part, two separate requirements must be checked: the size dimension and the virtual condition resulting from the straightness tolerance. Chapter 5 discussed how to verify the size dimension. Figure 10-15 shows how the straightness tolerance requirement could be verified.

A common verification method, for a straightness tolerance (at MMC) is to use a functional gage. The functional gage verifies that the feature of size fits into its virtual condition. The length of the gage would be at least as long as the feature of size it is verifying. The feature of size would have to fit through the gage for the straightness tolerance requirement to be accepted. In Figure 10-15, part #1 would fit through the gage and part #2 would not fit through the gage.

The size tolerance limits of the pin are verified separately.

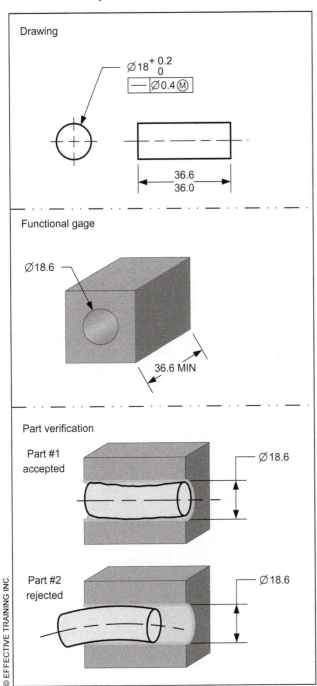

**FIGURE 10-15  Straightness Tolerance (at MMC) Applied to a Feature of Size Dimension**

## SUMMARY

### Key Points

- Straightness is the condition of a surface element or derived median line having all of its elements in one line.

- A derived median line is an imperfect line formed by the center points of all cross sections of a feature of size.

- A straightness tolerance is a geometric tolerance that defines the allowable straightness deviation permitted on surface elements or on a derived median line.

- A straightness tolerance is applied to a surface by directing its leader line to a surface or to an extension line of a surface.

- A straightness tolerance can only be applied to line elements of an individual surface or an individual feature of size.

- Where a straightness tolerance is applied to a surface element associated with a size dimension, and Rule #1 does not apply to the size dimension (i.e., overridden, independency, stock size, etc.), the straightness tolerance value may be larger than the size tolerance.

- Wherever Rule #1 applies to a cylindrical regular feature of size, it invokes an indirect straightness deviation requirement for all line elements.

- The "CARE" test can be used to determine if a straightness tolerance complies with the standard.

- Real-world applications for a straightness tolerance are:
  - o Support (equal load along the line elements)
  - o Path of motion (contact on a straight track or rail)
  - o Assembly (mating cylindrical features of size)
  - o Reducing manufacturing costs on size dimensions by allowing a diameter's straightness to vary more than its size limits

- Where a straightness tolerance is applied to surface elements, the tolerance zone is the space between two parallel lines.

- Where a straightness tolerance is applied to a surface element that is associated with a size dimension, and Rule #1 applies to the size dimension, the straightness tolerance value must be less than the size tolerance.

- Where a straightness tolerance is located beneath or beside a size dimension, it indicates that it applies to the feature of size.

- Where a straightness tolerance is applied to a regular cylindrical feature of size (at MMC)...
  - o Rule #1 is overridden
  - o The straightness tolerance value may be larger than the size tolerance
  - o The diameter symbol modifier must be specified
  - o The tolerance zone is the virtual condition boundary
  - o The virtual condition boundary size equals:
    External feature of size = MMC + straightness tolerance value
    Internal feature of size = MMC - straightness tolerance value
  - o A bonus tolerance is permissible
  - o The straightness of the derived median line is verified by passing the feature of size through or over its virtual condition boundary
  - o The feature of size must also be within its size limits

- Wherever a straightness tolerance with the MMC modifier is associated with a feature of size dimension, four things should come to mind:
  - o Rule #1 is overridden
  - o A bonus tolerance is permissible
  - o A functional gage may be used
  - o The function is probably assembly

## Additional Related Topics

*These topics are recommended for further study to improve your understanding of straightness.*

| Topic | Source |
|---|---|
| Straightness and independency | ETI's *Advanced Concepts of GD&T* textbook |
| Straightness per unit length | ASME Y14.5-2009, Para. 5.4.1.3 |
| Straightness and the CF modifier | ETI's *Advanced Concepts of GD&T* textbook |
| Calculating straightness tolerance values | ETI's *Advanced Concepts of GD&T* textbook |
| Using straightness in a tolerance stack | ETI's *Tolerance Stacks Using GD&T* textbook |
| Straightness of line elements of a planar feature of size | *ASME Y14.5-2009, Para 5.4.1.4* |

# QUESTIONS AND PROBLEMS

**Website Bonus Materials**
Additional questions are available at our website. To access bonus materials for this textbook, please visit:
www.etinews.com/textbookbonus

www.etinews.com

## True and False

*Indicate if each statement is true or false.*

T / F    1.    Straightness is where an axis is a straight line.

T / F    2.    A straightness tolerance may be applied to a width feature of size dimension.

T / F    3.    The placement of a straightness tolerance feature control frame affects the interpretation.

T / F    4.    A straightness tolerance (at MMC) may be verified with a functional gage.

T / F    5.    One real-world application of a straightness tolerance is to ensure assembly.

T / F    6.    A straightness tolerance cannot permit a bonus tolerance.

T / F    7.    A derived median line is the same as an axis.

T / F    8.    A straightness tolerance zone is two parallel planes.

## Multiple Choice

*Circle the best answer to each statement.*

1.    The tolerance zones for a straightness tolerance are:
   A.    Two parallel lines or a cylinder
   B.    Two lines and a cylinder
   C.    Two parallel planes or a cylinder
   D.    A parallel line or a cylinder

2.    Rule #1 affects the straightness deviation on a:
   A.    Planar feature of size
   B.    Cylindrical feature of size
   C.    Planar surface of a feature of size
   D.    All of the above

3.    What type of deviation is not limited by a straightness tolerance?
   A.    Waisting
   B.    Taper
   C.    Barreling
   D.    All of the above

4.    Where a straightness tolerance at MMC is applied to a cylindrical feature of size:
   A.    The function is probably assembly
   B.    A functional gage may be used to inspect the straightness
   C.    Rule #1 is overridden
   D.    All of the above

5.    Where a straightness tolerance is applied to surface elements that are associated with a size dimension, and Rule #1 applies to the size dimension, the specified tolerance value _____ the size tolerance.
   A.    Must be equal to
   B.    Must be less than
   C.    Must be greater than
   D.    May be less than or greater than

6.    A derived median line is...
   A.    The same as the axis of the actual mating envelope
   B.    A perfectly straight line established from the virtual condition boundary
   C.    An imperfect line formed by the center points of all cross sections of the feature of size
   D.    A spine formed by the center points of all cross section of the actual mating envelope

## Application Problems

*The application problems are designed to provide practice on applying the chapter concepts to situations that are similar to on-the-job conditions.*

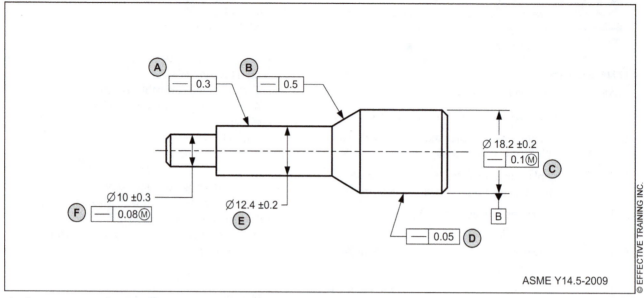

ASME Y14.5-2009

© EFFECTIVE TRAINING INC.

*Application questions 1–9 refer to the drawing above.*

1. Identify the straightness tolerances that apply to a feature of size. _____

2. Identify the straightness tolerances that apply to surface line elements. _____
   _____

3. Describe the tolerance zone for the straightness tolerance labeled "B." _____
   _____

4. How much bonus tolerance is permissible for the straightness tolerance labeled "C"? _____

5. How much bonus tolerance is permissible for the straightness tolerance labeled "F"? _____

6. Sketch a functional gage for verifying the straightness tolerance labeled "C."

7. Describe one method to inspect the straightness tolerance labeled "A." _____
   _____
   _____

8. Use the Significant Seven Questions to interpret the circularity tolerance labeled "D."

   1) Which dimensioning and tolerancing standard applies?
   _____

   2) What does the tolerance apply to? _____
   _____

   3) Is the specification standard-compliant? _____

   4) What are the shape and size of the tolerance zone?
   _____

   5) How much total tolerance is permitted? _____

   6) What are the shape and size of the datum simulators?
   _____

   7) Which geometry attributes are affected by this tolerance?
   _____

9. Use the CARE test to determine if the straightness tolerances in the chart below are standard-compliant. If a straightness tolerance is not standard-compliant, explain why.

| Straightness Tolerance | Standard-Compliant | Explanation |
|---|---|---|
| (A) | | |
| (B) | | |
| (C) | | |
| (D) | | |

# Circularity Tolerance

## Goal

Interpret the circularity tolerance

## Performance Objectives

Upon completing this chapter, you should be able to:

1. Describe the terms "circularity," "spine," "circularity tolerance," and minimum radial separation" (p.132)
2. Describe the tolerance zone for a circularity tolerance (p.132)
3. Describe how Rule #1 affects circularity deviation (p.133)
4. Describe the modifiers that may be used in a circularity tolerance (p.134)
5. Describe how to specify a circularity tolerance (p.134)
6. Evaluate if a circularity tolerance specification is standard-compliant (p.134)
7. Describe real-world applications for a circularity tolerance (p.136)
8. Interpret a circularity tolerance applied to cylindrical surface elements (p.136)
9. Interpret a circularity tolerance by using the "Significant Seven Questions" (p.137)
10. Understand the verification principles for a circularity tolerance (p.137)
11. Describe how to verify a circularity tolerance (p.137)

## New Terms

- Circularity
- Circularity tolerance
- Spine
- Minimum radial separation (MRS)
- Roundness (circularity) gage

## What This Chapter Is About

The circularity tolerance is one of the fourteen geometric characteristic symbols and is one of the four direct form tolerances. The circularity tolerance is used to limit deviations of circular elements of a part surface.

A circularity tolerance is often be used to ensure a good seal, rolling characteristics, bearing support, or appearance by defining the allowable deviation of circular elements.

This chapter explains how to interpret the circularity tolerance, where to use it, and how to inspect it.

# TERMS AND CONCEPTS

## Circularity

The definition of circularity in Y14.5 has two parts. **Circularity** is the condition of a surface where:

- For a feature other than a sphere (e.g., cylinder, cone, etc.), all points of the surface intersected by any plane perpendicular to an axis or spine (curved line) are equidistant from that axis or spine (curved line).

- For a sphere, all points of the surface intersected by any plane passing through a common center are equidistant from that center.

A **spine** is a simple (non self-intersecting) curve.

An example of circularity on a circular element of a cylinder is shown in Figure 11-1.

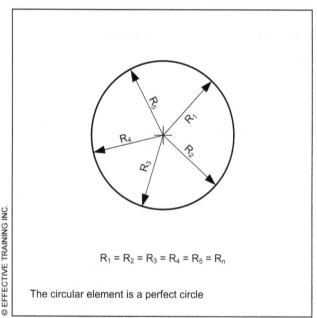

$$R_1 = R_2 = R_3 = R_4 = R_5 = R_n$$

The circular element is a perfect circle

*FIGURE 11-1  Circularity*

Circularity is a perfect condition. It results in perfect circular elements. Because circularity (the perfect condition) cannot be produced, it is necessary to specify an allowance for deviation from circularity.

## Circularity Tolerance

A **circularity tolerance** is a geometric tolerance that limits the circularity deviation (radial deviation between highest and lowest points) on individual circular elements. A circularity tolerance:

- Is one of the four direct form tolerances

- Never uses a datum feature reference

- Can only be applied to a single feature with circular cross sections (e.g., sphere, cylinder, cone, etc.)

- Applies to each circular element of a surface independently

An example of a circular tolerance is shown in Figure 11-2.

## Circularity Tolerance Zone

The tolerance zone for a circularity tolerance is the space between two coaxial circles on any cross section plane perpendicular to the axis or spine. The circles may be any size and are separated radially by the circularity tolerance value.

The tolerance zone circles may be best fit to the part circular element, or they may be established by a method that fits one of the tolerance zone circles to the part circular element. The outer circle may be the minimum circumscribed circle, or the inner circle may be the maximum inscribed circle. When the outer or inner circle is established, the second circle is coaxial with the first circle. The circularity tolerance value defines the radial distance between the circles.

For an external diameter, the size of the outer circle is determined by circumscribing a circle around the high points of a circular element. The radial distance to the inner circle is equal to the circularity tolerance value.

For an internal diameter, the size of the inner circle is determined by inscribing a circle inside the high points of a circular element. The radial distance to the outer circle is equal to the circularity tolerance value.

An example of a circularity tolerance zone is shown in Figure 11-2.

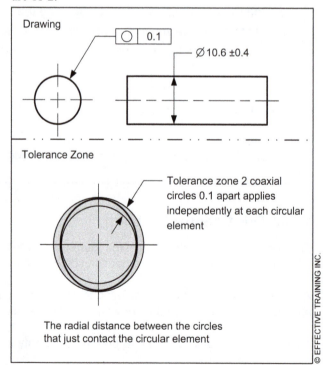

*FIGURE 11-2  Circularity Tolerance and Tolerance Zone*

The circular elements may have any type of form deviation within the tolerance zone. Typical form deviations for circular elements are lobing, ovality, and irregular (no specific shape). A circularity tolerance does not limit straightness or taper. Surface texture deviations are not included in a circularity tolerance measurement. Therefore, the surface texture deviations must be filtered out.

Figure 11-3 shows examples of circularity deviations.

***Design Tip***
Rule #1 is an indirect form tolerance Circularity effects from Rule #1 and a size dimension are often not inspected; they are a result of the relationship between the MMC boundary and size variations. If it is desired to have the circularity of the surface elements inspected, a circularity tolerance should be specified on the drawing.

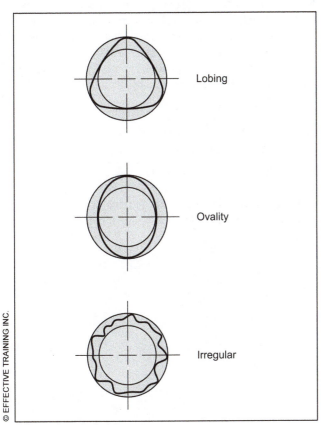

*FIGURE 11-3 Circularity Deviations*

## How Rule #1 Affects the Circularity Deviation of Surface Elements

Wherever Rule #1 applies to a regular feature of size (with an axis or spine), it restricts the circularity deviation on each circular element of the surface. This is a result of the relationship between Rule #1 (perfect form at MMC) and the size dimension.

When the feature of size is at MMC, all circular elements of the surface must be perfectly circular. As the size dimension departs from MMC, a circularity deviation equal to the amount of departure is permitted. The maximum circularity deviation occurs where the feature of size is at LMC. The maximum circularity deviation may not be larger than the size tolerance. An example is shown in Figure 11-4.

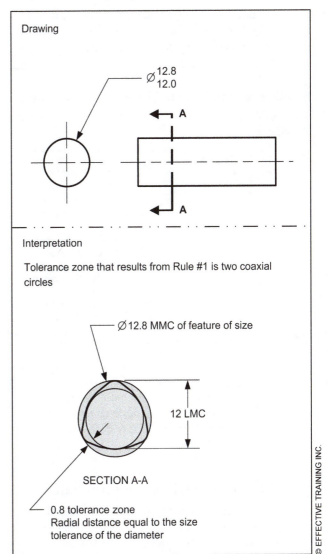

*FIGURE 11-4 Rule #1 Effects on Circularity Deviations*

© EFFECTIVE TRAINING INC.

---

**TECHNOTE 11-1**
**Rule #1 and Circularity**

- Where Rule #1 applies, a diameter dimension establishes a perfect form boundary at MMC.

- Rule #1 limits circularity deviations to within the size limits of the diameter.

- Where Rule #1 applies and a circularity tolerance is specified, the circularity tolerance must be a refinement of the size tolerance.

## Modifiers Used With Circularity Tolerances

The modifiers permitted with a circularity tolerance are shown in Figure 11-5.

| Tolerance Modifier | Can Be Applied To | Effect | Functional Application |
|---|---|---|---|
| ⓕ* | Feature | Release the restraint requirement | Non-rigid parts with restraint notes |
| ⟨ST⟩* | Feature | Requires statistical process controls | Statistically derived tolerances or tolerances used in statistical tolerance analyses |

\* These modifiers are only introduced in this text

**FIGURE 11-5 Modifiers Used With Circularity Tolerances**

**Author's Comment**
There are additional modifiers that can be used to invoke optional assessment methods for circularity described in ASME B89.3.1.

## How to Specify a Circularity Tolerance

A circularity tolerance is always applied to surface elements of a surface of revolution. Figure 11-6 shows examples of how to specify a circularity tolerance on a drawing. A circularity tolerance feature control frame may be attached to a leader directed to the surface or an extension line of the surface but not to the size dimension.

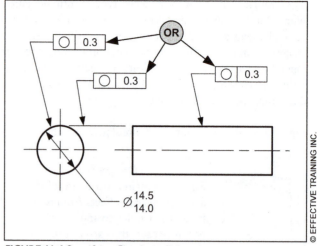

**FIGURE 11-6 Specifying Circularity Tolerances**

## Evaluate if a Circularity Tolerance Specification is Standard-Compliant

For a circularity tolerance to be a standard-compliant specification (legal), it must satisfy certain requirements. In order to make it easier to remember what to look for when evaluating the correctness of a geometric tolerance on a drawing, I created a mnemonic. The requirements are divided into four areas represented by the initials "CARE":

C - Check for datums

A - Assess the application

R - Review the modifiers

E - Evaluate the tolerance value

A flowchart that identifies the requirements for a standard-compliant circularity tolerance is shown in Figure 11-7.

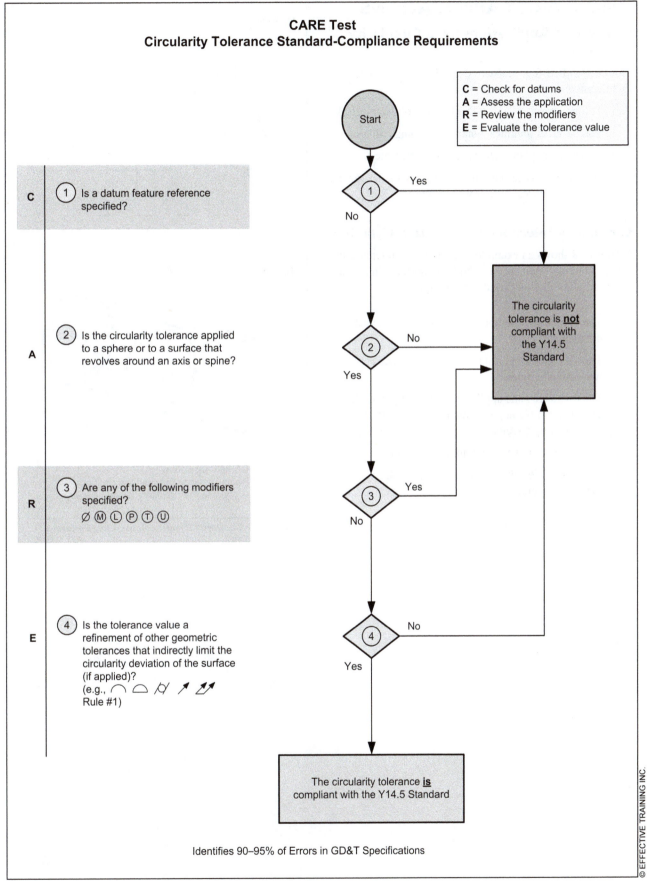

FIGURE 11-7 Test for a Standard-Compliant Circularity Specification

**CARE Test**
**Circularity Tolerance Standard-Compliance Requirements**

C = Check for datums
A = Assess the application
R = Review the modifiers
E = Evaluate the tolerance value

C  (1) Is a datum feature reference specified?

A  (2) Is the circularity tolerance applied to a sphere or to a surface that revolves around an axis or spine?

R  (3) Are any of the following modifiers specified?
Ø Ⓜ Ⓛ Ⓟ Ⓣ Ⓤ

E  (4) Is the tolerance value a refinement of other geometric tolerances that indirectly limit the circularity deviation of the surface (if applied)?
(e.g., ⌒ ⌓ ⌯ ⌰ ⌰ Rule #1)

Start

1 — Yes
No

2 — No
Yes

3 — Yes
No

4 — No
Yes

The circularity tolerance is **not** compliant with the Y14.5 Standard

The circularity tolerance **is** compliant with the Y14.5 Standard

Identifies 90–95% of Errors in GD&T Specifications

# CIRCULARITY APPLICATIONS

## Real-World Applications for a Circularity Tolerance

This section lists four real-world applications of a circularity tolerance:

- Rolling characteristics (wheels, cam followers, etc.)
- Seal surfaces (engines, pumps, plumbing, valves)
- Appearance (plates, clocks, cups, bottles, etc.)
- Performance (bearings in motors, saw blades, gears, sprockets, lenses, blower wheels, etc.)

## Circularity Tolerance Applied to a Cylinder

Figure 11-8 shows an example of a circularity tolerance applied to circular elements a cylindrical surface. In this application, the following conditions apply:

- The circularity tolerance applies to the full extent of the diameter.
- The tolerance zone is a 0.3 mm radial space between two coaxial circles.
- The tolerance zone applies independently at each cross section plane perpendicular to the axis of the actual mating envelope.
- The diameter must be within its size limits.
- Rule #1 applies, the diameter must fit within its MMC boundary.

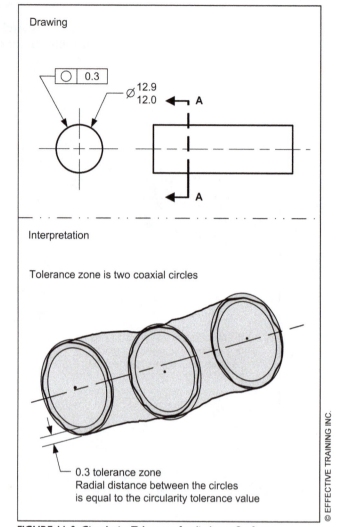

Drawing

⬜ 0.3

∅ 12.9 / 12.0 ← A

Interpretation

Tolerance zone is two coaxial circles

0.3 tolerance zone
Radial distance between the circles
is equal to the circularity tolerance value

© EFFECTIVE TRAINING INC.

**FIGURE 11-8  Circularity Tolerance Applied to a Surface**

## The Significant Seven Questions

This section uses the Significant Seven Questions to interpret a circularity tolerance application. The questions are explained in Appendix B.

Figure 11-9 contains a circularity tolerance application. The Significant Seven Questions are answered in the interpretation section of the figure. Notice how each answer increases your understanding of the geometric tolerance.

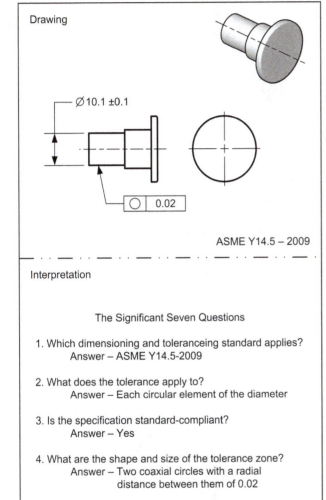

Drawing

Ø 10.1 ±0.1

⌀ 0.02

ASME Y14.5 – 2009

Interpretation

The Significant Seven Questions

1. Which dimensioning and toleranceing standard applies?
   Answer – ASME Y14.5-2009

2. What does the tolerance apply to?
   Answer – Each circular element of the diameter

3. Is the specification standard-compliant?
   Answer – Yes

4. What are the shape and size of the tolerance zone?
   Answer – Two coaxial circles with a radial distance between them of 0.02

5. How much total tolerance is permitted?
   Answer – 0.02

6. What are the shape and size of the datum simulators?
   Answer – Not applicable

7. Which geometry attributes are affected by this tolerance?
   Answer - Form

FIGURE 11-9 The Significant Seven Questions Applied to a Circularity Tolerance

# VERIFICATION METHODS AND PRINCIPLES

## Verification Principles for Circularity Tolerances

This section discusses verification principles and methods used to inspect a circularity tolerance applied to a cylindrical surface.

When verifying a circularity tolerance, the inspection method must be able to make a trace of a circular element (or collect a set of points that represent the circular element) and determine the radial distance between two coaxial circles that contain the trace (or points). Since a circularity tolerance is not related to any datums, care must be taken so that the location or orientation of the toleranced feature does not have an effect on the circularity inspection method.

Verifying a circularity tolerance has two parts:

1. Make a trace of a circular element (or collect a set of points that represent the circular element) on a plane normal to the axis (or spine)

2. Fit two coaxial circles to the circular trace using one of four assessment methods:

   • Minimum radial separation (MRS)

   • Least squares circle (LSC)

   • Minimum circumscribed circle (MCC)

   • Maximum inscribed circle (MIC)

***Minimum radial separation (MRS)*** is the radial distance between two concentric circles which just contain the measured polar profile as a minimum. Figure 11-10 shows an example of MRS. MRS is the default assessment method per B89.3.1 "Measurement of Out-of-Roundness."

There are several ways a circularity tolerance can be verified. We will look at one method in this section.

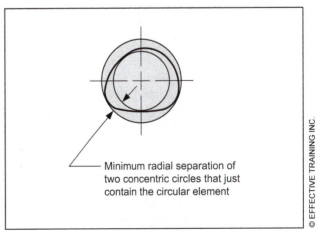

Minimum radial separation of two concentric circles that just contain the circular element

FIGURE 11-10 Minimum Radial Separation is the Default Tolerance Zone for Circularity Tolerances

## Verifying a Circularity Tolerance

Figure 11-8 shows a part with a circularity tolerance applied to a cylindrical surface. When verifying this part, three separate requirements must be checked: the size dimension, the Rule #1 boundary, and the circularity tolerance.

Circularity (roundness) is a difficult tolerance to measure. To accurately measure roundness, a measuring system that uses computer technology is required. One such device is a roundness gage.

A *roundness (circularity) gage* is a precision spindle machine used to measure the out of roundness (circularity deviations) of a surface of revolution. A roundness gage is shown in Figure 11-11. A roundness gage consists of four main components:

1. A rotating table on a precision spindle

2. A measuring probe

3. A computer control system

4. Measurement software

A part is placed on the turntable and a probe traces a circular path on the part surface. The software collects and analyzes the data and displays the circular element on a polar chart or other user selected format.

The tolerance zone circles may be best fit to the part circular element using the minimum radial separation (MRS) default assessment method. If one of three other assessment methods is wanted, least squares circle (LSC), minimum circumscribed circle (MCC), or maximum inscribed circle (MIC), they may be specified in the circularity tolerance feature control frame as defined in B89.3.1 or in a measurement plan. Each assessment method will yield different measurement results.

If the deviations of the circular element are within the tolerance zone (using the method and parameters as stated on the drawing or in a measurement plan), the circular element is acceptable. This process is repeated at several cross section locations until the inspector is confident the part meets its circularity requirements.

***Author's Comment***
Surface roughness deviations are not included in any of the form measurements; the surface roughness deviations must be filtered out when inspecting circularity. Therefore, in addition to the assessment method, the filter settings and probe tip radius will greatly affect the measurement results. If these variables are not specified on the drawing or in a measurement plan, then the default parameters for these variables apply as defined in ASME B89.3.1, "Measurement of Out of Roundness."

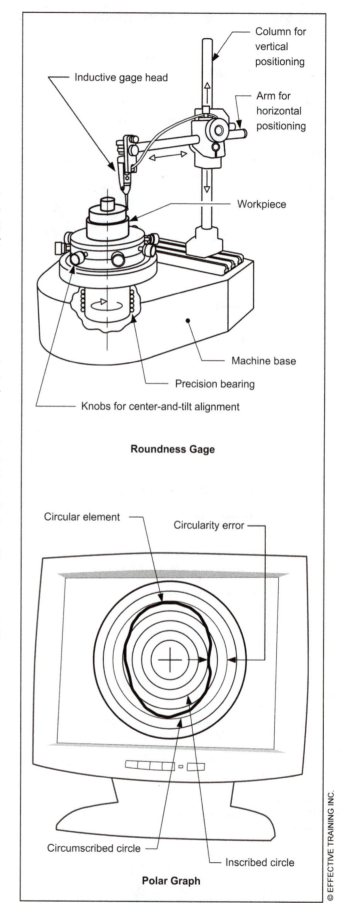

**Roundness Gage**

**Polar Graph**

© EFFECTIVE TRAINING INC.

**FIGURE 11-11** *Schematic Drawing of a Roundness Gage*

# SUMMARY

## Key Points

- Circularity is the condition of a surface where:
  - For a feature other than a sphere (e.g., cylinder, cone, etc.), all points of the surface intersected by any plane perpendicular to an axis or spine (curved line) are equidistant from that axis or spine.
  - For a sphere, all points of the surface intersected by any plane passing through a common center are equidistant from that center.

- A spine is a simple (non self-intersecting) curve.

- A circularity tolerance is a geometric tolerance that limits the circularity deviation (radial deviation between highest and lowest points) on individual circular elements.

- A circularity tolerance:
  - Is one of four direct form tolerances
  - Never uses a datum reference
  - Can only be applied to a single feature with a circular cross section (e.g., sphere, cylinder, cone, etc.).
  - Applies to each circular element of a surface independently.

- The circularity tolerance zone is the space between two coaxial circles.

- The two circles may be any size and are separated radially by the circularity tolerance value.

- For an internal diameter the inner circle may be a maximum inscribed circle.

- For an external diameter the outer circle may be a minimum circumscribed circle.

- The circular element may have any type of form deviation within the tolerance zone (e.g., lobing, ovality, irregular).

- Where Rule #1 applies, it indirectly limits the circularity deviations within the size limits.

- Where Rule #1 applies, a circularity tolerance value must always be less than the size tolerance limits.

- Where Rule #1 applies, the feature of size must fit within its MMC perfect form boundary.

- Where Rule #1 does not apply, a circularity tolerance value may be larger than the size tolerance.

- The associated feature of size must also be within its size limits.

- The free state modifier, and the statistical tolerance modifier are the only two modifiers that may be used with a circularity tolerance.

- A circularity tolerance may be applied to a surface by directing its leader line to the surface or extension line of the surface.

- The "CARE" test can be used to determine if a circularity tolerance complies with the standard.

- Four real-world applications of a circularity tolerance applied to surface elements on a drawing are to limit the circularity deviation of surface elements for:
  - Rolling characteristics (wheels, cam followers)
  - Seal surfaces (engines, pumps, plumbing, valves)
  - Appearance (plates, clocks, cups, bottles)
  - Performance (bearings in motors, saw blades, gears, sprockets, lenses, blower wheels, etc.)

- Verifying a circularity tolerance has three parts:
  1. Establish circular trace on plane normal to spine
  2. Fit two coaxial circles to the circular trace using one of four assessment methods
     a. Minimum radial separation (MRS)
     b. Least squares circle (LSC)
     c. Minimum circumscribed circle (MCC)
     d. Maximum inscribed circle (MIC)

- Minimum radial separation (MRS) is the radial distance between two concentric circles which just contain the measured polar profile is a minimum.

- A roundness (circularity) gage is a precision spindle machine used to measure the out of roundness (circularity deviations) of a surface of revolution.

## Additional Related Topics

*These topics are recommended for further study to improve your understanding of circularity.*

| Topic | Source |
|---|---|
| Circularity and independency | ETI's *Advanced Concepts of GD&T* textbook |
| Circularity and the free state modifier | ETI's *Advanced Concepts of GD&T* textbook |
| Circularity and the (CF) modifier | ETI's *Advanced Concepts of GD&T* textbook |
| Circularity and the "AVG" modifier | *ASME Y14.5, Para. 5.5.1* |
| Alternate circularity assessment methods based on B89.3.1 | ETI's *Functional Gaging and Measurement* textbook |
| Calculating circularity tolerance values | ETI's *Advanced Concepts of GD&T* textbook |
| Using circularity in a tolerance stack | ETI's *Tolerance Stacks Using GD&T* textbook |

## QUESTIONS AND PROBLEMS

***Website Bonus Materials***
Additional questions are available at our website. To access bonus materials for this textbook, please visit:
www.etinews.com/textbookbonus

www.etinews.com

### True and False

*Indicate if each statement is true or false.*

T / F      1. Rule #1 limits circularity deviations.

T / F      2. A circularity tolerance may be inspected with a roundness gage.

T / F      3. A circularity tolerance should be specified with the ⌀ modifier.

T / F      4. A circularity tolerance may use the Ⓜ modifier where a bonus tolerance is acceptable.

T / F      5. The tolerance zone of a circularity tolerance is a circle.

T / F      6. A circularity tolerance applies to the entire length, width, and depth of a feature simultaneously.

### Multiple Choice

*Circle the best answer to each statement.*

1. For a feature other than a sphere, circularity is where:
   A. The axis is a straight line
   B. The ⌀ modifier is specified with a size dimension
   C. All points of the surface intersected by any plane perpendicular to an axis or spine (curved line) are equidistant from that axis or spine
   D. All points of the surface intersected by any plane passing through a common center are equidistant from that center

2. What type of variation is limited by a circularity tolerance zone?
   A. Ovality          C. Bending
   B. Tapering         D. Warping

3. How does the Rule #1 boundary affect the application of a circularity tolerance?
   A. The ⌀ modifier must be used.
   B. The feature control frame must be placed next to the size dimension.
   C. The circularity tolerance value must be less than the limits of size tolerance.
   D. Circularity cannot be applied where a Rule #1 boundary exists.

4. A circularity tolerance may use a _____ modifier.
   A. ⌀
   B. Ⓕ
   C. Ⓜ
   D. Ⓟ

5. A real-world application for a circularity tolerance is:
   A. Assembly (i.e., shaft and hole)
   B. Sealing surface (i.e., engines, pumps, valves)
   C. Rotating clearance (i.e., shaft and housing)
   D. Support (equal load along a line element)

6. When verifying a circularity tolerance, the inspection method must be able to collect a set of points and determine the:
   A. Distance between two coaxial cylinders that contain the set of points
   B. Circle that circumscribes the set of points
   C. Circle that inscribes the set of points
   D. Distance between two coaxial circles that contain the set of points

## Application Problems

*The application problems are designed to provide practice on applying the chapter concepts to situations that are similar to on-the-job conditions.*

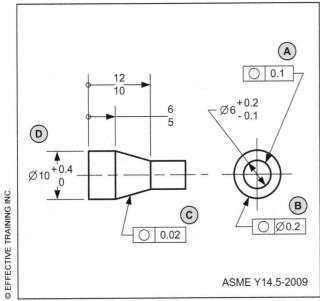

*Application questions 1–5 refer to the drawing above.*

1. Is the circularity tolerance labeled "A" standard-compliant? _____ If no, describe why. (Hint: use the CARE test.)_____

   _____

   _____

2. The allowable circularity deviation of the diameter labeled "D" is _____

3. What is the tolerance zone for the circularity tolerance labeled "C?"_____

   _____

4. Is the circularity tolerance labeled "B" standard-compliant? _____ If no, describe why. (Hint: use the CARE test.)_____

   _____

   _____

5. Use the Significant Seven Questions to interpret the circularity tolerance labeled "C."

   1) Which dimensioning and tolerancing standard applies?

   _____

   2) What does the tolerance apply to?_____

   _____

   3) Is the specification standard-compliant?_____

   4) What are the shape and size of the tolerance zone?

   _____

   5) How much total tolerance is permitted?_____

   6) What are the shape and size of the datum simulators?

   _____

   7) Which geometry attributes are affected by this tolerance?

   _____

# Cylindricity Tolerance

## Goal

Interpret the cylindricity tolerance

## Performance Objectives

Upon completing this chapter, you should be able to:

1.  Describe the terms "cylindricity" and "cylindricity tolerance" (p.144)
2.  Describe the tolerance zone for a cylindricity tolerance (p.144)
3.  Describe Rule #1's effects on cylindricity deviation (p.145)
4.  Describe how to specify a cylindricity tolerance (p.145)
5.  Describe the modifiers that may be used in a cylindricity tolerance (p.146)
6.  Evaluate if a cylindricity tolerance is standard-compliant (p.146)
7.  Describe real-world applications for a cylindricity tolerance (p.147)
8.  Interpret a cylindricity tolerance applied to a cylinder (p.147)
9.  Interpret a cylindricity tolerance by using the "Significant Seven Questions" (p.148)
10. Understand the verification principles for a cylindricity tolerance (p.148)
11. Describe how to verify a cylindricity tolerance applied to the surface of a diameter (p.148)

## New Terms

- Cylindricity
- Cylindricity gage
- Cylindricity tolerance

## What This Chapter Is About

The deviation of cylindrical surfaces is important because they are one of the most common shapes on many parts across all industries. Deviations in cylindricity should be considered when designing parts. A few examples where cylindricity tolerances are used are seals around shafts in an automobile, lip seal applications, and bearing applications.

This chapter explains how to interpret the cylindricity tolerance, where to use it, and how to inspect it. The cylindricity tolerance is one of the fourteen geometric characteristic symbols and is one of the four direct form tolerances. The cylindricity tolerance is used to limit the deviation of a cylindrical surface of a part.

# TERMS AND CONCEPTS

## Cylindricity

*Cylindricity* is the condition of a surface of revolution in which all points of the surface are equidistant from a common axis. An example of cylindricity is shown in Figure 12-1.

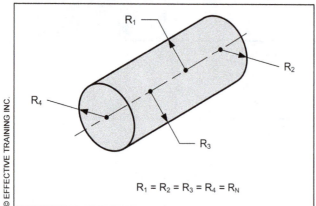

*FIGURE 12-1  Cylindricity Example*

Cylindricity is a perfect condition. It results in a perfect cylinder. Because cylindricity (the perfect condition) cannot be produced, it is necessary to specify an allowance for deviation from cylindricity. Figure 12-2 shows an example of a cylindricity deviation.

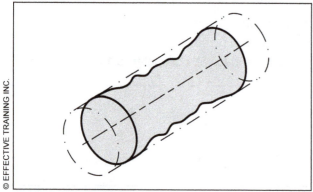

*FIGURE 12-2 Cylindricity Deviation Example*

## Cylindricity Tolerance

A *cylindricity tolerance* is a geometric tolerance that defines the deviation permitted on a cylindrical surface. A cylindricity tolerance applies along the entire cylindrical surface simultaneously.

A cylindricity tolerance:

- Is one of the four direct form tolerances
- Never uses a datum feature reference
- Can only be applied to a single cylindrical feature
- Applies to the entire surface simultaneously

An example of a cylindricity tolerance is shown in Figure 12-3.

## Cylindricity Tolerance Zone

The tolerance zone for a cylindricity tolerance is the space between two coaxial cylinders. The distance between the cylinders is equal to the cylindricity tolerance value. The tolerance zone applies to the entire cylindrical surface. All points of the cylindrical surface must be within the tolerance zone. The diameter of the tolerance zone is established from the toleranced cylinder. Figure 12-3 shows the cylindricity tolerance zone.

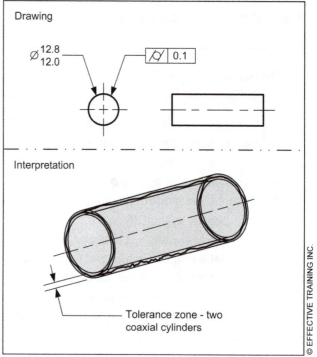

*FIGURE 12-3 Cylindricity Tolerance and Tolerance Zone*

For an external diameter, the size of the outer cylinder is determined by circumscribing a cylinder around the high points of the toleranced surface. The radial distance to the inner cylinder is equal to the cylindricity tolerance value.

For an internal diameter, the size of the inner cylinder is determined by inscribing a cylinder inside the high points of the toleranced surface. The radial distance to the outer cylinder is equal to the cylindricity tolerance value.

Where a cylindricity tolerance is specified, it applies to the entire surface. The surface may have any type of form deviation within the tolerance zone. Typical form deviations for a cylindrical surface are straightness, lobing, ovality, taper, and irregularities of no specific shape. A cylindricity tolerance simultaneously limits the straightness, roundness, and taper of a diameter. Surface texture deviations are not included in a cylindricity tolerance zone.

Figure 12-4 shows examples of types of cylindricity deviations.

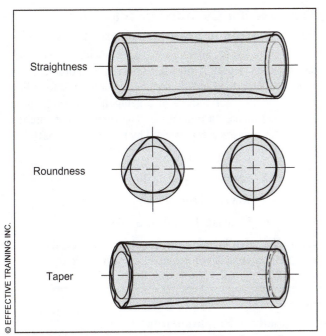

Straightness

Roundness

Taper

© EFFECTIVE TRAINING INC.

**FIGURE 12-4  A Cylindricity Tolerance Zone May Contain Any Combination of These Form Deviations**

## How Rule #1 Affects the Cylindricity Deviation of a Surface

Wherever Rule #1 applies to a cylindrical regular feature of size, it restricts the cylindricity deviation of the surface. This is a result of the relationship between Rule #1 (perfect form at MMC) and the size dimension. A cylindricity tolerance cannot override Rule #1.

When the feature of size is at MMC, the surface must be perfectly cylindrical. As the size dimension departs from MMC, a cylindricity deviation equal to the amount of departure is permitted.

In Figure 12-5, the Rule #1 boundary automatically limits the cylindricity deviations. The maximum cylindricity deviation occurs where the feature of size is at LMC. The maximum cylindricity deviation may not be larger than the difference between the size tolerance limits.

Since Rule #1 provides an automatic cylindricity tolerance, a cylindricity tolerance should not be specified on a surface, unless it is a refinement of the dimensional limits of the size dimension.

### Design Tip

Rule #1 is an indirect form control. Cylindricity effects from Rule #1 and a size dimension are often not inspected; they are a result of the relationship between the MMC boundary and the size variations. If it desired to have the cylindricity of a surface inspected, a cylindricity tolerance should be specified on the drawing.

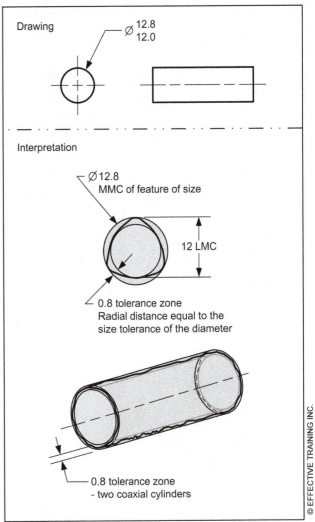

Drawing

Ø $\frac{12.8}{12.0}$

Interpretation

Ø12.8
MMC of feature of size

12 LMC

0.8 tolerance zone
Radial distance equal to the
size tolerance of the diameter

0.8 tolerance zone
- two coaxial cylinders

© EFFECTIVE TRAINING INC.

**FIGURE 12-5  Rule #1 Effects on Cylindricity Deviation of a Surface**

## How to Specify a Cylindricity Tolerance

A cylindricity tolerance is always applied to surface elements of a cylindrical shape. Figure 12-6 shows examples of how to specify a cylindricity tolerance on a drawing. A cylindricity tolerance feature control frame is attached to a leader directed to the surface or its extension line, but not to the size dimension.

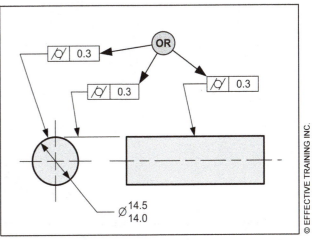

| ⌭ | 0.3 |

OR

| ⌭ | 0.3 |

| ⌭ | 0.3 |

Ø $\frac{14.5}{14.0}$

© EFFECTIVE TRAINING INC.

**FIGURE 12-6  Specifying a Cylindricity Tolerance**

## Modifiers Used With Cylindricity Tolerances

The modifiers permitted with a cylindricity tolerance application are shown in the chart in Figure 12-7.

| Tolerance Modifier | Can Be Applied To | Effect | Functional Application |
|---|---|---|---|
| Ⓕ* | Feature | Release the restraint requirement | Non-rigid parts with restraint notes |
| ⟨ST⟩* | Feature | Requires statistical process controls | Statistically derived tolerances or tolerances used in statistical tolerance analyses |

\* These modifiers are only introduced in this text

**FIGURE 12-7  Modifiers Used With Cylindricity Tolerances**

## Evaluate if a Cylindricity Tolerance Specification is Standard-Compliant

For a cylindricity tolerance to be a standard-compliant specification (legal), it must satisfy certain requirements. In order to make it easier to remember what to look for when evaluating the correctness of a geometric tolerance on a drawing, I created a mnemonic. The requirements are divided into four areas represented by the initials "CARE":

C - Check for datums

A - Assess the application

R - Review the modifiers

E - Evaluate the tolerance value

A flowchart that identifies the requirements for a standard-compliant cylindricity tolerance is shown in Figure 12-8.

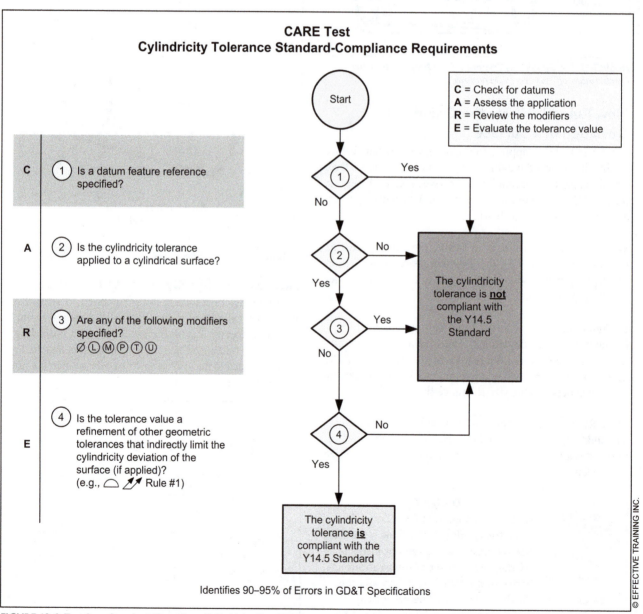

**FIGURE 12-8  Test for a Standard-Compliant Cylindricity Specification**

# CYLINDRICITY APPLICATIONS

## Real-World Applications for a Cylindricity Tolerance

Three real-world applications of a cylindricity tolerance are:

- Support (distribute loads or minimize wear)

- Sealing surface (lip seal of linear actuator shaft)

- Bearing applications

## Cylindricity Tolerance Applied to a Cylinder

A cylindricity tolerance is applied to a cylindrical surface by directing its leader line to a surface or to an extension line of a surface.

Where a cylindricity tolerance is applied to a cylindrical surface, the following conditions apply:

- The cylindricity tolerance does not override Rule #1.

- If Rule #1 applies, the cylindricity tolerance value must be less than the associated feature of size dimension tolerance.

- If Rule #1 does not apply, the cylindricity tolerance may be larger than the associated feature of size dimension tolerance.

- The tolerance zone is the space between two coaxial cylinders.

- The distance between the cylinders of the tolerance zone is equal to the cylindricity tolerance value.

- The tolerance zone applies along the entire length of the cylinder.

- The cylindricity tolerance limits straightness, taper, and circularity deviations of the surface.

- The associated feature of size must also be within its size limits.

- If Rule #1 applies, the diameter must also pass through its MMC boundary.

Figure 12-9 shows an example of a cylindricity tolerance applied to a cylindrical surface.

Where a cylindricity tolerance is applied to a surface that is associated with a size dimension, and Rule #1 applies to the size dimension, the cylindricity tolerance value must be less than the difference between the size tolerance limits. If Rule #1 does not apply (i.e., overridden, independency, stock size, etc.), the cylindricity tolerance value may be larger than the size tolerance limits.

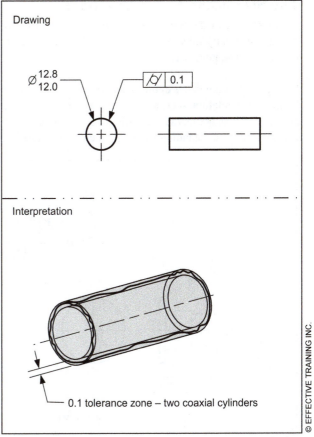

FIGURE 12-9 Cylindricity Tolerance Applied to a Cylindrical Surface

## The Significant Seven Questions

This section uses the Significant Seven Questions to interpret a cylindricity tolerance application. The questions are explained in Appendix B.

Figure 12-10 contains a cylindricity tolerance application. In the interpretation section of the figure the significant seven questions are answered. Notice how each answer increases your understanding of the geometric tolerance.

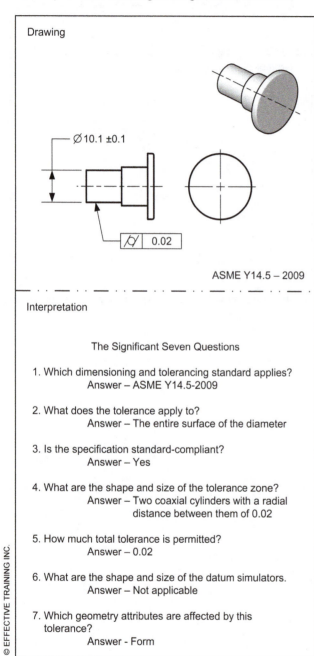

Drawing

Ø10.1 ±0.1

⌀ 0.02

ASME Y14.5 – 2009

Interpretation

The Significant Seven Questions

1. Which dimensioning and tolerancing standard applies?
   Answer – ASME Y14.5-2009

2. What does the tolerance apply to?
   Answer – The entire surface of the diameter

3. Is the specification standard-compliant?
   Answer – Yes

4. What are the shape and size of the tolerance zone?
   Answer – Two coaxial cylinders with a radial distance between them of 0.02

5. How much total tolerance is permitted?
   Answer – 0.02

6. What are the shape and size of the datum simulators.
   Answer – Not applicable

7. Which geometry attributes are affected by this tolerance?
   Answer - Form

**FIGURE 12-10 The Significant Seven Questions Applied to a Cylindricity Tolerance**

# VERIFICATION PRINCIPLES AND METHODS

## Verification Principles for Cylindricity Tolerances

This section contains a simplified explanation of the verification principles and methods that can be used to inspect a cylindricity tolerance applied to a cylindrical surface.

When a cylindricity tolerance is applied to a cylinder, the circularity deviation is verified by finding the distance between two coaxial cylinders that all of the points of the surface are located between. The inspection method must be able to collect a set of points and determine the distance between two cylinders that contains the set of points. Since location and orientation are not requirements of a cylindricity tolerance, the inspection method should not be influenced by these characteristics.

There are several ways cylindricity tolerances can be verified. In this section, we will look at one method.

## Verifying a Cylindricity Tolerance

Figure 12-9 shows a part with a cylindricity tolerance applied to the surface of a diameter. When verifying this part, three separate requirements must be checked, the size dimension, the Rule #1 boundary, and the cylindricity requirement. Verifying the cylindricity of a surface requires determining the distance of all points of a surface of revolution from the axis of that feature.

To accurately measure cylindricity deviations, a measuring system that uses computer technology is often used. One such device is a cylindricity gage (or precision spindle machine).

A *cylindricity gage* is an instrument used to measure cylindricity deviation. A cylindricity gage is similar to a roundness gage with the added ability to move axially as the turntable rotates. There are many different types of cylindricity gages used in industry today. A simplified description of a cylindricity gage follows.

A cylindricity gage consists of three main components: a rotating table with accurate spindle, a probe, and a computer with software that collects, analyzes, and displays the measurement results. The results are displayed graphically in several different user selected formats. A drawing of a cylindricity gage is shown in Figure 12-11.

A part is placed on the turntable, and a probe traces a spiral path on the part surface. The software analyzes the data and displays a map of the surface on a chart. The radial distance between two coaxial cylinders that circumscribe the high points of the surface and the cylinder that inscribes the low points of the surface is the cylindricity deviation.

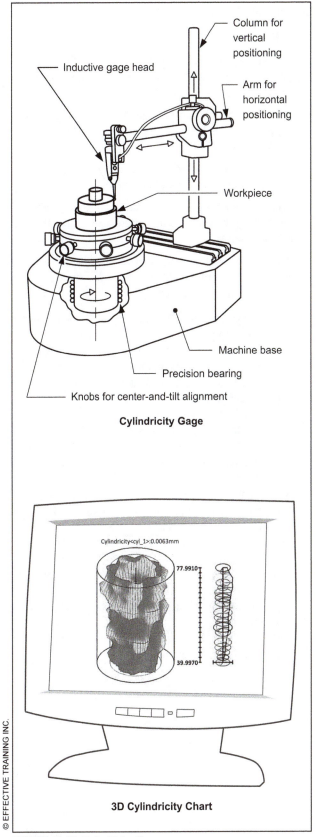

**Cylindricity Gage**

**3D Cylindricity Chart**

© EFFECTIVE TRAINING INC.

*FIGURE 12-11  Cylindricity Gage*

## SUMMARY

### Key Points

- Cylindricity is the condition of a surface of revolution in which all points of the surface are equidistant from a common axis.

- A cylindricity tolerance is a geometric tolerance that defines the cylindricity deviation permitted on a cylindrical surface.

- The tolerance zone for a cylindricity tolerance is the space between two coaxial cylinders.

- Where Rule #1 applies, a cylindricity tolerance value must always be less than the difference between the size tolerance limits.

- A cylindricity tolerance is applied to a cylindrical surface by directing its leader line to a surface or to an extension line of a surface.

- Where a cylindricity tolerance is applied, the following conditions apply:
  o The cylindricity tolerance does not override Rule #1.
  o If Rule #1 applies, the cylindricity tolerance value must be less than the associated feature of size dimension tolerance.
  o If Rule #1 does not apply, the cylindricity tolerance may be larger than the associated feature of size dimension tolerance.
  o The tolerance zone is the space between two coaxial cylinders.
  o The radial distance between the cylinders of the tolerance zone is equal to the cylindricity tolerance value.
  o The tolerance zone applies along the length of the cylinder.
  o The cylindricity tolerance limits roundness, straightness, and taper deviations of the surface.
  o The associated feature of size must also be within its size limits.
  o If Rule #1 applies, the diameter must also pass through its MMC boundary.

- Wherever Rule #1 applies to a cylindrical regular feature of size, it restricts the cylindricity deviation of the surface.

- Three real-world applications of a cylindricity tolerance applied to surface elements on a drawing are to limit the cylindricity deviation of surface elements for:
  o Support (distribute loads or minimize wear)
  o Sealing surface (lip seal of linear actuator shaft)
  o Bearing applications

- The "CARE" test can be used to determine if a cylindricity tolerance complies with the standard.

### Additional Related Topics

*These topics are recommended for further study to improve your understanding of cylindricity.*

| Topic | Source |
|---|---|
| Cylindricity and independency | ETI's *Advanced Concepts of GD&T* textbook |
| Cylindricity and the free state modifier | ETI's *Advanced Concepts of GD&T* textbook |
| Cylindricity and the (CF) modifier | ETI's *Advanced Concepts of GD&T* textbook |
| Cylindricity and the AVG modifier | ETI's *Advanced Concepts of GD&T* textbook |
| Calculating cylindricity tolerance values | ETI's *Advanced Concepts of GD&T* textbook |
| Using cylindricity in a tolerance stack | ETI's *Tolerance Stacks Using GD&T* textbook |

# QUESTIONS AND PROBLEMS

**Website Bonus Materials**
Additional questions are available at our website. To access bonus materials for this textbook, please visit:
www.etinews.com        www.etinews.com/textbookbonus

## True and False

*Indicate if each statement is true or false.*

T / F    1. A cylindricity tolerance can override Rule #1.

T / F    2. Cylindricity is the condition where all points of a circular element are equidistant from a common axis.

T / F    3. A cylindricity tolerance zone is a cylinder.

T / F    4. A cylindricity tolerance value must be less than half the size tolerance.

T / F    5. A cylindricity tolerance may contain the MMC modifier.

T / F    6. The specification of a cylindricity tolerance is incomplete without the Ø modifier.

T / F    7. A cylindricity tolerance may be verified with a fixed limit gage.

## Multiple Choice

*Circle the best answer to each statement.*

1. Which type of surface deviation is controlled by a cylindricity tolerance but not by a circularity tolerance?
   A. Ovality
   B. Taper
   C. Lobing
   D. None of the above

2. When verifying a cylindricity tolerance, the inspection method must be able to collect a set of points and determine the:
   A. Distance between two coaxial cylinders that contain the set of points
   B. Cylinder that circumscribes the set of points
   C. Cylinder that inscribes the set of points
   D. Distance between two coaxial circles that contain the set of points

3. Where Rule #1 applies to a cylindrical regular feature of size, the tolerance value of a cylindricity tolerance applied to the feature of size must be _____ the size tolerance.
   A. Less than
   B. Equal to
   C. Greater than
   D. None of the above

4. Which of the following modifiers may be applied with a cylindricity tolerance?
   A. Ⓜ                C. Ⓟ
   B. Ⓕ                D. ⌀

5. Which geometric tolerance can provide an indirect cylindricity control?
   A. ◎                C. ⫽
   B. ⌖                D. ⫽

6. A real-world application for a cylindricity tolerance is:
   A. Assembly (i.e., shaft and hole)
   B. Bearing journals on shafts
   C. Rotating clearance (i.e., shaft and housing)
   D. Sealing surface (between two shafts)

## Application Problems

*The application problems are designed to provide practice on applying the chapter concepts to situations that are similar to on-the-job conditions.*

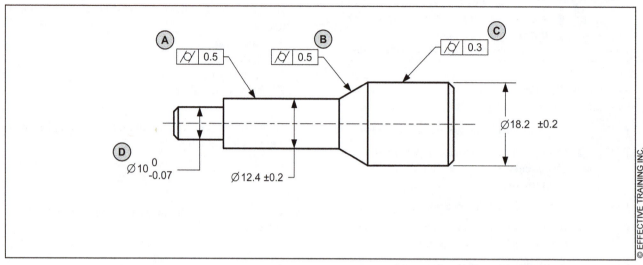

*Application questions 1-5 refer to the drawing above.*

1. Is the cylindricity tolerance labeled "B" standard-compliant? _____ If no, describe why. (Hint: use the CARE test.) _____

    _____

    _____

2. Could the circularity tolerance labeled "B" be replaced with cylindricity? _____ Why or why not? _____

    _____

    _____

    _____

3. What is the shape of the tolerance zone for the cylindricity control labeled "C"?
    A. Circular
    B. Cylindrical
    C. Two coaxial cylinders
    D. Two coaxial circles

4. What is the maximum cylindricity tolerance value that could be specified applied to diameter D?
    A. No limitation on tolerance value
    B. Less than 0
    C. Less than 10.07
    D. Less than 0.07

5. Use the Significant Seven Questions to interpret the circularity tolerance labeled "C."

    1) Which dimensioning and tolerancing standard applies?

        _____

    2) What does the tolerance apply to? _____

        _____

    3) Is the specification standard-compliant? _____

    4) What are the shape and size of the tolerance zone?

        _____

    5) How much total tolerance is permitted? _____

    6) What are the shape and size of the datum simulators?

        _____

    7) Which geometry attributes are affected by this tolerance?

        _____

# The Datum System

## Goal

Understand the datum system

## Performance Objectives

Upon completing this chapter, you should be able to:

1. Describe an implied datum (p.154)
2. Describe the shortcomings and consequences of implied datums (p.154)
3. Describe the datum system (p.155)
4. Describe three benefits of the datum system (p.155)
5. Define the terms "datum feature," "datum," "datum feature simulator," and "simulated datum" (p.155)
6. List seven types of datum feature simulators (p.156)
7. Describe six requirements of datum feature simulators (p.156)
8. Recognize the datum feature symbol (p.157)
9. Describe the datum reference frame symbol (p.157)
10. Describe four ways to specify a planar datum feature (p.158)
11. Explain how datum features are referenced in feature control frames (p.158)
12. Describe the six degrees of freedom (p.158)
13. Define "constraint" (p.159)
14. Describe a datum reference frame (p.160)
15. Describe what controls the relationships between datum features (p.160)
16. Describe the 3-2-1 Rule (p.161)
17. Explain the basis for selecting datum features (p.163)
18. Describe coplanar datum features (p.164)
19. Explain why multiple datum reference frames are used (p.164)
20. Explain the difference between datum-related and nondatum-related dimensions (p.165)

## New Terms

- Constraint
- Coplanar datum features
- Coplanarity
- Datum
- Datum feature
- Datum feature simulator (physical)
- Datum feature simulator (theoretical)
- Datum reference frame
- Datum system
- Degrees of freedom (DOF)
- Implied datum
- Partial datum reference frame
- Primary datum
- Rotation
- Secondary datum
- Simulated datum
- Tertiary datum
- 3-2-1 Rule
- Translation

## What This Chapter Is About

The datum system is one of the major components of the GD&T language; it provides the origin for part measurements, and it allows functional relationships to be communicated. The datum system lets the designer communicate which part surfaces are required to contact the datum feature simulators and in which sequence.

## TERMS AND CONCEPTS

The datum system is one of the major components of geometric dimensioning and tolerancing. Before we learn about datums, let's look at how we interpret dimensions using implied datums.

### Implied Datums

An *implied datum* is an assumed plane, axis, or point from which a dimensional measurement is made. Since the sequence in which they are used is not specified, each inspector must make assumptions about the setup for inspection. An example of implied datums is shown in Figure 13-1. The bottom and left sides of the block are considered implied datum features.

### Shortcomings of Implied Datums

Implied datums have two major shortcomings. First, they do not clearly communicate which surfaces should contact the inspection equipment, which means the inspector must make an assumption. Second, implied datums do not communicate the sequence in which each surface of the part should contact the inspection equipment. Figure 13-1 illustrates this problem.

### Consequences of Implied Datums

The use of implied datums results in two consequences:

- Functional parts are rejected

- Non-functional parts are accepted

The use of implied datums requires the inspector to assume which part surfaces should contact the inspection equipment and in what sequence. A part that meets specification when measured using the correct surfaces and oriented in the correct sequence may be rejected when measured from the wrong surfaces or when using the wrong sequence. Also, a part that does not meet its specification when measured while using the correct surfaces or in the correct sequence may be accepted when measured from an incorrect inspection setup.

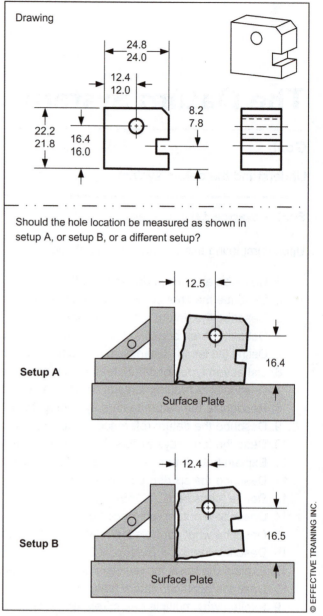

**FIGURE 13-1 Problem Using Implied Datums to Communicate the Sequence for Setting Up the Part for Measurement**

## The Datum System

The *datum system* is a set of symbols and rules on how to constrain a part to establish a relationship between the part and geometric tolerance zones.

From an engineering perspective, the datum system is used to define functional relationships between part features. From a measurement perspective, the datum system defines how a part is to be held for measurement. From a manufacturing perspective, the datum system communicates a clear description of the drawing requirement by indicating how the part will be held for inspecting the requirement. This allows manufacturing to evaluate options on how to hold the part in a manufacturing process and to compare different manufacturing processes.

The datum system is comprised of two major parts. The first component is identifying the features or features of size that are to be used to contact the measurement equipment. The second component is communicating the order in which the part surfaces contact the datums or measurement equipment.

The datum system constrains a part to restrict or remove some or all of the available degrees of freedom relative to a datum reference frame. This defines the relationship between the part, the geometric tolerance zones, and the datum reference frame. Figure 13-2 shows an example of the datum system.

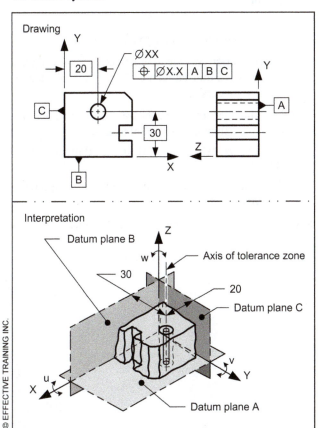

FIGURE 13-2  Datum System Example

## Benefits of the Datum System

The datum system provides three important benefits to industry. Each of these benefits helps to improve communications and reduce costs. The datum system benefits include:

- Precise communication of geometric relationships for proper fit and function
- Repeatable dimensional measurements
- Dimensional measurements that correlate to the fit and function requirements

**Design Tip**
Whenever dimensioning the location of a feature of size, do not use implied datums.

## Datum Terminology

A *datum feature* is a feature or feature of size that is identified with either a datum feature symbol or a datum target symbol.

A *datum* is a theoretically exact point, axis, line, plane, or combination thereof, derived from the theoretical datum feature simulator. It is the origin for dimensional measurements.

There are two types of datum feature simulators: theoretical and physical. A *datum feature simulator (theoretical)* is the theoretically perfect boundary used to establish a datum from a specified datum feature. A *datum feature simulator (physical)* is the physical boundary used to establish a simulated datum from a specified datum feature. A *simulated datum* is a datum established from a physical datum feature simulator.

The terms above describe both theoretical elements and physical entities used in the datum system. In industry, physical items, such as inspection equipment or gages (e.g., mandrels, chucks, collets, surface plates, etc., or digital data for mathematical simulation) of sufficient quality are used to establish simulated datums for dimensional measurements. Figure 13-3 shows the theoretical and physical datum feature simulators for a planar datum feature.

The datum feature symbol is shown in the Figure 13-4. Simply put, a datum feature is a part feature that is designated with a datum feature symbol.

**Author's Comment**
For clarity, in the interpretation section of the figures, the primary datum is colored in a light blue, and the secondary and tertiary datums are colored in darker shades of blue.

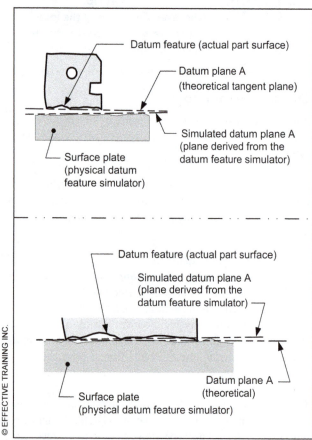

---

### TECHNOTE 13-1
#### Datum Features and Datums

- Datum features are features that exist on the part.
- Datums are theoretical reference planes, axes, lines, or points.
- A physical datum feature simulator is a gage element used to establish a simulated datum.
- For practical purposes, a simulated datum is considered a datum.

---

Datum feature (actual part surface)

Datum plane A (theoretical tangent plane)

Simulated datum plane A (plane derived from the datum feature simulator)

Surface plate (physical datum feature simulator)

---

Datum feature (actual part surface)

Simulated datum plane A (plane derived from the datum feature simulator)

Datum plane A (theoretical)

Surface plate (physical datum feature simulator)

© EFFECTIVE TRAINING INC.

**FIGURE 13-3  Datum Terminology**

## Datum Feature Simulators

Depending upon the type of datum feature (part geometry), a theoretical datum feature simulator can be:

- A tangent plane
- A maximum material boundary (MMB)
- A least material boundary (LMB) *
- A variable material boundary (RMB)
- A related actual mating envelope
- A datum target
- A mathematically defined contour *

* These theoretical datum feature simulator types are covered in ETI's *Advanced Concepts of GD&T* textbook.

---

The physical datum features simulators that correspond to the theoretical datum feature simulators listed above would be:

- A surface plate to simulate a tangent plane
- A fixed-size gage element (e.g., hole, width. pin, etc.) to simulate an MMB
- A fixed boundary established from points taken from the datum feature to simulate an LMB
- A variable boundary (e.g., precision chuck, collet, etc.) to simulate an RMB
- A gage element (e.g., gage pin, plate, etc.) to simulate a datum target
- A fixed-form gage element (shaped to match the math contour) to simulate a mathematically defined contour

---

#### Author's Comment
Whenever a planar surface is used as a primary datum feature, the form of the surface should be controlled. This will improve the stability of the part on its datum feature simulator.

---

## Datum Feature Simulator Requirements

Datum feature simulators (theoretical) have the following requirements:

- Perfect form
- Exact orientation relative to one another
- Exact location relative to other datum feature simulators, unless a translation modifier or movable datum target symbol is used
- Movable location where the translation modifier or movable datum target symbol is used
- Fixed in size where the MMB or LMB modifier is specified
- Adjustable in size where the datum feature applies at RMB

Physical datum feature simulators usually consist of precision gage elements that closely represent the theoretical datum feature simulator requirements.

## Datum Feature and Datum Reference Frame Symbols

Two symbols used in the specification of datums are the datum feature symbol and the datum reference frame symbol. The datum feature symbol is shown in Figure 13-4. Where a datum feature symbol is specified, it indicates that the part feature (surface or element) or feature of size is to be used to establish a datum. The datum feature symbol should be specified to be read from the bottom of a drawing.

When assigning datum letters, they should be assigned sequentially, without skipping any letters.

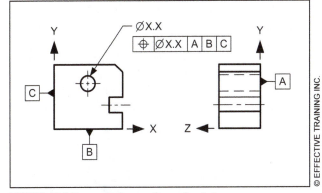

FIGURE 13-5 Datum Reference Frame Symbol

Where a datum feature symbol or a datum reference frame symbol is repeated on a drawing, it is not necessary to designate them as reference.

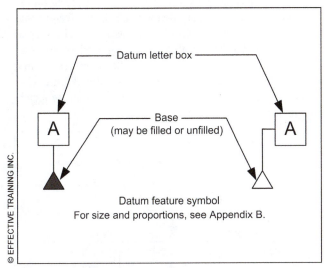

FIGURE 13-4 Datum Feature Symbol

### Author's Comment

Another instance where the datum reference frame symbol is required is where a customized datum reference frame is being specified. Customized datum reference frames are an advanced topic and covered in ETI's *Advanced Concepts of GD&T* textbook.

### Author's Comment

In the Y14.5 standard the term "datum feature" is used when referring to a "feature" or a "feature of size" that is used to establish a datum. This text will use the same convention.

### Author's Comment

If a datum feature is specified on a drawing but not referenced in a feature control frame, it should be removed from the drawing. Datum specifications that are not referenced in feature control frames are meaningless.

The second symbol used in the specification of datums is the datum reference frame symbol. The datum reference frame symbol is a trihedron with its axes labeled X, Y, and Z. The axis pointing out of the paper is not shown or labeled. The axes of the symbol often show coincident with surfaces of the part. The datum reference frame symbol identifies the X, Y, and Z axes of a datum reference frame. It also helps to associate the orthographic view to the 3D model coordinate system. An example is shown in Figure 13-5.

The datum reference frame symbol is optional on orthographic views, but it is required on axonometric views.

## Specifying Planar Datum Features

The method of indicating the datum feature symbol on a drawing determines the type of datum feature being specified. The placement can indicate a surface or a feature of size as the datum feature. For planar datum features, four common methods of attachment are shown in Figure 13-6.

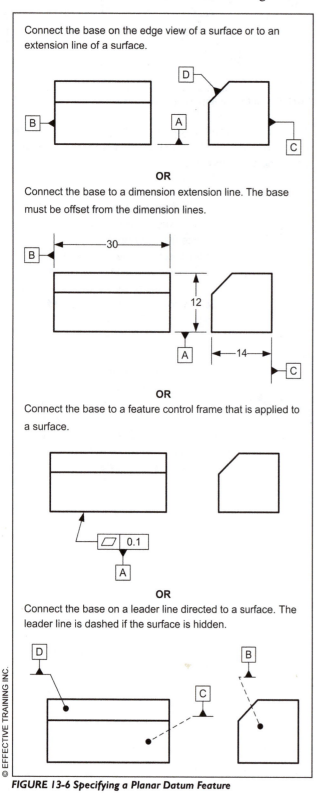

Connect the base on the edge view of a surface or to an extension line of a surface.

**OR**

Connect the base to a dimension extension line. The base must be offset from the dimension lines.

**OR**

Connect the base to a feature control frame that is applied to a surface.

**OR**

Connect the base on a leader line directed to a surface. The leader line is dashed if the surface is hidden.

*FIGURE 13-6 Specifying a Planar Datum Feature*

## Referencing Datum Features in Feature Control Frames

After datum features are specified, the drawing must also communicate when and how the datums should be used. This is typically done through the use of feature control frames. When feature control frames are specified, they reference the datums to be used for their measurement.

When feature control frames reference datums, they also specify the sequence for contacting the part to the datums or datum feature simulators (inspection equipment). The sequence is determined by reading the feature control frame datum feature references from left to right. Starting on the left, the first compartment denotes the datum feature that should contact the inspection equipment first. The second denotes the datum feature that should contact the inspection equipment second. The third denotes the datum feature that is to contact the inspection equipment third. Figure 13-7 shows an example of how to interpret the datum sequence in a feature control frame.

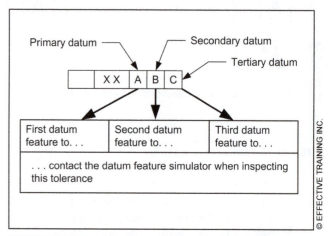

*FIGURE 13-7 Referencing Datums Features in Feature Control Frames*

## Degrees of Freedom

*Degrees of freedom (DOF)* – the movement of a part in space. A rigid part has six degrees of freedom: three translational degrees of freedom and three rotational degrees of freedom. *Translation* is the ability to move without rotating. *Rotation* is angular movement about an axis.

The constrained DOF are commonly shown using a 3D coordinate system symbol. The three translational DOF are labeled with the upper case letters X, Y, and Z. The three rotational DOF are labeled with the lower case letters u, v, and w. Figure 13-8 shows examples where the constrained DOF are labeled in the interpretation.

**Author's Comment**

The DOF that are constrained are not shown on an engineering drawing. They are used in this text to help you visualize the effects of the datum feature references in the feature control frame.

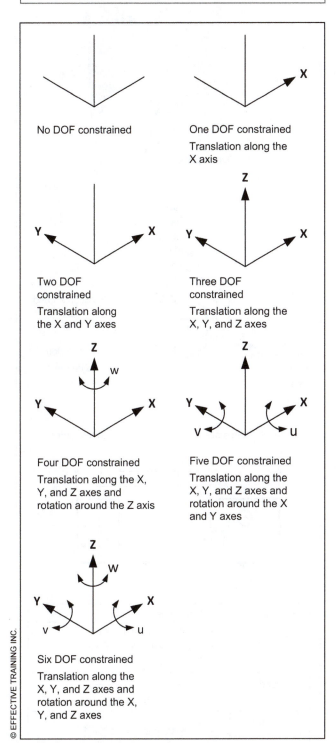

FIGURE 13-8  *Examples of Denoting Constrained Degrees of Freedom by Labeling the Axes of a 3D Coordinate System Symbol*

**TECHNOTE 13-2**
**Degrees of Freedom (DOF)**

Where DOF labels (X, Y, Z, u, v, w) are shown in the interpretation area of the figures of this text, the translational or rotational DOF that are constrained are labeled, and the DOF that are not labeled are not constrained.

## Constraint

*Constraint* is a limit to one or more DOF. As explained earlier in this chapter, when using the datum system, one or more DOF are being constrained. Constraint is necessary to define a relationship between the part, the datum reference frame, and the geometric tolerance zones. This relationship is necessary to achieve repeatable measurement of geometric tolerances.

Where a planar feature is referenced as a primary datum feature, three DOF are constrained: two rotational and one translational. See Figure 13-9.

**Author's Comment**

The use of the word "limit" in the definition of constraint means "reduce or remove." When using the datum system, you are reducing or removing DOF based on the datum features being referenced and boundary conditions.

| Feature Type | Planar (a) |
|---|---|
| On the Drawing | [A] |
| Datum and Datum Feature Simulator | Plane |
| Datum and Constraining DOF | |

FIGURE 13-9  *Degrees of Freedom Constrained*

## The Datum Reference Frame and Datum Feature Relationships

The primary purpose of the datum system is to constrain the movement of a part so that repeatable measurements can be made during inspection. When a part is free to move in space, it has six degrees of freedom. The six degrees of freedom are rotation around the X, Y, or Z axis and movement along the X, Y, or Z axis. They are shown in Figure 13-10.

In order to restrict the six degrees of freedom on a part with planar datums, it takes the use of three datum planes. When three datum planes are used, they are considered to be a datum reference frame. A *datum reference frame* is a set of three mutually perpendicular datum planes. The datum reference frame provides direction as well as an origin for dimensional measurements.

Simply put, a datum reference frame is a 3D coordinate system that the geometric tolerance zones are located and oriented within. The part is brought into contact with the coordinate system in a manner described by the datum references.

Some geometric tolerances require a part to have all six DOF constrained for their measurement. In these cases, all three planes of the datum reference frame are used.

Some geometric tolerances require fewer than six DOF constrained for their measurement. In these cases, a partial datum reference frame is used. A *partial datum reference frame* is where only one or two datum planes of the datum reference frame are used.

The planes of a datum reference frame have zero perpendicularity tolerance to each other by definition. Measurements are taken perpendicular from the datum planes. Figure 13-11 shows a datum reference frame for the part from Figure 13-10.

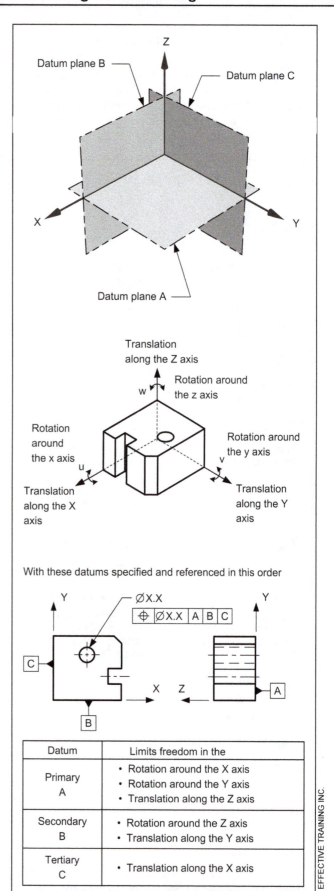

| Datum | Limits freedom in the |
|---|---|
| Primary A | • Rotation around the X axis<br>• Rotation around the Y axis<br>• Translation along the Z axis |
| Secondary B | • Rotation around the Z axis<br>• Translation along the Y axis |
| Tertiary C | • Translation along the X axis |

**FIGURE 13-10** *Planar Datum Reference Frame and Degrees of Freedom Constrained*

The 90° angles between datum planes have no tolerance because they are theoretical. The 90° angles between actual part surfaces have tolerance. The angular tolerance of the part surfaces may be specified on the drawing or in a general note. Figure 13-12 shows a part where the part surfaces (datum features) are not exactly 90° to each other. A measurement on this part will produce different results based on which part surface touches the datum reference frame first, second, and third. Therefore, the inspector must adhere to the datum sequence specified in the feature control frame.

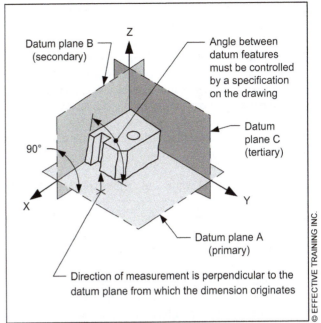

FIGURE 13-12 Datum Features That Are Not Perfect and Require Controls for Form and Orientation

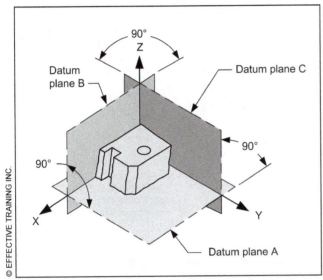

FIGURE 13-11 Planes of the Datum Reference Frame

When making a measurement for the location of a feature of size, the six DOF are constrained by using a datum reference frame. The method of bringing a part into contact with the planes of the datum reference frame has a significant impact on the measurement of the part dimensions.

---

### TECHNOTE 13-3
### Datum Feature Relationships

Datum features that are shown at 90° should be directly toleranced using geometric tolerances, otherwise their relationships will be governed by a general note or title block tolerances.

Feature control frames specify the order in which part surfaces are to contact the datum reference frame. A *primary datum* is the first datum referenced in a feature control frame. It the first datum that the part contacts in a dimensional measurement. The *secondary datum* is the second datum referenced in a feature control frame. It is the second datum that the part contacts in a dimensional measurement. The *tertiary datum* is the third datum referenced in a feature control frame. It is the third datum that the part contacts in a dimensional measurement.

Refer to Figure 13-13 while reading the explanation below.

A feature control frame is read from left to right. In this case, datum A is referenced first, datum B is referenced second, and datum C is referenced third.

The feature control frame indicates that the part should contact datum A first. The primary datum plane establishes the orientation of the part (stabilizes the part) to the datum reference frame. The part contacts the datum plane with at least three points of contact.

This primary datum plane constrains three degrees of freedom: translation along the Z axis, rotation around the X axis, and rotation around the Y axis. Three degrees of freedom remain unconstrained.

> *For more info...*
> See Paragraph 4.4.1 of Y14.5.

The feature control frame indicates that the part should contact datum B second. In this example, the secondary datum plane locates the part (constrains translation) and constrains the remaining rotational degree of freedom within the datum reference frame.

Where angular deviations exist between the primary and secondary datum features, the part may have a line contact with the secondary datum plane; therefore, it requires a minimum of two points of contact with the secondary datum plane.

The secondary datum restricts two additional DOF: rotation around the Z axis and translation along the Y axis. One degree of freedom remains unconstrained.

The feature control frame indicates that the part should contact datum C third. The tertiary datum plane locates the part (restricts part movement) within the datum reference frame.

The part may have a single point of contact with the tertiary datum; therefore, it requires a minimum of one point of contact with the tertiary datum plane. This tertiary datum restricts the last remaining degree of freedom: translation along the X axis.

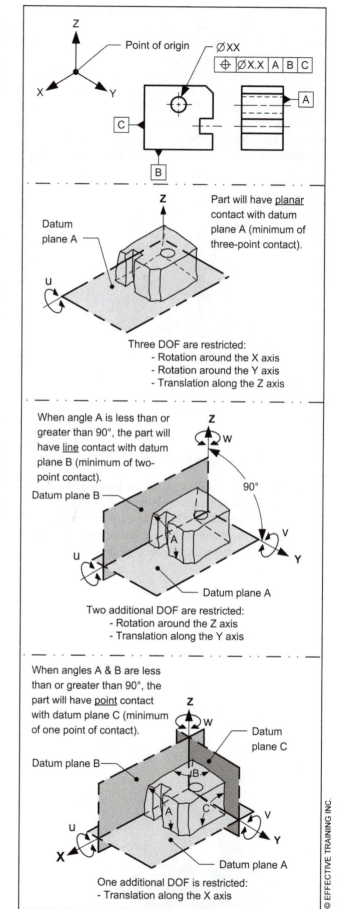

*FIGURE 13-13  Primary, Secondary, and Tertiary Datums*

## The 3-2-1 Rule

The 3-2-1 Rule defines the minimum number of points of contact for a part with its primary, secondary, and tertiary datum planes.

> The **3-2-1 Rule** – A part will have a minimum of three points of contact with its primary datum plane, two points minimum contact with its secondary datum plane, and a minimum of one point of contact with its tertiary datum plane.

The 3-2-1 Rule only applies to a part with all planar datums.

---

### TECHNOTE 13-4
### Datum Feature Optimization

The three-point minimum contact between a datum feature and its datum plane is the ideal condition. Form deviation of the datum feature can cause a one point contact which results in a part "rocking" on its primary datum plane. In these cases, the part must be stabilized/optimized on its primary datum plane. For additional information about this condition, see Y14.5, paragraph 4.11.2.

---

## DATUM APPLICATIONS

### Datum Feature Selection

Datum features are selected on the basis of part function and assembly conditions. The datum features are often the features that orient (stabilize) and locate the part in its assembly. For example, the cover shown in Figure 13-14 mounts on surface A and is located by diameter B. For assembly, the holes need to be located relative to the features that mount and locate the part to the mating part. Therefore, surface A and diameter B are designated as datum features.

After designating the datum features, the drawing must also communicate when and how the datum features should be used to control the geometric relationships required for function and assembly. For example, in Figure 13-14, the bolt holes are toleranced with a position tolerance that references the orienting and locating features to ensure that they will be aligned with the tapped holes in the mating part. Since the part is mounted on the surface (datum feature A), it establishes the orientation and becomes the primary datum feature reference. After the part is oriented by datum feature A, it is located by the large diameter (datum feature B). It establishes the location and becomes the secondary datum.

Since the bolt holes are equally spaced around the bolt circle diameter and there are no other features required to further orient the part in the assembly, a tertiary datum feature reference is not required.

Depending upon the complexity and functions of a part, it may have many datum features. Most simple parts will have at least three datum features and one datum reference frame. However, complex parts often have multiple datum reference frames and many datum features.

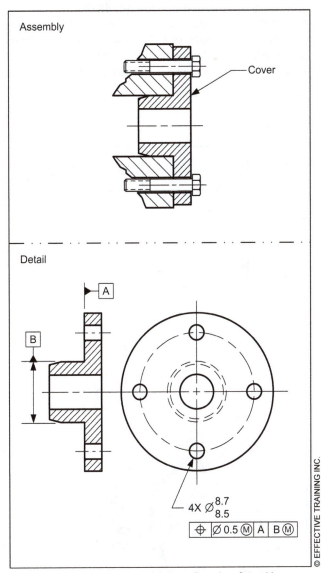

FIGURE 13-14 *Datum Feature Selection Based on Assembly Conditions*

---

### TECHNOTE 13-5
### Datum Selection

Datum features are selected on the basis of part function and assembly conditions; they are often the features that orient (stabilize) and locate the part in its assembly.

---

## Coplanar Datum Features

*Coplanarity* is two or more surfaces having all elements on one plane. *Coplanar datum features* are two or more datum features that are on the same plane. A single datum plane can be established from multiple surfaces. Figure 13-15 illustrates coplanar datum features. In this case, a datum feature symbol is attached to a profile control. The profile control limits the flatness and coplanarity of the surfaces. The note following this profile control—"two surfaces"—also denotes that datum feature A is comprised of two surfaces.

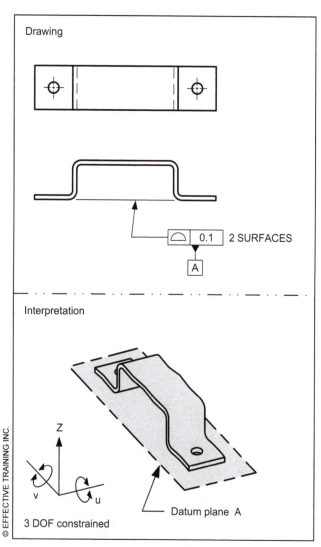

FIGURE 13-15 Coplanar Datum Features

## Multiple Datum Reference Frames

In certain cases, the functional requirements of a part call for the part to contain more than one datum reference frame. A part may have as many datum reference frames as needed to define its functional relationships. The datum reference frames may be at right angles or at angles other than 90°. Also, a datum plane may be used in more than one datum reference frame. Figure 13-16 shows a part with four datum reference frames.

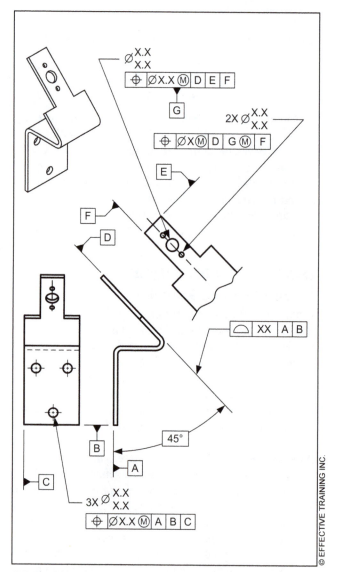

FIGURE 13-16 Multiple Datum Reference Frames

### Author's Comment

On complex parts, it is common to have multiple datum reference frames. I have seen drawings with as many as thirty different datum reference frames.

## Datum-Related Versus Nondatum-Related Dimensions

Only dimensions that are related to a datum reference frame through geometric tolerances should be measured in a datum reference frame. If a dimension is not associated to a datum reference frame with a geometric tolerance and basic dimensions relating the feature or feature of size to the datums, then there is no specification on how to locate the part in the datum reference frame.

In Figure 13-17, the hole locations are related to the datum reference frame, D primary, E secondary, and F tertiary. During inspection of the hole location dimensions, the part should be mounted in datum reference frame D-E-F, but the length, width, and height dimensions are not related to the datum reference frame. The length, width, and height dimensions are feature of size dimensions. During inspection of feature of size dimensions, corner radii, and chamfers, the part should not be mounted in the datum reference frame.

---

### TECHNOTE 13-6
### Datum-Related Dimensions

Only dimensions related to a datum reference frame through the use of geometric tolerances are measured relative to a datum reference frame.

---

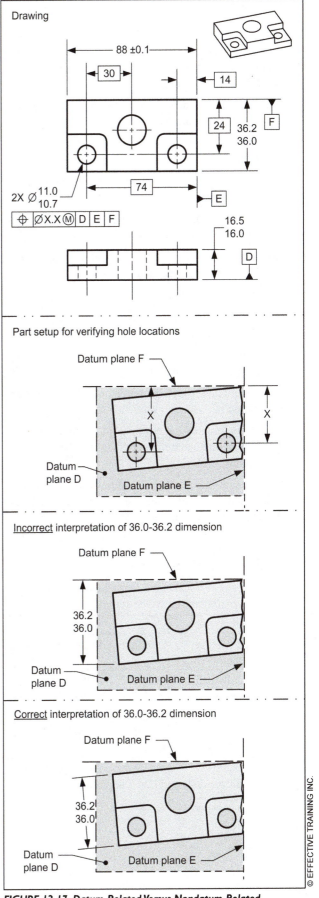

**FIGURE 13-17  Datum-Related Versus Nondatum-Related Dimension**

## SUMMARY

### Key Points

- The purpose of the datum system is to constrain a part relative to one or more datums to be able to make repeatable measurements.

- All parts have six degrees of freedom: three translational and three rotational.

- The datum system aids in communicating functional relationships.

- There are two types of datum feature simulators: theoretical and physical.

- Theoretical datum feature simulators have perfect form, location, and orientation.

- Datum features should be selected on the basis of the part functional requirements.

- Constraint is a limit to one or more DOF.

- The planes of a datum reference frame have perfect orientation to each other.

- The orientation deviation between datum features must be defined on the drawing.

- The variation between datum features affects the part orientation in the datum reference frame.

- Due to communicating multiple functional relationships, parts often contain more than one datum reference frame.

- Only dimensions related to a datum reference frame (with feature control frames) should be measured from the datum reference frame.

## Additional Related Topics

*These topics are recommended for further study to improve your understanding of the datum system.*

| Topic | Source |
|---|---|
| Customized datum reference frame | *ASME Y14.5-2009, Para. 4.22* |
| Mathematically defined contours used as datum features | *ASME Y14.5-2009, Para. 4.13* and ETI's *Advanced Concepts of GD&T* textbook |
| Least material boundary | *ASME Y14.5-2009, Para. 4.11.7* |
| Interrelationships between multiple datum reference frames | ETI's *Advanced Concepts of GD&T* textbook |

## QUESTIONS AND PROBLEMS

**Website Bonus Materials**
Additional questions are available at our website. To access bonus materials for this textbook, please visit:
www.etinews.com    www.etinews.com/textbookbonus

### True and False

*Indicate if each statement is true or false.*

T / F    1. An implied datum is an assumed plane, axis, or point from which a dimensional measurement is made.

T / F    2. A datum is a part surface, axis, or center plane.

T / F    3. The datum system constrains a part to restrict or remove some or all of the available DOF relative to a datum reference frame.

T / F    4. Datum sequence is usually communicated through feature control frames.

T / F    5. A part may have no more than three datum features.

T / F    6. Datum features should be selected based on part functional requirements.

T / F    7. Implied datums communicate which surfaces should contact the inspection equipment.

T / F    8. One benefit of using the datum system is the ability to make repeatable dimensional measurements.

T / F    9. A datum feature is theoretical.

## Multiple Choice

*Circle the best answer to each statement.*

1. A _____ is a gage element used to establish a simulated datum:
   A. Datum feature
   B. Datum feature simulator (physical)
   C. Datum feature simulator (theoretical)
   D. Datum reference frame

2. _____ is a limit to one or more DOF.
   A. Constraint
   B. Restraint
   C. Simulated datum
   D. Datum feature simulator

3. All parts have ___ DOF.
   A. 2
   B. 4
   C. 6
   D. 8

4. How many datum reference frames can a part have?
   A. 1
   B. 2
   C. 3
   D. No limit (theoretically)

5. On a part with all planar datums, what is the minimum number of points of contact with the datum feature simulator for a tertiary datum?
   A. 3
   B. 2
   C. 1
   D. None (theoretically)

6. Datum feature simulators (theoretical) must:
   A. Have perfect form
   B. Have location tolerance relative to one another
   C. Be variable in size where the MMB or LMB modifier is specified
   D. All of the above

7. A datum feature simulator may be:
   A. A tangent plane
   B. A variable material boundary
   C. A datum target
   D. All of the above

## Application Problems

*The application problems are designed to provide practice on applying the chapter concepts to situations that are similar to on-the-job conditions.*

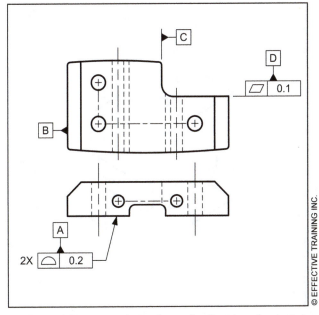

© EFFECTIVE TRAINING INC.

*Application questions 1–5 refer to the drawing above.*

1. Which datum is the primary datum?_____
   How did you determine your answer? _____
   _____

2. Which datum involves coplanar datum features?
   _____

3. If datum A was the primary datum, what is the minimum number of points of contact (theoretical) between the datum feature(s) and the simulated datum?
   _____

4. If datum A was the primary datum, how many DOF would it constrain?_____

5. What controls the orientation deviation between datum features A and B?_____
   _____

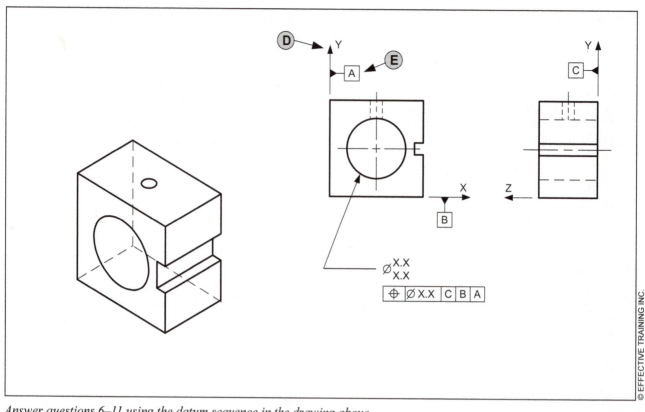

*Answer questions 6–11 using the datum sequence in the drawing above.*

6.  List the six degrees of freedom. _____

    _____

    _____

7.  List the minimum number of points of contact that the part must have with . . . .
    a. Datum plane C _____

    b. Datum plane B _____

    c. Datum plane A _____

8.  List the degrees of freedom that contact with datum plane E will restrict. _____

    _____

    _____

9.  List the degrees of freedom that contact with datum plane B will restrict (after the part contacts the datum plane E).

    _____

    _____

    _____

10. The name of the symbol labeled "D" is _____

11. The name of the symbol labeled "E" is _____

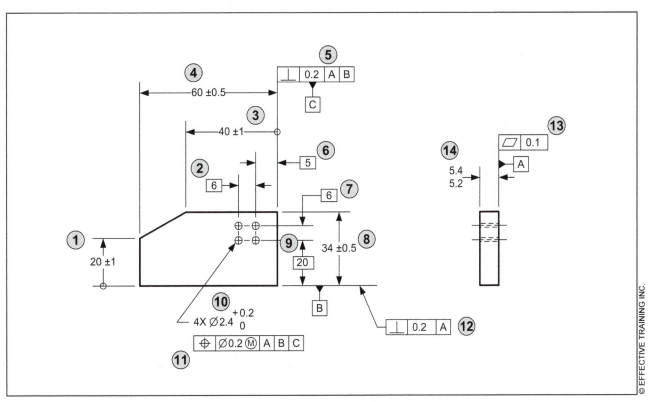

12. Fill in the chart based on the figure above.

| Dimension Number | Dimension measured from datum reference frame ABC | Dimension is a feature of size; not related to datums | Describe how the part is held to measure the dimension |
|---|---|---|---|
| 1 | | | |
| 2 | | | |
| 3 | | | |
| 4 | | | |
| 5 | | | |
| 6 | | | |
| 7 | | | |
| 8 | | | |
| 9 | | | |
| 10 | | | |
| 11 | | | |
| 12 | | | |
| 13 | | | |
| 14 | | | |

# Datum Targets

## Goal

Interpret applications of datum targets

## Performance Objectives

Upon completing this chapter, you should be able to:

1. Describe datum targets (p.172)
2. Recognize the datum target symbols (datum target, movable datum target, datum target leader line, point, line, and area) (p.172)
3. Explain where datum targets should be used (p.172)
4. Describe how to establish a complete datum reference frame using datum targets (p.173)
5. Describe a datum feature simulator for a point datum target application (p.174)
6. Describe a datum feature simulator for a line datum target application (p.175)
7. Describe a datum feature simulator for an area datum target application (p.176)
8. Describe the datum feature simulator for datum targets applied to a non-planar surface (p.177)
9. Interpret applications of datum targets applied to a partial surface, to offset parallel surfaces, on irregular surfaces, and using the movable datum target symbol (p.177)

## New Terms

- Datum target
- Datum target area
- Datum target line
- Datum target point
- Datum target symbol
- Movable datum target symbol

## What This Chapter Is About

Datum targets are part of the datum system. They enable repeatable datums to be established from uneven or warped surfaces and from partial surfaces. Datum targets are symbols that describe datum feature simulators that are used to establish datums.

This chapter explains where to use and how to interpret datum targets.

## TERMS AND CONCEPTS

### Datum Targets

*Datum targets* are a set of symbols that describe the shape, size, and location of datum feature simulators that are used to establish datum planes, axes, or points. Datum targets are shown on an engineering drawing and describe datum feature simulators. The use of datum targets enables a stable and repeatable relationship of a part with its datum feature simulator.

### Datum Target Symbols

A datum target application includes two types of symbols: a datum target symbol and symbols that describe the datum feature simulator elements.

A *datum target symbol* is a circular symbol with a horizontal line dividing the symbol into two halves. The bottom half denotes the datum letter and the target's number. The top half denotes the datum feature simulator size, when applicable. The datum target symbol is shown in Figure 14-1.

A *movable datum target symbol* consists of the datum target symbol with a pointed end as illustrated in Figure 14-1. The movable datum target symbol indicates that the datum feature simulator is movable.

The leader line from the datum target symbol indicates whether the datum feature simulator for the target contacts the part on the surface shown (near side) or the surface that is hidden (the opposite side). If the leader line is solid, the datum feature simulator for the target contacts the part on the surface shown.

Figure 14-1 shows the solid and hidden datum target symbol leader lines. Examples of solid and dashed datum target leader lines are shown in Figure 14-2. If the leader line is dashed, the datum feature simulator contacts the part on the hidden surface of the part in the view shown. See datum targets A1, A2, and A3 in Figure 14-2.

A datum target leader line connects the datum target symbol with the datum target point, line, or area symbol. The datum target leader line does not terminate with an arrow. It simply touches the datum target point, line, or area symbol.

There are three different datum target types: a datum target point, datum target line and datum target area. The symbol for a datum target point is an X-shaped symbol, consisting of a pair of lines intersecting at 90°. The symbol for a datum target line is a phantom line. The symbol for a datum target area is a phantom outline of the area that is crosshatched.

The symbols for a datum target point, line, and area are shown in Figure 14-1. These symbols are used to indicate the shape of the datum feature simulator and the contact with the part surface.

| Symbol | Description |
|---|---|
| ⊖ A1 | **Datum Target**<br>The bottom half of the symbol contains the datum letter and target number |
| ⊖ A1 ▷ | **Movable Datum Target**<br>The bottom half of the symbol contains the datum letter and target number. The pointed end indicates the datum feature simulator is movable. |
| (solid leader line) | **Datum Target Solid Leader Line**<br>When the datum target identification symbol leader line is solid, it denotes that the datum target exists on the visible side of the part in the view shown. |
| (dashed leader line) | **Datum Target Dashed Leader Line**<br>When the datum target identification symbol leader line is dashed, it denotes that the target exists in the back side of the part in the view shown. |
| ✕ | **Datum Target Point**<br>A datum target point symbol denotes that the datum feature simulator has a theoretical point contact with the theoretical part surface. |
| – – – | **Datum Target Line**<br>A datum target symbol denotes that the theoretical datum feature simulator contacts the theoretical line element on the theoretical part surface. |
| ⊘ (crosshatched area) | **Datum Target Area**<br>A datum target area denotes that the shape of the theoretical datum feature simulator contacts that area on the theoretical part surface. |

**FIGURE 14-1** *Symbols Used with Datum Targets*

© EFFECTIVE TRAINING INC.

### Where to Use Datum Targets

Datum targets should be specified on parts where it is not practical (or possible) to use an entire surface as a datum feature. A few examples are castings, forgings, irregular shaped parts, plastic parts, large parts, and weldments. These types of parts often do not have a planar surface to use as a datum feature, or the datum feature is likely to be warped or bowed resulting in an unstable contact with a datum feature simulator.

Where a datum feature is warped or bowed, the part will rock or wobble on its datum feature simulator. Where datum targets are specified, the part will have only one orientation with the datum feature simulator and repeatable measurements can be made.

When using planar datums, the datum feature references in a feature control frame indicate which datum is primary, secondary, and tertiary (as explained on page 158). The same practice applies to datum targets. The datum feature references indicate which datum is primary, secondary, and tertiary.

---

**TECHNOTE 14-1**
**Datum Target Usage**

Datum targets should be used wherever:

- It is not practical to use the entire surface as a datum feature.
- The designer suspects the part may rock or wobble when the datum feature contacts the datum feature simulator.
- Only a portion of the feature is used in the function of the part.

---

## Guidelines for Establishing a Complete Datum Reference Frame With Datum Targets

The guidelines below are for establishing and using a complete datum reference frame (a datum reference frame that constrains all six DOF):

1. The datum targets should be located and defined with basic dimensions to ensure repeatability between the part and the datum feature simulators. Standard gage tolerances apply to the basic dimensions (see text conventions).

2. The datum reference frame must constrain all six degrees of freedom.

3. The part dimensioning must ensure that the part will have only one orientation and location in the datum feature simulator.

---

*Author's Comment*
The Y14.5 standard allows toleranced dimensions to be used to locate datum targets; however, the most common practice in industry is to use basic dimensions.

Basic dimensions should be used to locate datum targets relative to each other and to the additional datums of the datum reference frame. Using basic dimensions assures that there will be minimum variation between gages. Where basic dimensions are used, gage tolerances apply.

---

A datum target application is shown in Figure 14-2. This application uses datum target points, lines, and areas. When a datum target application is indicated on a drawing, the datum feature identification symbol may also be used, as shown in Figure 14-2B.

A  Datum target identification symbol

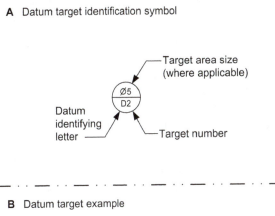

B  Datum target example

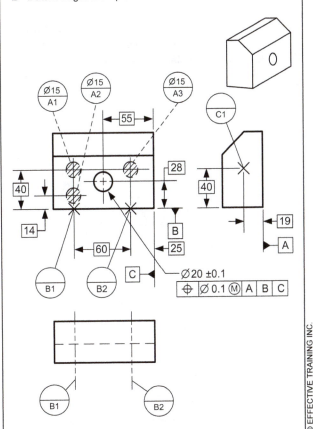

**FIGURE 14-2** *Datum Target Application*

---

*Author's Comment*
When interpreting a drawing, it is important to be able to identify which DOF are constrained for the datums referenced in each geometric tolerance. This also applies when datum targets are used.

For each datum target application in this chapter, examine the datums referenced and list the DOF constrained.

## Datum Target Points, Lines, and Areas

### Datum Target Points

A *datum target point* is a designated point used in establishing a datum. A datum target point is indicated by the datum target point symbol. A datum target point symbol may be shown two ways on a drawing. The symbol may be shown and dimensioned on a plan view of the surface to which it is being applied, as shown in Figure 14-3A.

Where a plan view is not available, the symbol can be shown and dimensioned in two adjacent views as shown in Figure 14-3B. Basic dimensions are used to locate datum target points relative to each other and to the other datums on the part. Using basic dimensions assures that there will be minimum variation between gages. A spherical-tipped gage pin is often used as the datum feature simulator for a datum target point, as shown in Figure 14-4.

Figure 14-4 shows an application of a datum target point used to establish a tertiary datum. The datum target point is simulated with a spherical-tipped gage pin. The simulated datum exists at the top of the gage pin and is perpendicular to both datum planes A and B. Portions of the actual part surface may be above or below the simulated datum plane.

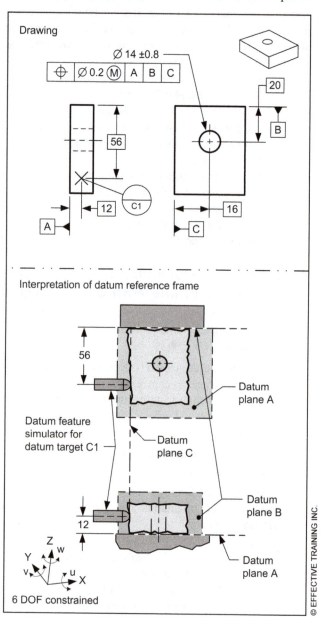

FIGURE 14-4 *Datum Target Point Example*

Where datum target points are used to establish a primary datum from a planar surface, three target are used. Where datum target points are used to establish a secondary datum, two target points are used. Where datum target points are used to establish a tertiary datum, one target point is used.

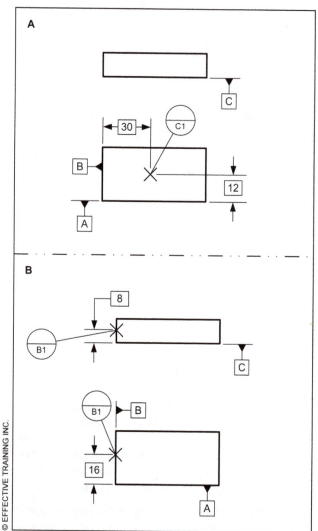

FIGURE 14-3 *Datum Target Point Specifications*

## Datum Target Line

A *datum target line* is a designated line used in establishing a datum. A datum target line symbol may be specified three different ways on a drawing.

The symbol may be shown and dimensioned on a plan view of the surface to which it is being applied, as shown in Figure 14-5A. Where a plan view is not available, the target symbol can be shown and dimensioned in an edge view, as shown in Figure 14-5B. Since the edge view of a line is a point, where a line target is shown in an edge view, the point symbol is used. The third method is to show and dimension the line symbol in the plan view and the point symbol in the edge view, as shown in Figure 14-5C.

Basic dimensions should be used to locate datum target lines relative to each other and to the additional datums on the part. Using basic dimensions assures that there will be minimum variation between gages.

**Author's Comment**
Using the method from Panel B in Figure 14-5 is not recommended. This method could be mistaken for a point target.

The side of a gage pin is often used as the datum feature simulator for a datum target line, as shown in Figure 14-6.

Figure 14-6 shows an application of a datum target line used to establish a tertiary datum. The datum target line is simulated with the side of a cylindrical gage pin. The simulated datum exists at the tangent of the gage pin and is perpendicular to both datum planes D and E. Portions of the actual part surface may be above or below the simulated datum plane.

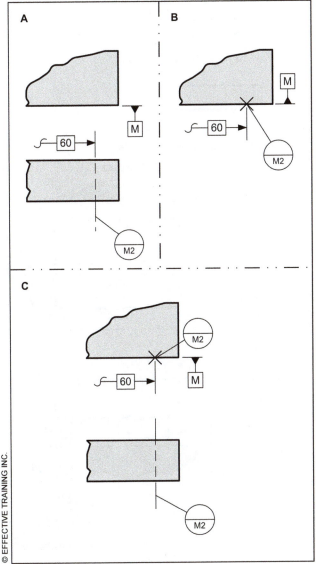

**FIGURE 14-5 Datum Target Line Specification**

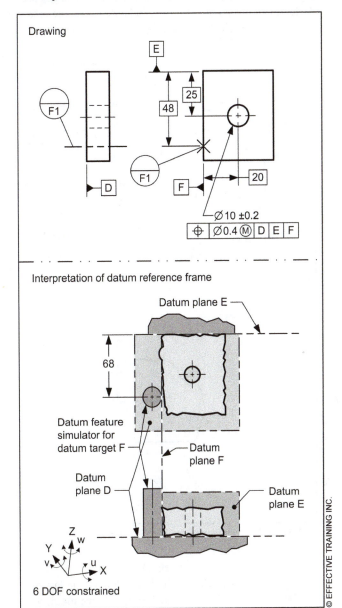

**FIGURE 14-6 Datum Target Line Example**

## Datum Target Area

A *datum target area* is a designated area used in establishing a datum. A datum target area symbol may be specified various ways on a drawing. The symbol may be shown and dimensioned on a plan view of the surface to which it is being applied as shown in Figure 14-7A. In this figure, a rectangular datum target area is specified. The area is indicated with a crosshatched phantom outline. The size of the area is defined with basic dimensions.

In Figure 14-7B, a circular datum target area is specified. The area is indicated with a crosshatched phantom outline. The size of the area is defined with basic dimensions. The size of a circular area may be shown as a separate dimension or indicated inside the top half of the datum target symbol.

Figure 14-7C also indicates a circular datum target area. A datum target point symbol is indicated, and the size of the datum is indicated in the top half of the datum target symbol.

A flat-tipped gage element is often used as the datum feature simulator for a datum target area, as shown in Figure 14-8. The figure shows an application of datum target areas used to establish a primary datum. The datum target areas are simulated with three cylindrical gage pins with the same shape as the nominal part contour. The simulated datum exists at the tangent of the gage pins. Portions of the actual part surface may be above or below the simulated datum plane.

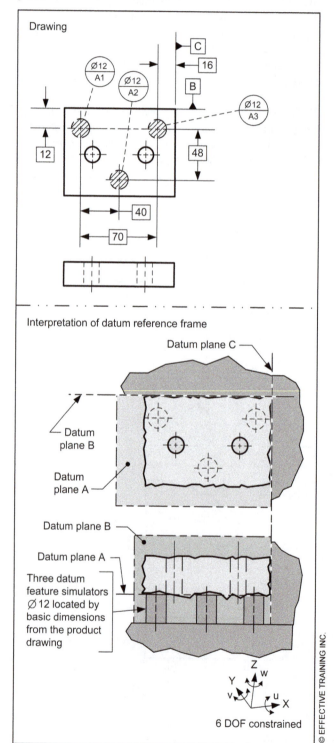

FIGURE 14-8 Datum Target Area Example

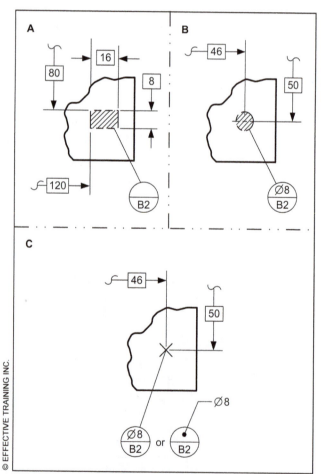

FIGURE 14-7 Datum Target Area Specification

---

### Author's Comment

Using the method from Panel C in Figure 14-7 is not recommended. This method could be mistaken for a point target.

© EFFECTIVE TRAINING INC.

## Datum Targets on Non-Planar Surfaces

Where a datum target line or area is indicated on a non-planar surface, the datum feature simulator follows the contour of the surface. Figure 14-9 shows an application of a datum target line and datum target area used on a cylindrical surface. The datum target line may be simulated with a knife-edged contracting cylindrical gage element. The simulated datum is a point that exists at the center of the gage element. The datum target area may be simulated with a contracting cylindrical gage element that is 5mm wide. The simulated datum is an axis that exists at the axis of the cylindrical gage element.

# DATUM TARGET APPLICATIONS

## Datum Target Application – Partial Surface

In certain applications, part function requires a datum to be established from a portion of a surface or a feature of size.

Datum targets can be used to establish a simulated datum from a feature of size. Figure 14-10 shows an application that uses datum targets to define gage elements that contact partial surfaces. In this figure, datum targets A1 and B1 designate gage elements that contact only a portion of the datum features. The datum feature simulator for datum target A1 is an annulus-shaped flat area. Where a datum target area (or line) is shown on a cylindrical surface like datum target B1, it wraps around the cylindrical surface. The datum feature simulator for datum target B1 is a adjustable cylinder (e.g., a collet) 5mm wide.

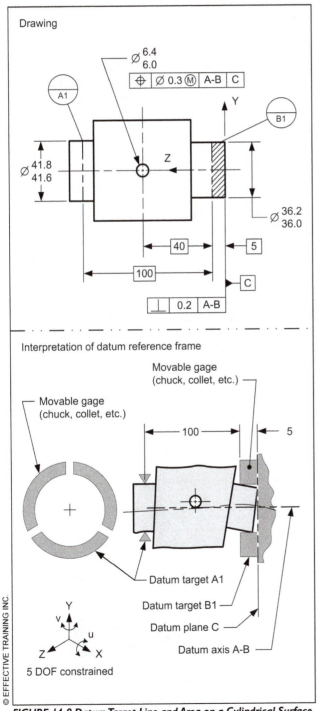

**FIGURE 14-9 Datum Target Line and Area on a Cylindrical Surface**

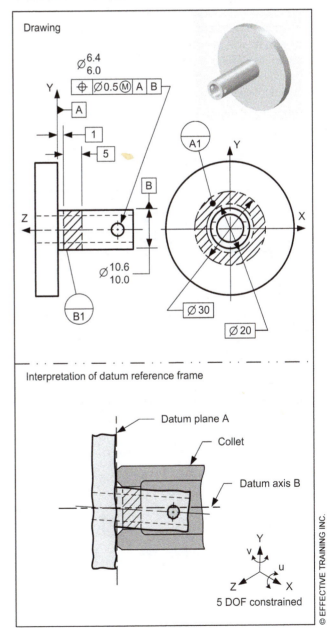

**FIGURE 14-10 Datum Target Area Application**

## Datum Target Application – Offset Parallel Surfaces

In certain applications, part function may require establishing a datum from two offset parallel surfaces. Datum targets can be used to establish a simulated datum from two offset parallel surfaces. Figure 14-11 shows an application that uses datum targets to define gage elements that contact multiple surfaces. In this figure, datum target A1 is on one surface and datum targets A2 and A3 are on a second offset parallel surface. Since all three datum targets are point targets, the datum feature simulators are spherical-tipped gage elements.

Since the datum targets are on offset surfaces, a basic dimension is used to define the offset between the datum feature simulators. The datum plane should contain at least one of the datum targets. Although not required, a datum feature symbol may be associated with either of the surfaces where the datum targets are shown. The datum feature symbol may also be shown on a plane offset from the part by basic dimensions. Where targets are used to establish a primary datum, all size degrees of freedom usually need to be constrained to achieve repeatability.

### Author's Comment

Beware, the drawing in Figure 14-11 does not define a repeatable relationship between the part and the datum feature simulators. In most cases, secondary and tertiary datum references are required when the primary datum is established with datum targets.

Constraint of all six degrees of freedom is usually needed to achieve repeatability between the part and the datum feature simulators.

## Datum Target Application – Irregular Surfaces

In certain applications, part function requires establishing a datum from irregular surfaces. Datum targets can be used to establish a simulated datum from irregular surfaces. Figure 14-12 shows an application that uses datum targets.

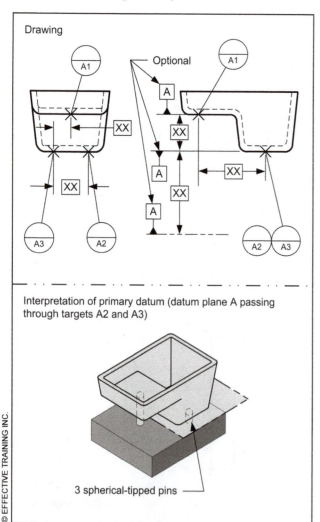

Interpretation of primary datum (datum plane A passing through targets A2 and A3)

3 spherical-tipped pins

*FIGURE 14-11 Datum Targets on Offset Parallel Surfaces*

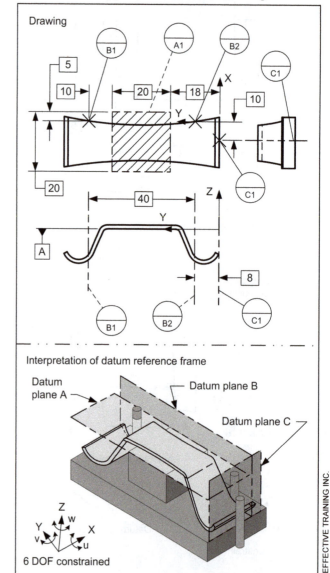

*FIGURE 14-12 Datum Targets on Irregular Surfaces*

This application contains two interesting uses of datum targets. The first is the specification of a datum target area that is larger than the datum feature. Datum target A1 is a 20mm square, which is larger than the datum feature. Wherever a target area is specified, the datum feature simulator is made to the same shape as the target area. A square target area was specified to keep the gage construction costs lower. If a target area was specified to match the part contour, the sides of the datum feature simulator would have to be produced with the contour shape.

The length of the target area was also specified to 20mm, so it will be less than the minimum length of surface A.

The second use of datum targets is to establish datums from irregular-shaped surfaces. Datum target lines B1, B2 and C1 are all contacting irregular-shaped surfaces. These targets are simulated with the side of gage pins located by the basic dimensions indicated.

## Datum Target Application – Movable Target Symbol

Datum targets can be used to establish a simulated datum axis or center plane on a part. When specifying datum targets on a drawing, the datum targets may indicate stationary or movable datum feature simulators.

Datum targets require their location to be defined with basic dimensions relative to other datum features simulators for all the datum feature references in a feature control frame. The basic dimensions establish a stationary location for the datum feature simulators unless one of the following conditions exist:

- The targets are applied to a feature of size at RMB (see page 177 and Chapter 15)
- A movable target symbol is specified
- A datum translation symbol is specified (see Additional Related Topics on page 191)

Where the movable target symbol is specified, it indicates that the datum feature simulator is movable in location. Unless otherwise indicated, the direction of movement is normal to the part surface. If it desired to have the direction of movement of a datum feature simulator that is not normal to the part surface, the movable target symbol and direction of movement must both be specified.

Figure 14-13 shows datum targets B1 and B2 that define a 12mm high v-block, and C1 and C2 that defines datum target points that are movable. The part is located and centered by the v-block. Datum targets B1 and B2 are area targets that are 90° to each other. The datum feature simulator for datum targets B1 and B2 is a v-block. Datum targets C1 and C2 are point targets. In this case, the movable datum target symbol is indicated.

Drawing

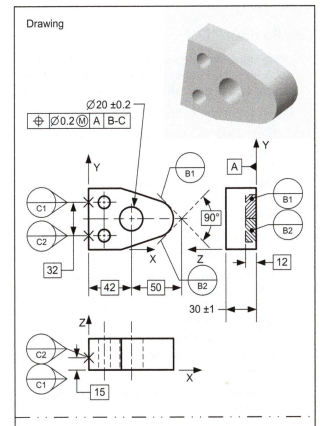

Interpretation of datum references frame

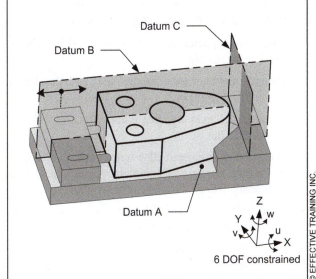

FIGURE 14-13 Movable Datum Target Application

**Author's Comment**
The movable datum target symbol is new in the Y14.5 2009 Standard. When to use the movable datum target symbol is a bit confusing. When specifying datum targets on a cylindrical surface as shown in Figure 14-9, the movable datum target symbol is optional. When specifying datum targets as shown in Figure 14-13, the movable datum target symbol is required.

## SUMMARY

### Key Points

- Datum targets are a set of symbols that describe the shape, size, and location of datum feature simulators that are used to establish datum planes, axes, or points.

- Datum targets should be specified where:

  o It is not practical (or possible) to use an entire surface as a datum feature.

  o The part may rock or wobble when the datum feature contacts the datum feature simulator.

  o Only a portion of a part surface is used in the function of the part.

- Guidelines for establishing a complete datum reference frame datum targets:

  o The datum targets should be located and defined with basic dimensions.

  o The datum reference frame must constrain all six degrees of freedom.

  o The part dimensions should ensure one orientation and location between the part and the datum feature simulator.

- There are three different datum target types: points, lines, and areas.

- Where the leader line from a datum target symbol is dashed, it denotes that the target exists on the back (hidden) side of the part in the view shown.

- Datum target points are often simulated with a spherical-tipped gage pin.

- Datum target lines are often simulated with the side of a cylindrical gage pin.

- Datum target areas are often simulated with a flat-tipped gage pin.

- Unless otherwise specified, where a datum target line or area is indicated on a non-planar surface, the datum feature simulator follows the contour of the surface.

- A movable datum feature simulator is used where the movable datum target symbol is specified.

- Unless otherwise specified, where a movable datum feature simulator is used, the direction of movement is normal to the part surface.

## Additional Related Topics

*These topics are recommended for further study to improve your understanding of datum targets.*

| Topic | Source |
|---|---|
| Chain line for partial surface datum features | *ASME Y14.5-2009, Para. 4.12.5* |
| Mathematically defined surfaces as a datum feature | *ASME Y14.5-2009, Para. 4.13* and ETI's *Advanced Concepts of GD&T* textbook |
| NOTE to invoke all three datums where only one is referenced | *ASME Y14.5-2009, Para. 4.24.14* |
| Datum targets on non-rigid parts | ETI's *Advanced Concepts of GD&T* textbook |
| Candidate datum set | *ASME Y14.5.1M-1994* |

## QUESTIONS AND PROBLEMS

***Website Bonus Materials***
Additional questions are available at our website. To access bonus materials for this textbook, please visit: www.etinews.com/textbookbonus

### True and False

*Indicate if each statement is true or false.*

T / F   1. Datum target lines are often simulated with a conical shaped gage pin.

T / F   2. Datum targets describe the shape, size, and location of datum feature simulators.

T / F   3. Datum targets enable a stable relationship between the part and gage elements.

T / F   4. The datum target leader line may be a solid or phantom line.

T / F   5. The datum target leader line may be solid or dashed line.

T / F   6. A datum target point should be simulated with a sharp gage element, like a cone tip or needle point.

## Multiple Choice

*Circle the best answer to each statement.*

1. Basic dimensions are used to locate datum targets because:
   A. They are exempt from title block tolerances
   B. To ensure repeatability between the part and the datum feature simulator.
   C. They define the nominal for the gagemaker.
   D. All of the above.

2. When using datum target points to establish a primary datum from a planar surface, _____ targets must be specified.
   A. 6       C. 2
   B. 3       D. 1

3. One place where datum targets should be used is on
   _____ .
   A. Features of size
   B. Simulated gages
   C. An irregular-shaped surface
   D. A planar surface

4. How is a datum target line typically simulated?
   A. With the edge of a gage block
   B. With a blade
   C. With the side of a cylindrical gage pin
   D. With a line of gage balls

5. How are the basic dimensions defining the size, location, and orientation between datum targets toleranced?
   A. The designer must apply geometric tolerances to the datum target
   B. With general or title block tolerances
   C. With a class of gage tolerances
   D. None of the above; they must be perfect

## Application Problems

*The application problems are designed to provide practice on applying the chapter concepts to situations that are similar to on-the-job conditions.*

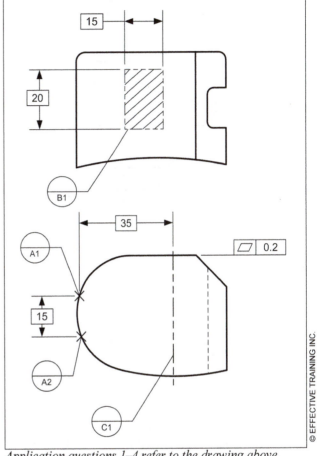

© EFFECTIVE TRAINING INC.

*Application questions 1–4 refer to the drawing above.*

1. What is missing in the datum target specification for target B1?
   A. The target area should be crosshatched.
   B. The height of the target area needs a dimension.
   C. The target area should be located relative to other targets.
   D. All of the above

2. Datum targets A1 and A2 indicate...
   A. Two target points     C. Two target areas
   B. Two target lines      D. None of the above

3. Datum target C1 is _____ target.
   A. A line
   B. A point
   C. An area
   D. Cannot determine from the drawing specifications

4. Datum _____ is the primary datum.
   A. A
   B. B
   C. C
   D. Cannot determine from the drawing specifications

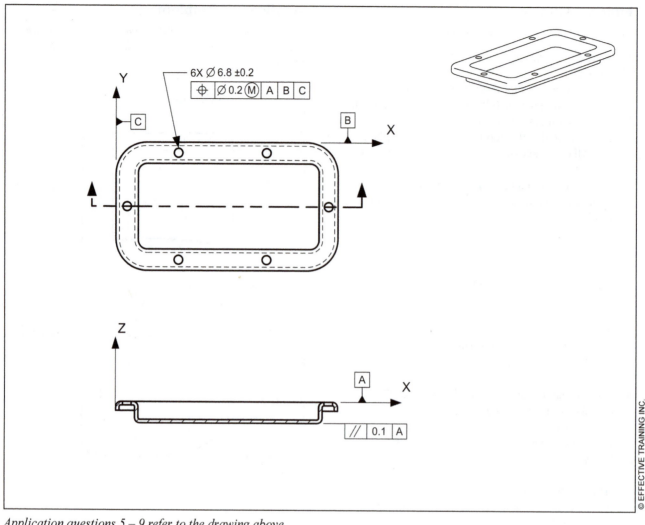

*Application questions 5 – 9 refer to the drawing above.*

5. Draw and dimension the datum target according to the following instructions.
   A. Use three 15mm dia area targets to establish datum plane A.
   B. Use two-line datum targets to establish datum plane B.
   C. Use a one-line datum target to establish datum plane C.

6. List the DOF constrained by the position tolerance.

   _____

   _____

   _____

7. List the DOF constrained by the parallelism tolerance.

   _____

   _____

   _____

8. Is the relationship between the part and the datum feature simulator repeatable for the verifying position tolerance? _____ If no, explain why not.

   _____

   _____

9. Is the relationship between the part and the datum feature simulator repeatable for the verifying parallelism tolerance? _____ If no, explain why not.

   _____

   _____

# Size Datum Features – RMB

## Goal

Interpret size datum features (RMB)

---

## Performance Objectives

Upon completing this chapter, you should be able to:

1. Describe the terms, "regardless of material boundary," "datum axis," "datum center plane," "datum center point," and "coaxial datum features" (p.184)
2. Describe common methods used to specify a feature of size as a datum feature (p.184)
3. Describe how to reference a feature of size datum feature at RMB (p.184)
4. Describe two effects of referencing a feature of size datum feature at regardless of material boundary (RMB) (p.185)
5. Describe the datum feature simulator and list the DOF constrained for external and internal cylindrical feature of size datum features (RMB primary) (p.186)
6. Describe the datum feature simulator and list the DOF constrained for external and internal planar feature of size datum features (RMB primary) (p.187)
7. Describe the datum feature simulator and list the DOF constrained for a planar surface primary and internal feature of size (RMB secondary) (p.188)
8. Describe the datum feature simulator and list the DOF constrained for a planar surface primary and internal feature of size (RMB secondary and tertiary) (p.189)
9. Describe the datum established from coaxial datum features of size referenced at RMB (p.190)
10. Describe the datum feature simulator for coaxial datum features of size referenced at RMB (p.190)

---

## New Terms

- Coaxial datum features
- Datum axis
- Datum center plane
- Datum center point
- Regardless of material boundary (RMB)

---

## What This Chapter Is About

Often, during the tolerancing of a part, toleranced features need to be related to an axis, center plane, or center point. Features of size (at RMB) are used to establish an axis, center plane, or center point as a datum. Where a feature of size is used to establish a datum, the type of datum feature simulator must also be defined. The datum feature simulator may be a fixed boundary, or it may be adjustable. Where the datum feature simulator is to be a fixed boundary, it is referenced in a feature control frame at MMB or LMB. Where a datum feature simulator is to be adjustable, it is referenced in a feature control frame at RMB.

This chapter explains where to use and how to interpret feature of size datum features referenced at RMB.

## TERMS AND CONCEPTS

When a feature of size is specified as a datum feature, and referenced in a feature control frame, two conditions exist:

1. The resulting datum is an axis, center plane, or center point.

2. The feature control frame must communicate the material condition

There are three material conditions used with datum feature references: maximum material boundary, least material boundary, and regardless of material boundary. The feature control frame indicates which material condition applies. This chapter covers regardless of material boundary. Lets begin by looking at how regardless of material boundary is specified.

## Regardless of Material Boundary (RMB)

Where no modifiers are specified in the datum portion of a feature control frame, RMB automatically applies to datum features per Rule #2. See Chapter 7, page 82, for a review of Rule #2. *Regardless of material boundary* indicates that a datum feature simulator is adjustable (or movable) from MMB towards LMB until it makes maximum contact with the extremities of the datum feature(s).

## Datum Axis, Datum Center Plane, Datum Center Point

When a cylindrical feature of size is specified as a datum feature, the resulting datum is an axis. A *datum axis* is the axis of a datum feature simulator established from the datum feature. An example is shown in Figure 15-2.

When a width feature of size is specified as a datum feature, the resulting datum is a center plane. A *datum center plane* is the center plane of a datum feature simulator established from the datum feature. An example is shown in Figure 15-3.

When a spherical feature of size is specified as a datum feature, the resulting datum is a center point. A *datum center point* is the center point of a datum feature simulator established from the datum feature.

## Coaxial Datum Features

One term that you will need to be familiar with that is new in this chapter is coaxial datum features. *Coaxial datum features* are coaxial features of size used to create a single datum axis.

## Common Methods to Specify a Feature of Size as a Datum Feature

Where a feature of size is used as a datum feature, the resulting datum is an axis, a center plane, or a point. A feature of size is specified as a datum feature by associating the datum feature symbol, as described below and shown in Figure 15-1.

The base of the datum feature symbol is attached:

1. On a dimension line or as an extension of a dimension line of a feature of size, as illustrated in Figure 15-1A, F, & H

2. On a dimension line of a feature of size dimension, as illustrated in Figure 15-1B & G

3. Touching the surface of a diameter, as illustrated in Figure 15-1C

4. To the horizontal portion of a leader line of a size dimension as illustrated in 15-1D & I

5. To the top or bottom of a feature control frame associated with a size dimension, as illustrated in Figure 15-1E & J

**Author's Comment**

A common drawing error is to attach the datum feature symbol to a center line. The datum feature symbol should never be attached to a center line. The symbol should always be associated with a feature of size to designate the axis or center plane as a datum.

## Referencing Feature of Size Datum Features at RMB

After indicating the datum features, the drawing must also communicate when and how the datum features should be used to control the geometric relationships required for function and assembly. This is done through the use of a feature control frame.

Where a size datum feature is referenced, its boundary condition must also be communicated. The feature control frame not only communicates the datum sequence, but also the boundary condition for size datum features. A size datum feature may be referenced at RMB (regardless of material boundary), MMB (maximum material boundary), or LMB (least material boundary). Where MMB or LMB are not indicated, RMB is the default condition per Rule #2. We will discuss size datum features referenced at MMB and LMB in Chapter 16.

| Methods for specifying an axis as a datum | Methods for specifying a center plane as a datum |
|---|---|

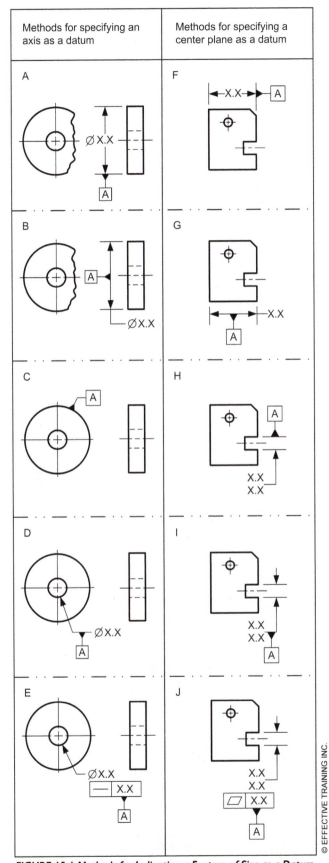

FIGURE 15-1 Methods for Indicating a Feature of Size as a Datum Feature

## Effects of Referencing a Feature of Size as a Datum Feature at RMB

Where a size datum feature is referenced at RMB, two conditions exist:

- The datum feature simulator must always contact the datum feature.
- The datum feature simulator is adjustable (movable) to allow for maximum contact with the datum feature.

The following sections describe various size datum features referenced at RMB and their resulting datums, datum feature simulators, datum reference frames, and degrees of freedom constrained.

### Author's Comment

There is a difference between "restraint" (as discussed in Fundamental Rule "m" in Chapter 7) and constraining a datum feature at RMB.

When constraining a datum feature, the datum feature simulator contacts the datum feature with the minimum force required to orient and locate the datum feature to the datum(s), In contrast, "restraining" forces are often significant enough to alter the form of a part between its free state and its restrained state.

## FEATURE OF SIZE DATUM FEATURE APPLICATIONS

Datum feature selection and datum feature references are directly related to how a part functions in its assembly. Often, the features that orient and locate a part in its assembly become the datum features for the part.

Where a feature of size is used as a datum feature on a drawing, the boundary condition for establishing the datum must be communicated. This is typically done through the feature control frames that reference the datum features. The feature control frames communicate both the boundary condition (RMB, MMB, or LMB) and the datum sequence.

Where a feature of size datum feature is referenced at RMB, two conditions exist. The datum feature simulator is adjustable (or movable), and there is no looseness between the part and the datum feature simulator.

The following sections describe various feature of size datum features referenced at (RMB) and the resulting datum and datum feature simulators.

## Cylindrical Datum Feature RMB Primary

In certain applications, the function of a part requires using a cylindrical feature of size as a datum feature referenced primary at RMB.

Where a cylindrical feature of size is designated as a datum feature and referenced in a feature control frame at RMB, the resulting datum is an axis. The datum axis is established through physical contact between the datum feature simulator and the datum feature. The feature of size is surrounded (or filled) by the datum feature simulator. The datum axis is the axis of the unrelated actual mating envelope.

Devices that are adjustable in size, such as a precision chuck, a collet, a mandrel, or other centering device, may be used as the datum feature simulator. The datum feature simulator contacts the high points of the datum feature. The axis of the datum feature simulator becomes the datum axis and establishes the orientation of the part. Figure 15-2 A & B shows examples of external and internal features of size used to establish a primary datum axis RMB.

Where a cylindrical feature of size is used as a primary datum feature at RMB, four degrees of freedom are constrained: two translational and two rotational. The degrees of freedom constrained are shown on the model coordinate system in Figure 15-2.

---

### TECHNOTE 15-1
#### Cylindrical Datum Feature – Primary RMB

Where a feature of size is used as a datum feature and referenced as a primary datum feature RMB, the following applies:

- An adjustable datum feature simulator is used.
- The datum is an axis (line).
- The feature of size must be of sufficient length to provide repeatable orientation of the part in the datum feature simulator.
- Four degrees of freedom are constrained (two translational and two rotational).

---

### Design Tip
When a cylindrical or planar feature of size is used to establish a datum axis (or center plane) primary RMB, its length must be sufficient to establish a repeatable relationship (orientation) with the datum feature simulator. For example, a hole in a stamped part, like the one shown in Figure 13-16 (datum G) would not have sufficient length to be used as a primary datum feature.

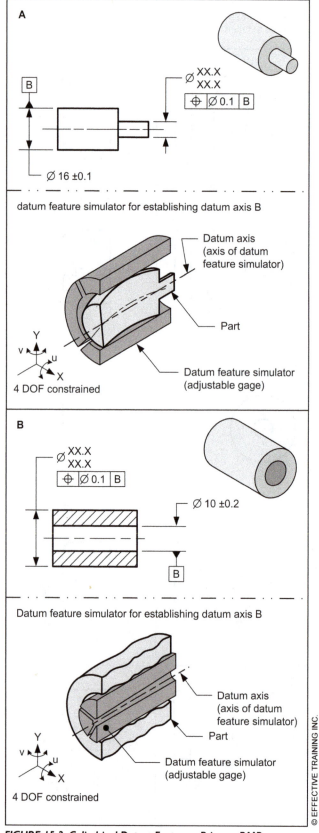

*FIGURE 15-2  Cylindrical Datum Feature – Primary RMB*

## Planar Width Datum Feature RMB Primary

In certain applications, the function of a part requires using a planar feature of size (width) as a datum feature referenced primary at RMB.

Where a planar feature of size (width) is designated as a datum feature and referenced in a feature control frame at RMB, the resulting datum is a center plane. The datum center plane may be established through physical contact with datum feature simulators or through a set of data points obtained from the physical surface. To make it easier to visualize the conditions, I will use physical datum feature simulators to describe the datums. The datum center plane is established through physical contact between the datum feature simulator and the datum feature. The width is surrounded (or filled) by the datum feature simulator.

Devices that are adjustable in size, such as contracting or expanding parallel plates, are used as the datum feature simulator. The datum feature simulator contacts the high points of the datum feature. The center plane of the datum feature simulator becomes the datum center plane and establishes the orientation of the part. Figure 15-3 A & B shows examples of external and internal widths used to establish a primary datum center plane RMB, the resulting datum reference frame, and the degrees of freedom constrained.

Where a width is used as a primary datum feature at RMB, three degrees of freedom are constrained: one translational and two rotational.

---

**TECHNOTE 15-2**
**Width Datum Feature – Primary RMB**

Where a width feature of size is used as a datum feature and referenced as a primary datum feature RMB, the following applies:

- An adjustable datum feature simulator is used.
- The datum is a center plane.
- Width must be of sufficient length to provide repeatable orientation of the part in the datum feature simulator.
- Three degrees of freedom are constrained (one translational and two rotational).

---

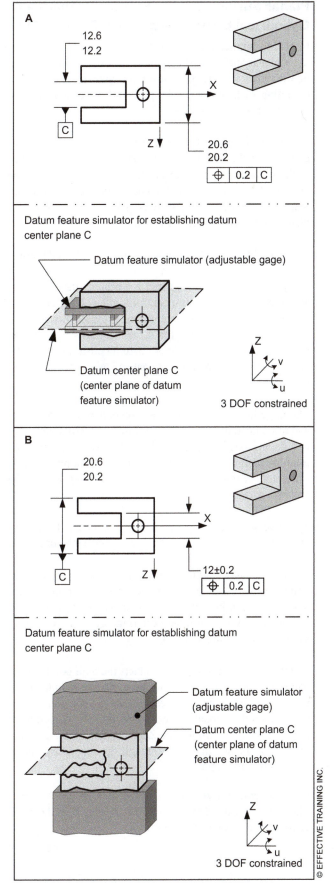

FIGURE 15-3 Width Datum Feature – Primary RMB

© EFFECTIVE TRAINING INC.

## Planar Surface Datum Feature Primary and Cylindrical Datum Feature RMB Secondary

In certain applications, the function of a part requires using a planar surface and a cylindrical feature of size as datum features. For the part shown in Figure 15-4, the planar mounting surface orients the part, and the center hole locates the part in its assembly.

The feature that orients the part is referenced as the primary datum, and the feature that locates the part is referenced as the secondary datum. In the case where a surface is the primary datum feature and the hole is the secondary datum feature at RMB, the datum reference frame consists of a plane and an axis 90° to the plane.

An example is shown in Figure 15-4 and explained below.

The planar surface is designated as a primary datum feature and referenced in a feature control frame at RMB, a datum plane is established. The datum plane is established through physical contact between the datum feature simulator and the datum feature. The primary datum constrains three degrees of freedom: one translational and two rotational. Now let's discuss the secondary datum feature.

The large hole is designated as a secondary datum feature and referenced in a feature control frame at RMB, resulting in a datum axis being established. The datum axis may be established through physical contact with datum feature simulators or through a set of data points obtained from the physical surface.

To make it easier to visualize the conditions, I will use physical datum feature simulators to describe the datums.

The datum feature simulator is expanded to contact the high points of the related actual mating envelope of the hole. A device that is adjustable in size, such as a precision chuck, a collet, a mandrel, or other centering device, may be used as the datum feature simulator.

The datum feature simulator contacts the high points of the datum feature. The axis of the datum feature simulator becomes the datum axis and establishes the location of the part. Two translational degrees of freedom are constrained. Since the rotation around the Z axis is not constrained, the two datum planes that pass through the secondary datum axis are free to rotate.

In this example, five degrees of freedom are constrained. The degrees of freedom constrained are shown on the model coordinate system in Figure 15-4.

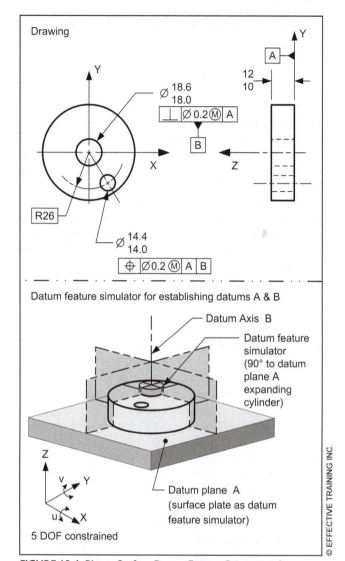

**FIGURE 15-4 Planar Surface Datum Feature Primary and Cylindrical Datum Feature RMB Secondary**

In summary, when referencing a planar surface as a primary datum feature and a cylindrical feature of size as a secondary datum feature RMB, the following conditions apply:

- The part will have a minimum of three-point contact with the primary datum feature simulator.

- The datum feature simulator for the secondary datum is perpendicular to the primary datum and adjustable.

- The datum reference frame consists of a plane and an axis 90° to the plane.

- A second and third datum plane is associated with the datum axis.

- Five degrees of freedom are constrained (three translational and two rotational).

## Planar Surface Datum Feature Primary, Cylindrical Datum Feature RMB Secondary, and Width Datum Feature Tertiary

In certain applications, the function of a part requires using three datum features to establish a datum reference frame. For the part shown in Figure 15-5, the planar mounting surface orients the part, the center hole locates the part, and the width stops the rotation about the center hole.

In the case where a surface is the primary datum feature, a center hole is the secondary datum feature at RMB, and a width is the tertiary datum feature at RMB, the datum reference frame consists of a plane and an axis with two planes passing through it. The rotation of the two datum planes is constrained by the tertiary datum's center plane. An example is shown in Figure 15-5 and explained below.

The planar surface is designated as a datum feature and referenced in a feature control frame as primary. Datum plane A is established through physical contact between the datum feature simulator and the datum feature. The primary datum constrains three degrees of freedom: one translational and two rotational.

The center hole is designated as a secondary datum feature and referenced in a feature control frame at RMB. Datum axis B is established. The datum axis may be established through physical contact with datum feature simulators or through a set of data points obtained from the physical surface. To make it easier to visualize the conditions, I will use physical datum feature simulators to describe the datums.

The datum feature simulator is expanded to contact the high points of the cylinder. A device that is adjustable in size, such as a precision mandrel or other centering device, may be used as the datum feature simulator that contacts the high points of the related actual mating envelope of datum feature. The axis of the datum feature simulator becomes the datum axis and establishes the location of the part. Two translational degrees of freedom are constrained. A second and third datum plane exists through the axis, but they are free to rotate about the axis. The rotation of the datum planes is constrained by invoking a tertiary datum.

A width is designated as a tertiary datum feature, and referenced at RMB. This constrains the rotation of the second and third datum planes. The datum center plane is constrained through physical contact between the datum feature simulator and the datum feature. The width is filled by the datum feature simulator.

A device that is perpendicular to the primary and its center plane, passes through the secondary datum feature simulator axis, and is adjustable in size (e.g., expanding parallel plates) is used as the datum feature simulator. The device contacts the high points of the related actual mating envelope of the datum feature.

The center plane of the datum feature simulator becomes the datum center plane and establishes the angular orientation of the part.

In this example, all six degrees of freedom are constrained. The degrees of freedom constrained are shown on the model coordinate system in Figure 15-5.

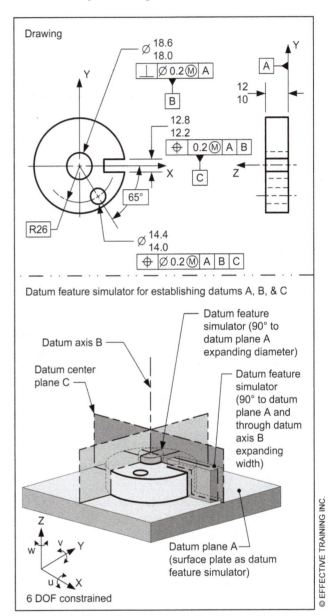

FIGURE 15-5 *Planar Surface Datum Feature Primary, Cylindrical Datum Feature RMB Secondary, and Width Datum Feature RMB Tertiary*

**Author's Comment**
A datum feature simulator for a cylindrical feature of size at RMB should simulate the minimum circumscribed cylinder or maximum inscribed cylinder. Therefore, a v-block or three points of contact are not good representations of a datum feature simulator.

In summary, when referencing a planar surface as a primary datum feature, a cylindrical feature of size as a secondary datum feature RMB, and a width as a tertiary datum feature RMB, the following conditions apply:

- The part will have a minimum of three-point contact with the primary datum feature simulator.
- The datum feature simulator for the secondary datum is perpendicular to the primary and secondary datums and adjustable.
- The datum reference frame consists of a plane and an axis 90° to the plane.
- A second and third datum plane is associated with the datum axis.
- The tertiary datum constrains the rotation of the second and third datum planes.
- All six degrees of freedom are constrained (three translational and three rotational).

## Coaxial Features of Size as a Datum Feature Referenced at RMB

In certain applications, the function of a part requires establishing a single datum axis from two (or more) coaxial features of size. For the part shown in Figure 15-6, the two coaxial cylinders (at each end of the part) work together to orient and locate the part in its assembly. Individually, each cylinder is too short to orient the part in the datum feature simulator. Therefore, the cylinders are referenced together in a feature control frame as multiple datum features "A-B." Where coaxial features of size are designated as a datum feature and referenced in a feature control frame at RMB, a datum axis is established.

## Datum Feature Simulators for Coaxial Features of Size datum Features Referenced at RMB

Where coaxial features of size are referenced as datum features at RMB, the datum axis may be established through physical contact with datum feature simulators or though a set of data points obtained from the physical surface. To make it easier to visualize the conditions, I will use physical datum feature simulators to describe the datums. The datum axis is established through physical contact between the datum feature simulator and both datum features. The datum features are surrounded (or filled) by the datum feature simulator.

Devices that are adjustable in size, such as precision chucks, collets, mandrels, or other centering devices, may be used as the datum feature simulators. The datum feature simulators expand and contract about a common center simultaneously and contact both datum features to establish a common datum axis. The axis of the datum feature simulators becomes the datum axis and establishes the orientation of the part. Figure 15-6 shows examples of two external coaxial features of size used to establish a primary datum axis RMB.

---

> ### TECHNOTE 15-3
> ### Coaxial Datum Feature – Primary RMB
> Where a coaxial datum feature is referenced at RMB:
> - An adjustable datum feature simulator is used to contact both cylindrical features of size simultaneously.
> - The datum is an axis (line).
> - The part is oriented in the gage.
> - Four degrees of freedom are constrained (two translational and two rotational).

Where a coaxial datum feature is referenced at RMB, four degrees of freedom are constrained: two translational and two rotational. The degrees of freedom constrained are shown on the model coordinate system in Figure 15-6.

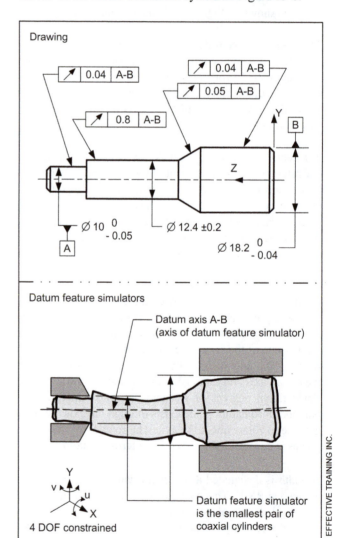

**FIGURE 15-6 Coaxial Datum Features – Primary RMB**

© EFFECTIVE TRAINING INC.

## SUMMARY

### Key Points

- Common methods used to specify a feature of size as a datum feature are:
  - o The datum feature symbol is placed touching the surface of a diameter.
  - o The datum feature symbol is placed on the horizontal portion of a leader line of the size dimension of a feature of size.
  - o The datum feature symbol is placed touching a feature control frame (that applies to the feature of size dimension).
  - o The datum feature symbol is placed on a dimension line or as an extension of the dimension line of a feature of size.

- Where a feature of size datum feature is referenced at RMB, two conditions exist:
  - o The datum feature simulator is adjustable (or movable).
  - o There is no looseness between the part and the datum feature simulator.

- Where a cylindrical feature of size datum feature is referenced at RMB, the following conditions exist:
  - o An adjustable datum feature simulator is used.
  - o The datum is an axis (line).
  - o The part is oriented by the datum feature simulator.
  - o Four degrees of freedom are constrained (two translational and two rotational).

- Where a width feature of size datum feature is referenced at RMB, the following conditions exist:
  - o An adjustable datum feature simulator is used.
  - o The datum is a center plane.
  - o The part is oriented by the datum feature simulator.
  - o Three degrees of freedom are constrained (one translational and two rotational).

- Coaxial datum features are where coaxial features of size are used simultaneously as a datum feature.

### Additional Related Topics

*These topics are recommended for further study to improve your understanding of size datum features (RMB).*

| Topic | Source |
|---|---|
| Customized datum reference frames | *ASME Y14.5-2009, Para. 4.22* |
| Translation modifier | *ASME Y14.5-2009, Para. 4.11.10* |

## QUESTIONS AND PROBLEMS

### True and False

*Indicate if each statement is true or false.*

T / F    1. Where a cylindrical feature of size is designated as a datum feature and referenced in a feature control frame as primary at RMB, the resulting datum is an axis.

T / F    2. Where a cylindrical feature of size is designated as a datum feature and referenced in a feature control frame as primary at RMB, three DOF are constrained.

T / F    3. Where a width is designated as a datum feature and referenced in a feature control frame as primary at RMB, four DOF are constrained.

T / F    4. RMB indicates that a datum feature simulator is adjustable (or movable).

T / F    5. Two cylindrical features of size shown on the drawing with one center line are considered coaxial datum features.

T / F    6. A datum axis is the axis of a datum feature simulator established from the datum feature.

## Multiple Choice

*Circle the best answer to each statement.*

1.  Where a width is referenced as a primary datum at RMB, the datum is...
    A. An axis
    B. A center plane
    C. An area
    D. A point

2.  When referencing a datum feature of size at RMB primary, the datum feature simulator is...
    A. Fixed at the RMB size
    B. Adjustable in size
    C. Fixed at the MMB size
    D. Adjustable in location

3.  A datum axis may be established from _____.
    A. A hole
    B. A shaft
    C. Coaxial features of size
    D. All of the above

4.  If an external diameter of a shaft was a datum feature referenced as primary at RMB, the best choice for a datum feature simulator would be_____.
    A. A v-block
    B. A precision chuck or collet
    C. A functional gage
    D. None of the above

5.  Identifying two coaxial features of size as a datum feature establishes_____.
    A. An axis as the datum
    B. Two axes as the datum
    C. A datum target
    D. An axis and a point as the datums

## Application Problems

*The application problems are designed to provide practice on applying the chapter concepts to situations that are similar to on-the-job conditions.*

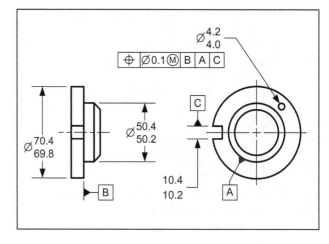

*Application questions 1–5 refer to the drawing above.*

1.  Which datum features are feature of size datum features?
    A. A & B
    B. A & C
    C. B & C
    D. All of the above

2.  How is datum feature A to be simulated?
    A. A 50.4 fixed diameter cylinder
    B. A v-block
    C. An adjustable cylinder
    D. All of the above

3.  How is datum feature C to be simulated?
    A. A 10.2 fixed-width gage element
    B. A wedge
    C. An adjustable width
    D. All of the above

4.  When verifying the position tolerance, how many degrees of freedom are constrained?
    A. 3
    B. 4
    C. 5
    D. 6

5.  The datum reference frame for the position tolerance is_____ primary, _____ secondary, _____ tertiary.
    A. plane, axis, line
    B. plane, axis, axis
    C. plane, axis, center plane
    D. plane, center plane, axis

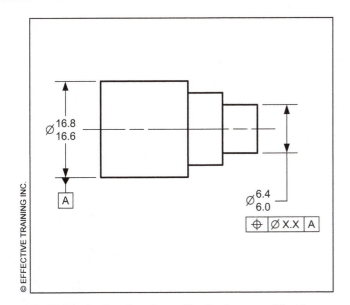

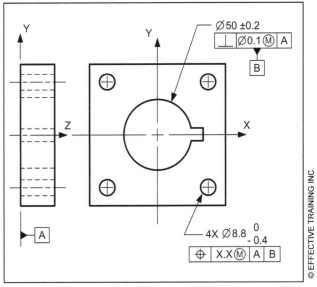

6. Using the drawing above, list the degrees of freedom constrained by the datum feature references in the position tolerance. _____

_____

_____

7. Using the drawing above, list the degrees of freedom constrained by the datum feature references in the position tolerance. _____

_____

_____

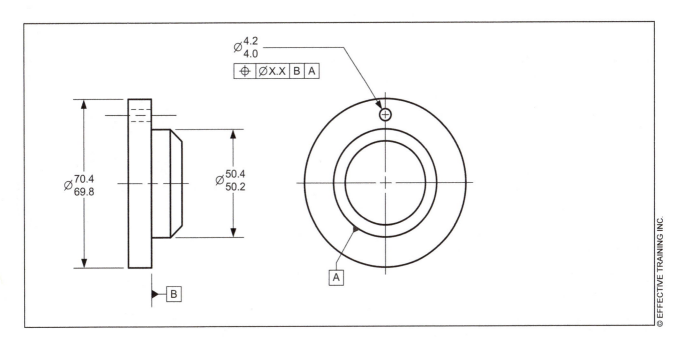

8. Using the drawing above, list the degrees of freedom constrained by the datum feature references in the position tolerance.

_____

_____

# Size Datum Features – MMB

## Goal

Interpret size datum features (MMB)

## Performance Objectives

Upon completing this chapter, you should be able to:

1. Describe the term "maximum material boundary" (MMB) (p.196)
2. Describe two effects of referencing a datum feature at MMB (p.196)
3. Calculate the size of the datum feature simulator for a primary datum MMB (p.196)
4. Calculate the size of the datum feature simulator for a secondary datum feature at MMB (p.197)
5. Calculate the size of the datum feature simulator for a tertiary datum feature at MMB (p.198)
6. Describe the concept of datum feature shift (p.199)
7. Explain where datum feature shift is permissible (p.199)
8. Calculate the amount of datum feature shift permissible (p.199)
9. Determine when datum feature shift is or is not additive to a geometric tolerance (p.200)
10. Calculate the MMB of the datum feature simulator for a planar surface datum feature at MMB (p.201)
11. Describe the datum feature simulator and the DOF constrained for internal and external cylindrical feature of size datum features (MMB primary) (p.202)
12. Describe the datum feature simulator and the DOF constrained for internal and external width feature of size datum features (MMB primary) (p.203)
13. Describe the datum feature simulator for an internal feature of size (MMB secondary) (p.204)
14. Describe the datum feature simulator for an internal feature of size (MMB tertiary) (p.205)
15. Describe the datum feature simulator and the DOF constrained for coaxial datum features (MMB) (p.206)
16. Describe the datum feature simulator and the DOF constrained for a pattern of features of size (MMB) (p.207)
17. Analyze the effects that changing the datum feature reference sequence in a feature control frame has on the part and datum feature simulator (p.208)

## New Terms

- Datum feature shift
- Maximum material boundary (MMB)

## What This Chapter Is About

Where a feature of size is used as a datum feature, it may be referenced in a feature control frame at MMB, LMB or RMB. In Chapter 15, we looked at datum features referenced at RMB. In this chapter, we will focus on datum features referenced at maximum material boundary.

This chapter explains where to use and how to interpret feature of size datum features referenced at MMB.

# TERMS AND CONCEPTS

## Maximum Material Boundary (MMB)

The *maximum material boundary* (MMB) is the boundary established by the collective effects of the MMC of a datum feature and any applicable tolerances.

## Effects of Referencing a Datum Feature at MMB

Where a datum feature is referenced at MMB, it has two effects:

- A fixed-size datum feature simulator (gage element) must be used.

- A clearance may exist between the datum feature and the datum feature simulator.

An MMB datum feature simulator is to be fixed in size and, in some cases, may be fixed in location and orientation, as well. The MMB concept is important because it ensures that:

- The datum feature simulator will accept all datum features that are within their drawing specifications.

- The dimensioning represents the function of assembly.

To explain how the MMB concept affects drawing interpretation, we will look at how MMB applies to features of size as datum features. Using the functional dimensioning approach, the MMB modifier is commonly used where the function of the datum feature is assembly. MMB can also be applied to non-size datum features. An example is provided later in this chapter.

## MMB Size Calculations for Primary Datum Feature

Where a feature of size datum feature is referenced as a primary datum feature at MMB, the datum feature simulator is a fixed size. In applications where Rule #1 applies, the MMB of a feature of size datum feature is equal to the MMC Rule #1 boundary of the datum feature.

In applications where Rule #1 is overridden, the MMB of a feature of size datum feature will involve combining the MMC of the datum feature and the effects from any flatness or straightness tolerance applied to the feature of size. In each application, the size of the datum feature simulator is equal to the MMB of the datum feature.

Primary datum feature MMB size calculations:
- Applications where Rule #1 applies to the size dimension of the datum feature)
  o  MMB = MMC of datum feature

- Applications where Rule #1 is overridden by a straightness or flatness tolerance)
  o  MMB = MMC + form tolerance (for external feature of size)
  o  MMB = MMC - form tolerance (for internal feature of size)

An example of each application is shown in Figure 16-1.

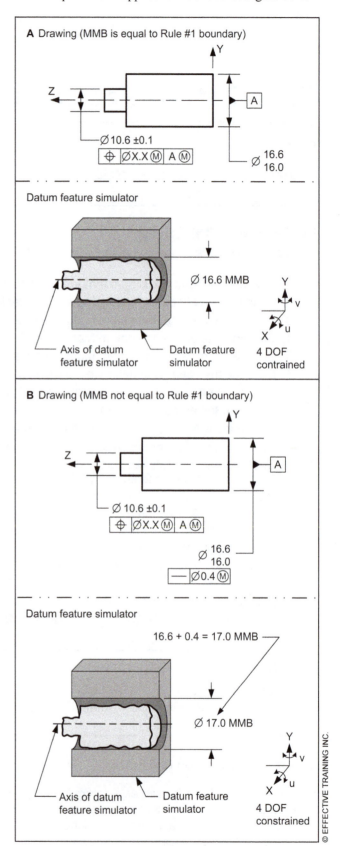

**FIGURE 16-1 Calculating the MMB Size of a Datum Feature Simulator for a Primary Datum Feature**

---

### TECHNOTE 16-1
### MMB Size Calculations for Primary Datum Features

Where a datum feature is referenced at MMB as a primary datum feature reference, the following applies:

- The datum feature simulator is fixed at the MMB size.

- If Rule #1 applies to the feature of size dimension, the MMB is equal to the Rule #1 boundary of the datum feature.

- If a straightness or flatness tolerance is applied to an external feature of size, the MMB is equal to the MMC plus the effects of the geometric tolerance.

- If a straightness or flatness tolerance is applied to an internal feature of size, the MMB is equal to the MMC minus the effects of the geometric tolerance.

## MMB Size Calculations for Secondary Datum Feature

Where a secondary datum feature is referenced at MMB, it indicates the datum simulator is fixed in size. The datum simulator is also fixed at the specified orientation (and sometimes location) relative to the primary datum. The datum simulator is made to the MMB size of the datum feature.

The formula for calculating the MMB of a secondary datum feature is shown below.

Secondary datum feature MMB size calculation:

- For an external feature of size
  MMB = MMC + applicable geometric tolerance*

- For an internal feature of size
  MMB = MMC - applicable geometric tolerance*

\* The effects of any orientation or location tolerances relative to the primary datum

Figure 16-2 shows an example of a secondary datum feature referenced at MMB and its resulting MMB size datum feature simulator.

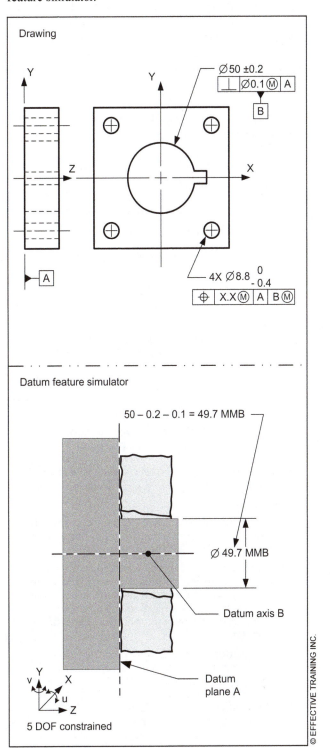

*FIGURE 16-2 Calculating the MMB Size of a Datum Feature Simulator for a Secondary Datum Feature*

## MMB Size Calculation for Tertiary Datum Feature

Where a tertiary datum feature is referenced at MMB, it indicates that the datum simulator is fixed in size. The datum simulator is also fixed at the specified orientation and location relative to the primary and secondary datums. The datum simulator is made to the MMB size of the datum feature.

The formula for calculating the MMB of a tertiary datum feature is shown below:

Tertiary datum feature MMB size calculation:

- For an external feature of size

  MMB = MMC + applicable geometric tolerance*

- For an internal feature of size:

  MMB = MMC - applicable geometric tolerance*

\* The effects of any orientation or location tolerances relative to the primary and secondary datums

Figure 16-3 shows an example of a tertiary datum feature referenced at MMB and its resulting MMB size datum feature simulator.

---

### TECHNOTE 16-2
### MMB Size Calculations for Secondary or Tertiary Datum Features

Where a datum feature is referenced at MMB as a secondary (or tertiary) datum feature reference, the following applies:

- The datum feature simulator is fixed at the MMB size, which includes orientation (and sometimes location), relative to the primary datum.

- If an orientation control is applied to the feature of size dimension, relating the secondary (or tertiary) datum feature back to the primary datum feature, the MMB is equal to the MMC of the datum feature combined with the effects of the orientation control.

- If a location control is applied to the feature of size dimension, relating the secondary (or tertiary) datum feature back to the primary or secondary datum feature, the MMB is equal to the MMC of the datum feature combined with the effects of the location control.

**Author's Comment**

In applications where the MMB modifier is specified in a secondary or tertiary datum feature reference, there must be geometric tolerances relating the datum feature(s) to the higher ranking datums.

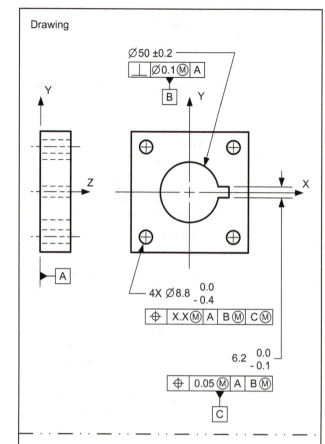

Drawing

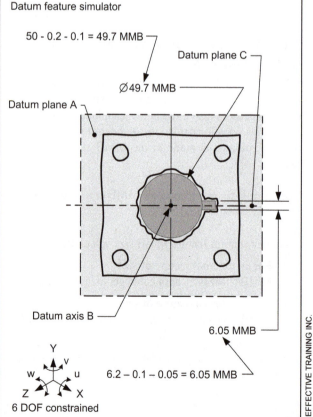

Datum feature simulator

50 - 0.2 - 0.1 = 49.7 MMB

Ø 49.7 MMB

Datum plane C

Datum plane A

Datum axis B

6.05 MMB

6.2 - 0.1 - 0.05 = 6.05 MMB

6 DOF constrained

© EFFECTIVE TRAINING INC.

**FIGURE 16-3 Calculating the MMB Size of a Datum Feature Simulator for Secondary and Tertiary Datum Features**

## Datum Feature Shift

Where a feature of size is referenced as a datum feature at MMB, the datum feature simulator is a fixed size. If the datum feature on the actual part is departs from its MMB, looseness (clearance) will exist between the part and the datum feature simulator. This looseness is called datum feature shift.

*Datum feature shift* is the allowable movement or looseness between the part datum feature and the datum feature simulator.

## Where Datum Feature Shift is Permissible

A datum feature shift is permissible when two conditions exist:

- The MMB modifier is specified in the datum portion of a feature control frame
- The datum feature being referenced is a feature of size

## Datum Feature Shift Calculations

Figure 16-4 shows a part with a position tolerance that references a primary datum feature at MMB, and the part is located in the datum feature simulator. When the datum feature is at MMB, there is no looseness between the part and the datum simulator. The part cannot shift in the datum simulator.

As the datum feature departs from MMB towards LMB, the looseness between the part and the datum simulator increases. This looseness is the datum feature shift. The maximum amount of datum feature shift possible is the difference between the datum simulator and the LMB of the datum feature (the maximum clearance between the part and the datum simulator). The chart in Figure 16-4 shows the amount of datum feature shift for various datum feature sizes.

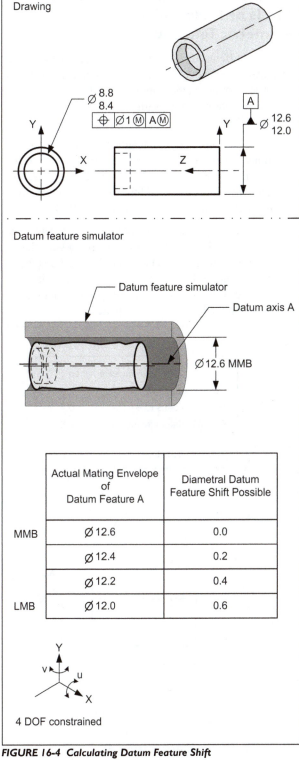

| | Actual Mating Envelope of Datum Feature A | Diametral Datum Feature Shift Possible |
|---|---|---|
| MMB | ⌀ 12.6 | 0.0 |
| | ⌀ 12.4 | 0.2 |
| | ⌀ 12.2 | 0.4 |
| LMB | ⌀ 12.0 | 0.6 |

4 DOF constrained

*FIGURE 16-4 Calculating Datum Feature Shift*

## Recognizing Whether Datum Feature Shift Is or Is Not Additive to a Geometric Tolerance

One major difference between bonus and datum feature shift is that datum feature shift is not always additive to the stated geometric tolerance in the feature control frame. An explanation of where datum feature shift is additive or is not additive is shown below:

- Coaxial features of size toleranced relative to each other—one is the toleranced feature; the other is the datum feature referenced at MMB. In this case, the datum feature shift is additive to the geometric tolerance. An example is shown in Figure 16-5.

- A pattern of holes toleranced relative to a datum feature at MMB (as shown in Figure 16-6)—the datum feature shift only affects the location of the pattern (as a group) relative to the datum axis. The datum feature shift does not affect the spacing tolerance between the holes within the pattern.

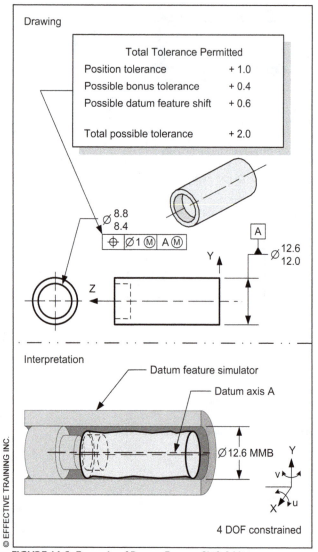

Drawing

| Total Tolerance Permitted | |
|---|---|
| Position tolerance | + 1.0 |
| Possible bonus tolerance | + 0.4 |
| Possible datum feature shift | + 0.6 |
| Total possible tolerance | + 2.0 |

Interpretation

4 DOF constrained

*FIGURE 16-5 Example of Datum Feature Shift Additive to a Geometric Tolerance*

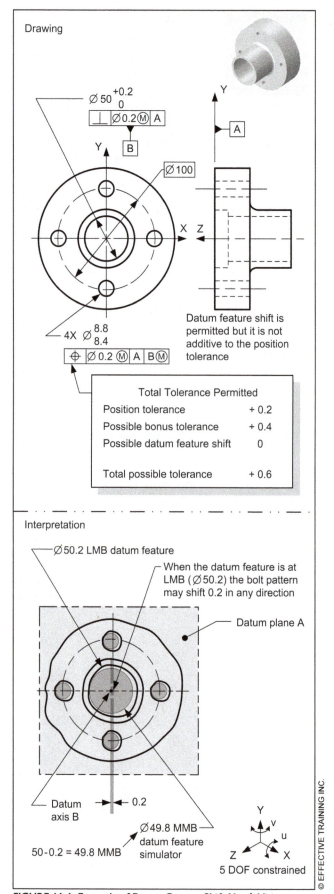

Drawing

Datum feature shift is permitted but it is not additive to the position tolerance

| Total Tolerance Permitted | |
|---|---|
| Position tolerance | + 0.2 |
| Possible bonus tolerance | + 0.4 |
| Possible datum feature shift | 0 |
| Total possible tolerance | + 0.6 |

Interpretation

Ø 50.2 LMB datum feature

When the datum feature is at LMB (Ø 50.2) the bolt pattern may shift 0.2 in any direction

Datum plane A

Datum axis B

0.2

50 - 0.2 = 49.8 MMB

Ø 49.8 MMB datum feature simulator

5 DOF constrained

*FIGURE 16-6 Example of Datum Feature Shift Not Additive to a Geometric Tolerance*

## MMB for a Surface as a Datum Feature

**Author's Comment**
Using the MMB modifier on planar and contoured datum features is a complex topic and, therefore, only introduced in this text.

The MMB modifier may also be applied to surface (planar or contoured) datum features. Where a surface datum feature is referenced as a secondary or tertiary datum at MMB, the datum feature simulator is fixed in location.

The location of the datum feature simulator is established at the MMB of the datum feature. The MMB of the datum feature is a established by the basic location plus the effects of geometric tolerances applied to the datum feature.

Figure 16-7 shows an example of a planar surface used as a datum feature referenced as a secondary datum at MMB. The MMB boundary is equal to the basic dimension plus the effects of the profile tolerance. (For a complete explanation of profile tolerances, see Chapters 26 & 27). In this example, the secondary datum is constraining the rotational degree of freedom of the planes of the datum reference frame around the axis of the datum feature simulator from datum feature A.

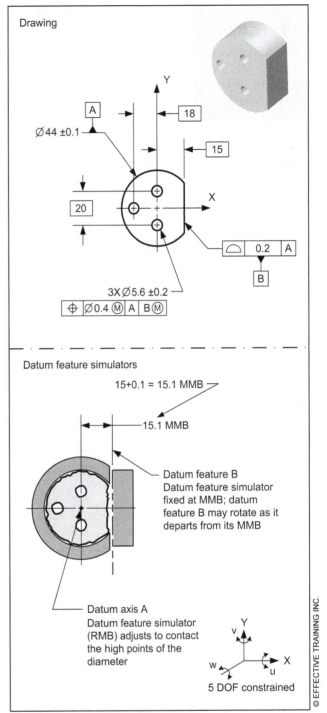

**FIGURE 16-7 Planar Surface Used as a Datum Feature Referenced as a Secondary Datum at MMB**

## DATUM FEATURE REFERENCE MMB APPLICATIONS

The following sections describe various datum features referenced at MMB and the resulting datum(s), datum feature simulators, datum reference frames, and degrees of freedom constrained.

### Cylindrical Datum Feature MMB Primary

In certain applications, the function of a part requires using a cylindrical feature of size as a datum feature referenced primary at MMB.

Where a cylindrical feature of size is designated as a datum feature and referenced in a feature control frame at MMB, a datum axis (line) is established. The datum axis is established by the datum feature simulator. The cylinder is surrounded (or filled) by the datum feature simulator.

Devices that are fixed in size, such as a precision hole, width, or pin, may be used as the datum feature simulator. The axis of the datum feature simulator becomes the datum axis and establishes the orientation of the part within the amount of clearance between the part and the datum feature simulator. This looseness is the datum feature shift. Figures 16-8 A & B show examples of datum axis MMB.

Where a cylindrical feature of size is used as a primary datum feature at MMB, four degrees of freedom are constrained: two translational and two rotational.

The degrees of freedom constrained are shown on the model coordinate system in Figure 16-8.

---

**TECHNOTE 16-3**
**Cylindrical Datum Feature – Primary MMB**

Where a feature of size is referenced as a primary datum feature MMB, the following applies:

- A fixed-size (MMB) datum feature simulator is used.

- The datum is an axis (line).

- The part is oriented in the datum feature simulator.

- Four degrees of freedom are constrained (two translational and two rotational).

- The datum (or datum reference frame) is established from the datum feature simulator.

- There may be looseness between the datum feature and the datum feature simulator.

- A datum feature shift may be available.

---

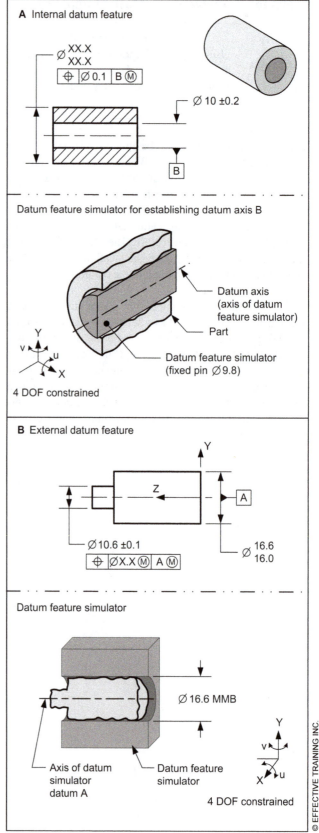

**FIGURE 16-8 Internal and External Cylindrical Features of Size as Primary Datum Features at MMB**

# Planar Feature of Size Datum Feature MMB Primary

In certain applications, the function of a part requires that a planar feature of size (width) is used as a datum feature and is referenced primary at MMB.

Where a planar feature of size (width) is designated as a datum feature and referenced in a feature control frame at MMB, a datum center plane is established. The datum center plane is established as the center plane of the datum feature simulator. The width is surrounded (or filled) by the datum feature simulator. Devices that are fixed in size, such as a set of parallel plates, are used as the datum feature simulator. The center plane of the datum feature simulator becomes the datum center plane and establishes the orientation of the part.

Figure 16-9 A & B shows examples of internal and external widths used to establish a primary datum center plane MMB.

Where a width is used as a datum feature at MMB, three degrees of freedom are constrained: one translational and two rotational. The degrees of freedom constrained are shown on the model coordinate system in Figures 16-7 and 16-8.

---

**TECHNOTE 16-4**
**Width Datum Feature – Primary MMB**

Where a width is referenced as a primary datum feature MMB, the following applies:

- A fixed-size (MMB) datum feature simulator is used.

- The datum is a center plane.

- The part is oriented in the datum feature simulator.

- Three degrees of freedom are constrained (one translational and two rotational).

- The datum feature must be of sufficient length to provide repeatable orientation of the part in the datum simulator.

- There may be looseness between the part and the datum feature simulator.

- A datum feature shift may be available.

---

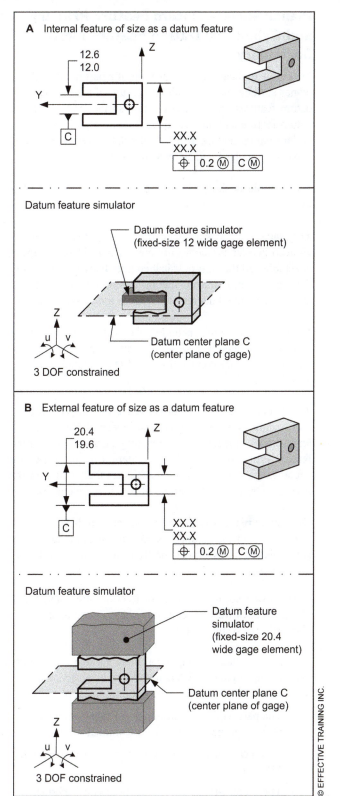

FIGURE 16-9 *Internal and External Width Features of Size as Primary Datum Features at MMB*

## Planar Surface Datum Feature Primary and Cylindrical Datum Feature MMB Secondary

In certain applications, the function of a part requires that a planar surface and a cylindrical feature of size are used as datum features. In the case where a surface is the primary datum feature and the cylindrical feature of size is the secondary datum feature at MMB, the datum reference frame consists of a plane and an axis 90° to the plane.

An example is shown in Figure 16-10 and explained below.

The planar surface is designated as a datum feature and referenced in a feature control frame at RMB. Datum plane A is established through physical contact between the datum feature simulator and the datum feature. One translational and two rotational degrees of freedom are constrained.

Now let's discuss the secondary datum feature. The center hole is designated as a secondary datum feature and referenced in a feature control frame at MMB. Datum axis B is established.

The datum feature simulator is fixed at the MMB of the datum feature. A device that is fixed in size (17.8mm diameter), such as a precision gage pin, may be used as the datum feature simulator. The axis of the datum feature simulator becomes the datum axis and establishes the location of the part. Two translational degrees of freedom are constrained.

Two datum planes exist through the datum axis, but are free to rotate about the datum axis. Due to the geometry of the part, it is not necessary to constrain the rotation of the datum planes.

In this example, five degrees of freedom are constrained. The degrees of freedom constrained are shown on the model coordinate system in Figure 16-9.

In summary, when referencing a planar surface as a primary datum and a cylindrical feature of size as a secondary datum feature MMB, the following conditions apply:

- The part will have a minimum of three-point contact with its primary datum feature simulator.
- The secondary datum feature simulator is fixed at the MMB size.
- The secondary datum feature simulator is perpendicular to the primary datum plane.
- The datum reference frame consists of a primary plane and two planes passing through the secondary datum axis 90° to the primary plane.
- Five degrees of freedom are constrained (three translational and two rotational).
- Datum feature shift may be available.

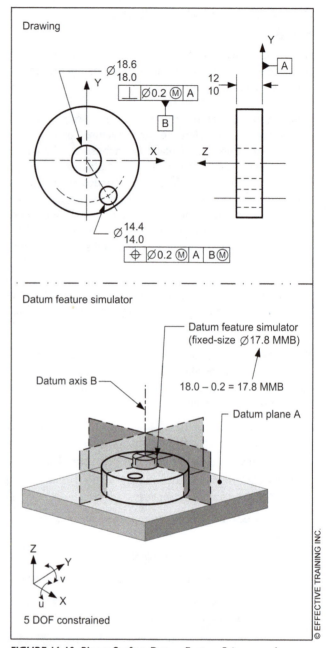

**FIGURE 16-10** *Planar Surface Datum Feature Primary and Cylindrical Datum Feature MMB Secondary*

# Planar Surface Datum Feature Primary, Cylindrical Datum Feature MMB Secondary, and Width Datum Feature MMB Tertiary

In certain applications, the function of a part requires using three datum features to establish a datum reference frame.

In the case where a surface is the primary datum feature, a cylindrical feature of size is the secondary datum feature at MMB, and a width is the tertiary datum feature at MMB, the datum reference frame consists of a plane and an axis with two planes passing through it. The rotation of the two datum planes is constrained by the tertiary datum.

An example using the datum reference frame from the position tolerance is shown in Figure 16-11 and explained below.

The planar surface is designated as a datum feature and referenced in a feature control frame at RMB. Datum plane A is established through physical contact between the datum feature simulator and the datum feature. One translational and two rotational degrees of freedom are constrained.

The internal center hole is designated as a secondary datum feature and referenced in a feature control frame at MMB. Datum axis B is established.

The datum feature simulator is fixed at the MMB of the datum feature. A device that is fixed in size, such as a precision gage pin, may be used as the datum feature simulator. The axis of the datum feature simulator becomes the datum axis and establishes the location of the part: two translational degrees of freedom are constrained.

For the tertiary datum feature, the datum feature simulator is fixed at the MMB size of the width. A device that is fixed in size, such as a precision gage block, may be used as the datum feature simulator.

The center plane of the datum feature simulator becomes the datum center plane and establishes the angular relationship of the part relative to the datum reference frame. One translational degree of freedom is constrained. The third datum plane exists through the datum axis and is constrained by the datum feature simulator for the tertiary datum.

In this example, all six degrees of freedom are constrained. The degrees of freedom constrained are shown on the model coordinate system in Figure 16-11.

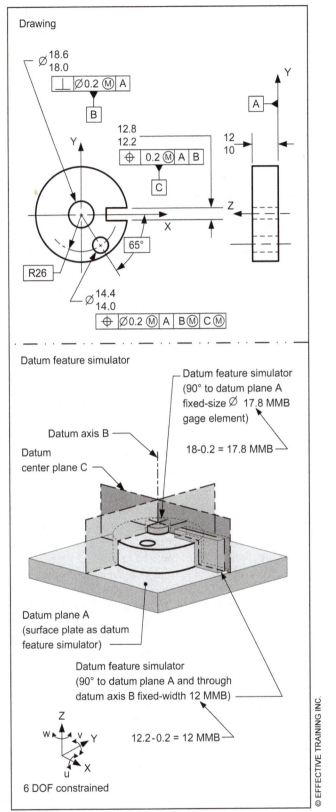

FIGURE 16-11 Planar Surface Datum Feature Primary, Cylindrical Datum Feature MMB Secondary, and Width Datum Feature Tertiary

In summary, when referencing a planar surface as a primary datum feature, a cylindrical feature of size as a secondary datum feature MMB, and a width as a tertiary datum feature MMB, the following conditions apply:

- The part will have a minimum of three-point contact with the primary datum feature simulator.

- The secondary datum feature simulator is fixed at the MMB size.

- The secondary datum feature simulator is perpendicular to the primary datum plane.

- The datum reference frame consists of a plane primary and two planes passing through the secondary datum axis 90° to the plane.

- A second and third datum plane is associated with the datum axis.

- The tertiary datum constrains the rotation of the second and third datum planes.

- All six degrees of freedom are constrained (three translational and three rotational).

- Datum feature shift is permissible.

- The part is allowed to shift within the looseness between the datum features and the datum feature simulators.

## Coaxial Datum Features as Primary Datum Feature MMB

For the part shown in Figure 16-12, the two coaxial features of size at each end of the part work together to properly orient and locate the part in its assembly. Separately, each is too short to establish repeatable orientation of the part. Therefore, the diameters must be referenced together in a feature control frame as primary co-datums A-B.

Because the primary design concern is assembly, and to help reduce costs, the co-datums are referenced at MMB in a feature control frame. The resulting datum reference frame consists of two datum planes passing through the primary datum axis.

The two fixed-size datum feature simulators (made to the MMB size of each datum feature) must fit around each datum feature simultaneously to orient and locate the part in its datum reference frame and constrain four degrees of freedom (two translational and two rotational).

There may be looseness (datum feature shift) between the datum features and their datum feature simulators. The axis of the datum feature simulators becomes the datum axis, the origin for part measurement.

In summary, when referencing coaxial features of size as primary co-datums MMB, the following conditions apply:

- The datum is an axis (line).

- The axis of the datum feature simulators is the datum axis.

- The datum feature simulators are fixed at the MMB size.

- The part is oriented by the datum feature simulator.

- The datum reference frame consists of two planes passing through the datum axis.

- Four degrees of freedom are constrained (two translational and two rotational).

Where a coaxial datum feature is referenced at MMB, four degrees of freedom are constrained: two translational and two rotational. The degrees of freedom constrained are shown on the model coordinate system in Figure 16-12.

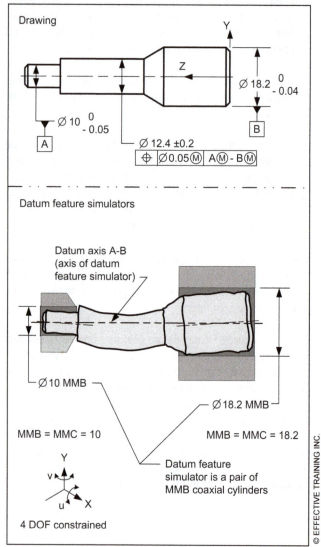

**FIGURE 16-12 Coaxial Datum Features – MMB Primary**

## Pattern of Holes as a Datum Feature MMB Secondary

For the part shown in Figure 16-13, the planar mounting surface orients the part, and the four hole pattern (clearance holes for fasteners) locates the part in its assembly. The center hole does not locate the part. The feature that orients the part is referenced as the primary datum, and the feature that locates the part is referenced as the secondary datum. A mounting surface is always referenced at RMB.

Because there is assembly clearance between the holes and fasteners, the part is allowed to be moved to align the holes for assembly without affecting the function of the part. Therefore, the secondary datum feature (pattern of holes) may be referenced at MMB. For proper fit and function, a position tolerance is used to control the orientation and location of the 12.2-12.6mm center hole relative to primary datum A at RMB and secondary datum B at MMB. The resulting datum reference frame consists of a primary datum plane and two datum planes passing through the secondary datum axis (the axis of the pattern of holes) that is 90° to the primary plane.

To inspect the position tolerance of the center hole, the primary datum feature (mounting surface) is brought into physical contact with a planar datum feature simulator (gage surface) that orients the part in its datum reference frame and constrains three degrees of freedom (one translational and two rotational). Four fixed-size datum feature simulators (4.2mm precision gage pins) perpendicular to the primary datum feature simulator must fit inside the holes to locate the part in its datum reference frame and to constrain three more degrees of freedom (two translational and one rotational). There may be looseness (datum feature shift) between the holes and their gage pins.

Since the primary datum constrains three degrees of freedom and the secondary constrains three degrees of freedom, all six degrees of freedom are constrained.

When referencing a planar surface as a primary datum feature and a pattern of holes as a secondary datum feature MMB, the following conditions apply:

- The part will have a minimum of three-point contact with the primary datum feature simulator.

- The datum feature simulator for the secondary datum will be fixed-size gage elements equal to the MMB of each hole at the MMB size.

- The datum reference frame consists of a plane and an axis 90° to the plane.

- All six degrees of freedom are constrained (three translational and three rotational).

- Datum feature shift is permissible.

- The part is allowed to shift within the looseness between the datum features and the datum feature simulators.

Where a pattern of holes are used as a datum feature, the spacing between the holes should be defined. This is often done by using basic dimensions and a tolerance of position control with a single datum reference. Any of the four fixed-size pins may be used as the datum axis.

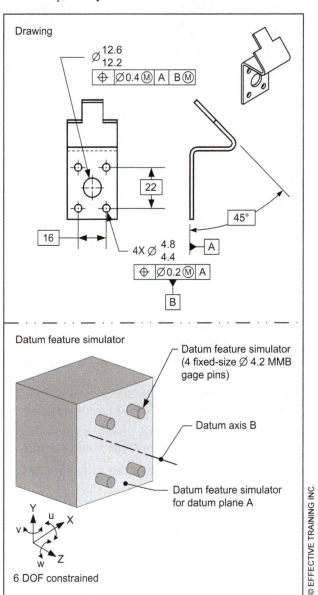

FIGURE 16-13 Planar Surface Datum Feature Primary, Pattern of Holes Datum Feature MMB Secondary

**Author's Comment**
Using a pattern of features of size is a common practice when functionally dimensioning a part. However, this concept is not well understood and often controversial.

© EFFECTIVE TRAINING INC.

## Effects of Changing the Datum Feature Reference Sequence in a Feature Control Frame

Where datum reference frames involve both feature of size datum features and planar datum features, the sequence of the datum feature references plays an important role in the part tolerances. Changing the order (sequence) of datum feature references in a feature control frame changes the relationship between part features and results in a different interpretation.

An example is shown in Figure 16-14. In the top section of the figure, the datum portion of the feature control frame is left blank. In the lower portion, the datum references of the feature control frame are completed using three different datum feature sequences.

In panel A, the datum feature sequence is A primary at RMB, and B secondary at RMB. The following applies:

- An adjustable datum feature simulator is required.

- No datum feature shift is permitted.

- The part is oriented in the datum feature simulator by datum feature A.

- Datum feature B will have a minimum of one point of contact with the datum feature simulator.

- The orientation of the holes is relative to datum axis A.

In panel B, the datum feature sequence is revised to B primary at RMB, and A secondary at RMB. The following applies:

- Datum feature B will have a minimum of three points of contact with the datum feature simulator.

- The part is oriented in the datum feature simulator by datum feature B.

- An adjustable datum feature simulator is required.

- No datum feature shift is permitted.

- The orientation of the holes is relative to datum plane B.

In panel C, the datum feature sequence is revised to B primary at RMB, and A secondary at MMB. The following applies:

- Datum feature B will have a minimum of three points of contact with the datum feature simulator.

- The part is oriented in the datum feature simulator by datum feature B.

- A fixed-size datum feature simulator is used.

- Datum feature shift is permitted.

- The orientation of the holes is relative to datum plane B.

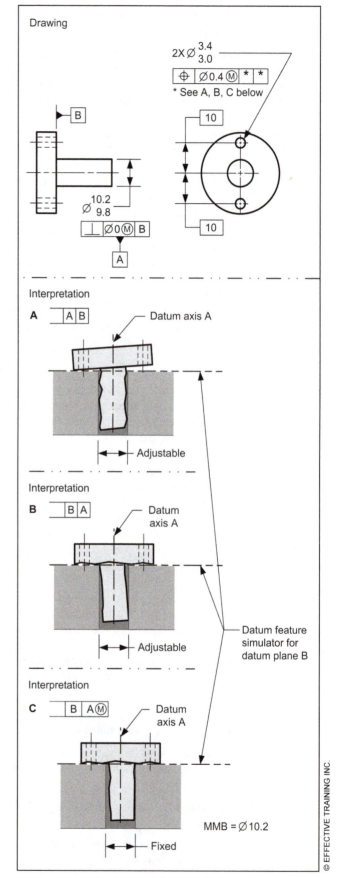

**FIGURE 16-14 Effects of Datum Feature Reference Sequence**

## SUMMARY

### Key Points

- Maximum material boundary (MMB) is the boundary established by the collective effects of the MMC of a datum feature and any applicable tolerances.

- Where a feature of size datum feature is referenced as a primary datum at MMB, the following applies:
  - o The size of datum feature simulator must accept all datum features that are within their drawing specifications.
  - o The datum feature simulator is fixed at the MMB size.
  - o The datum (or datum reference frame) is established from the datum feature simulator.
  - o There may be looseness between the part and the datum feature simulator.
  - o A datum feature shift is permitted.

- Where a feature of size is referenced as a secondary (or tertiary) datum feature at MMB on a drawing, the MMB (datum feature simulator size) is determined as follows:
  - o If an orientation control is applied to the feature of size dimension, relating the secondary (or tertiary) datum feature back to the higher ranking datums, the MMB is equal to the MMC of the datum feature combined with the effects of the orientation control.
  - o If a location control is applied to the feature of size dimension, relating the secondary (or tertiary) datum feature back to the higher ranking datums, the MMB is equal to the MMC of the datum feature combined with the effects of the location control.

- Datum feature shift is the allowable movement or looseness between the datum feature and the datum feature simulator.

- It is important to be able to recognize when a datum feature shift is additive to a geometric tolerance value and when it is not. Generally, there are three common cases:
  - o Where coaxial features of size are being controlled relative to each other, one feature of size being the toleranced feature and the other being the datum feature referenced at MMB, the datum feature shift is additive to the geometric tolerance.
  - o Where a pattern of holes are toleranced relative to a datum feature, the datum feature shift does not affect the spacing tolerance between the holes of the pattern.
  - o Where several features or patterns of features are a simultaneous requirement, the datum feature shift does not affect the tolerance between the features that are gaged simultaneously.

- Where a cylindrical feature of size datum feature is referenced at MMB as a primary datum feature, the following applies:
  - o A fixed-size (MMB) datum feature simulator is used.
  - o The datum is an axis (line).
  - o The part is oriented by the datum feature simulator.
  - o Four degrees of freedom are constrained (two translational and two rotational).
  - o The datum (or datum reference frame) is established from the datum feature simulator.
  - o There may be looseness between the part and the datum feature simulator.
  - o A datum feature shift is permitted.

- Where a width feature of size datum feature is referenced at MMB as a primary datum feature, the following applies:
  - o A fixed-size (MMB) datum feature simulator is used.
  - o The datum is a center plane.
  - o The part is oriented by the datum feature simulator.
  - o Three degrees of freedom are constrained (one translational and two rotational).
  - o The datum (or datum reference frame) is established from the datum feature simulator.
  - o There may be looseness between the part and the datum feature simulator.
  - o A datum feature shift is permitted.

- Where a coaxial datum feature is referenced at MMB, the following applies:
  - o A fixed-size datum feature simulator is used to surround both diameters simultaneously.
  - o The datum is an axis (line).
  - o The part is oriented in the gage.
  - o Four degrees of freedom are constrained (two translational and two rotational).
  - o A datum feature shift is permissible.

## Additional Related Topics

*These topics are recommended for further study to improve your understanding of size datum features (MMB).*

| Topic | Source |
|---|---|
| LMB | *ASME Y14.5-2009, Para. 4.11.7* |
| Translation modifier | *ASME Y14.5-2009, Para. 4.11.10* |
| MMB calculations | *ASME Y14.5-2009, Para. 4.11.6* |
| MMB applied to planar datum features | *ASME Y14.5-2009, Para. 4.16.7* |
| Screw threads, gears, and splines as datum features | ETI's *Advanced Concepts of GD&T* textbook |
| Customized datum reference frames | *ASME Y14.5-2009, Para. 4.22* |
| Simultaneous requirements | ETI's *Advanced Concepts of GD&T* textbook |
| Qualifying datum features in a datum reference frame | ETI's *Advanced Concepts of GD&T* textbook |
| Additional datum feature types | ETI's *Advanced Concepts of GD&T* textbook |
| Datum targets referenced at MMB | *ASME Y14.5-2009, Para. 4.24.10* |

## QUESTIONS AND PROBLEMS

**Website Bonus Materials**
Additional discussion questions are available at our website. To access bonus materials for this textbook, please visit: www.etinews.com/textbookbonus

## True and False

*Indicate if each statement is true or false.*

T / F    1. An MMB is always equal to a Rule #1 boundary.

T / F    2. An MMB datum feature simulator is a fixed-size boundary.

T / F    3. When a datum feature of size is referenced at MMB, a datum feature shift will always be available.

T / F    4. Where a datum feature is referenced at MMB, the datum simulator size is adjustable.

T / F    5. Datum feature shift is the allowable looseness between the datum feature and the datum feature simulator.

T / F    6. When a datum feature is referenced at MMB, a fixed size datum feature simulator must be used.

T / F    7. Changing the order (sequence) of datum feature references in a feature control frame changes the relationships between part features and results in a different interpretation.

T / F    8. When two coaxial features of size are designated datum features and referenced as primary co-datums, five DOF are constrained.

## Multiple Choice

*Circle the best answer to each statement.*

1. If a 10.2-10.6 diameter hole is designated as a datum feature and is referenced as a primary datum at MMB, the size of the datum feature simulator would be_____ .
   A. 10.2
   B. 10.4
   C. 10.6
   D. Variable between 10.2 and 10.6

2. Where a datum feature is referenced MMB primary, the simulated datum is_____.
   A. The center of the datum feature
   B. The axis or center plane of the datum feature simulator
   C. Derived from the related actual mating envelope
   D. All of the above

3. When clearance (looseness) exists between a datum feature and its datum feature simulator...
   A. The looseness may be used to decrease the part feature's acceptance criteria.
   B. The looseness may be used to increase the part feature's acceptance criteria.
   C. The looseness must be ignored by centering the part on the gage.
   D. The part is not acceptable.

4. When a hole is a datum feature referenced at MMB, the datum feature simulator should be...
   A. A fixed-size ball
   B. A fixed-size pin
   C. A tapered pin
   D. A variable-size pin

5. Where a hole is a secondary datum feature referenced at MMB the datum feature is simulated at its _____ size.
   A. Basic
   B. MMB
   C. VC
   D. MMC

## Application Problems

*The application problems are designed to provide practice on applying the chapter concepts to situations that are similar to on-the-job conditions.*

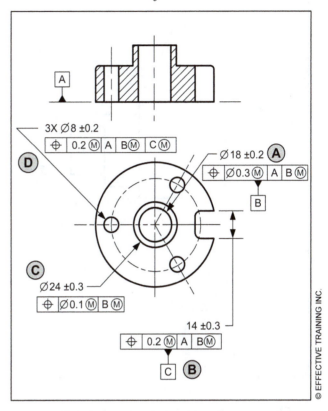

*Application questions 1–3 refer to the drawing above.*

1. Fill in the amount of datum feature shift permissible for each feature control frame.

   (A) _____

   (B) _____

   (C) _____

   (D) _____

2. Describe the datum feature simulator for datum feature B for the feature control frame labeled C. _____

   _____

3. Describe the datum feature simulator for datum feature B for the feature control frame labeled B. _____

   _____

4.  Using the drawing below, draw and dimension the datum feature simulator for establishing datum axis A.

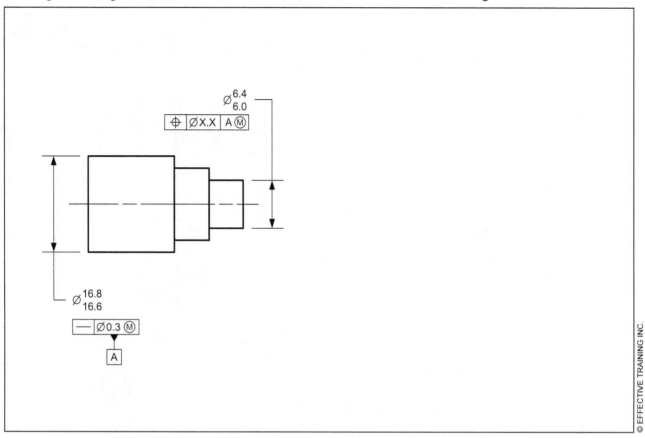

5.  For the feature control frame shown, draw and dimension the datum feature simulator for establishing datum axis A.

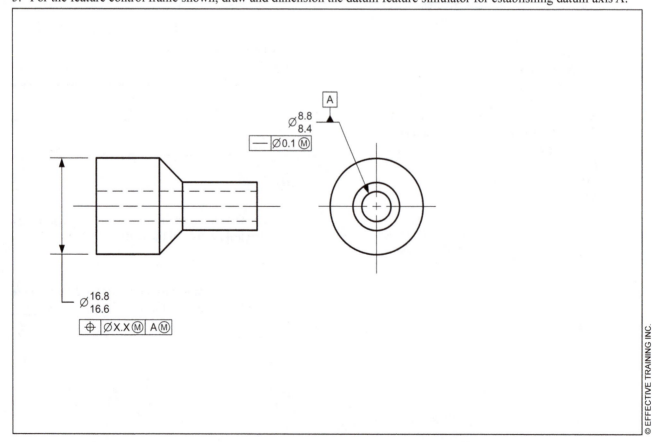

# Perpendicularity Tolerance

## Goal

Interpret the perpendicularity tolerance

## Performance Objectives

Upon completing this chapter, you should be able to:

1. Describe what controls the tolerance on implied 90° angles (p.214)
2. Describe the terms "perpendicularity" and "perpendicularity tolerance" (p.214)
3. Describe two common tolerance zone shapes for a perpendicularity tolerance (p.215)
4. List three indirect perpendicularity tolerances (p.216)
5. Describe the modifiers that may be used in a perpendicularity tolerance (p.216)
6. Recognize when a perpendicularity tolerance is applied to a planar surface or a feature of size (p.217)
7. Evaluate if a perpendicularity tolerance is standard-compliant (p.218)
8. Describe real-world applications for a perpendicularity tolerance (p.219)
9. Interpret a perpendicularity tolerance applied to a planar surface (p.219)
10. Interpret a perpendicularity tolerance with multiple datum feature references applied to a planar surface (p.220)
11. Interpret a perpendicularity application that uses the tangent plane modifier (p.221)
12. Interpret a perpendicularity tolerance at RFS applied to a width (p.222)
13. Interpret a perpendicularity tolerance at MMC applied to a width (p.223)
14. Interpret a perpendicularity tolerance at MMC applied to a cylindrical feature of size (p.224)
15. Interpret a perpendicularity tolerance by using the "Significant Seven Questions" (p.225)
16. Understand the verification principles for perpendicularity (p.225)
17. Describe how a perpendicularity tolerance applied to a surface is verified (p.226)
18. Draw a gage for inspecting perpendicularity applied to a feature of size (MMC) (p.226)

## New Terms

- Height gage
- Implied 90° angles
- Perpendicularity
- Perpendicularity tolerance
- Precision square
- Tangent plane modifier

## What This Chapter Is About

This is the first of three chapters that explain the orientation tolerances. Orientation tolerances define angular deviations (and sometimes form deviations) on surfaces or features of size, but cannot control location. In this chapter, we'll look at how to control right angles (90° angles) on a part.

This chapter explains where to use and how to interpret perpendicularity tolerances and how to interpret implied 90° angles. We'll cover the fundamental uses of the perpendicularity tolerance on surfaces and features of size.

The perpendicularity tolerance is important because it allows the drawing to clearly communicate functional requirements for 90° angles on parts. Perpendicularity tolerance zones are easier to visualize and verify consistently than angular dimensions or general angular tolerances.

# TERMS AND CONCEPTS

## Implied 90° Angles

An *implied 90° angle* is where center lines and lines depicting features are shown on a 2D orthographic drawing at right angles, and no angle is specified (as defined by Fundamental Dimensioning Rule "i" in Y14.5; see Chapter 7).

Since an angle is not specified, there is no tolerance shown for the angle. If the angular tolerance is not controlled with a geometric tolerance, the general tolerance applies and can be in the title block or in the general notes of the drawing.

A general angular tolerance or an angular dimension invokes a wedge-shaped tolerance zone. All of the points (high and low points of the surface) of the toleranced surface must be within the tolerance zone.

Figure 17-1 shows an example of a common note used to specify a tolerance on implied 90° angles.

The general note for implied 90° angles may work for some drawings, but it does have four shortcomings:

1. The tolerance zone is wedge-shaped. It increases as the surface gets farther from the origin.

2. There is no datum sequence specified, which means there are at least two ways the part can be measured. Figure 17-1 shows two ways the angle could be verified, using the long side as the implied datum, or using the short side as the implied datum.

3. The lack of a specified dimension results in the requirement often being overlooked.

4. When general angular tolerances are inspected, common inspection methods often do not comply with the requirement for an angular dimension in the standard.

### Author's Comment
My experience indicates that in many cases implied 90° angles are not verified during inspection. This leads me to believe that it is better to specify the tolerance on implied 90° angles wherever possible.

## Perpendicularity

*Perpendicularity* is the condition where a surface, axis, or center plane is exactly 90° to a datum plane or axis. For a part surface, axis, or center plane to achieve the condition known as perpendicularity is to say the surface, axis, or center plane is to be perfectly perpendicular. Since perfect perpendicularity cannot be manufactured, it is necessary to specify allowances for perpendicular relationships. This is often done with a perpendicularity tolerance.

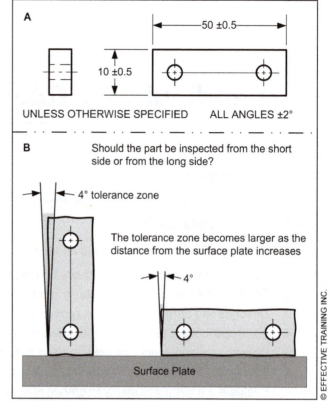

**A**

50 ±0.5

10 ±0.5

UNLESS OTHERWISE SPECIFIED    ALL ANGLES ±2°

**B**    Should the part be inspected from the short side or from the long side?

4° tolerance zone

The tolerance zone becomes larger as the distance from the surface plate increases

4°

Surface Plate

© EFFECTIVE TRAINING INC.

**FIGURE 17-1  Shortcomings of Using General Tolerances for Implied 90° Angles**

## Perpendicularity Tolerance

A *perpendicularity tolerance* is a geometric tolerance that limits the amount a surface axis, or center plane is permitted to deviate from being perpendicular to a datum. A perpendicularity tolerance can only constrain rotational degrees of freedom relative to the datum referenced. Even in cases where the datum features may constrain all degrees of freedom, the tolerance zone only orients to the datum reference frame.

A perpendicularity tolerance:

- Is one of the three direct orientation tolerances

- Must always use a datum reference

- May be applied to a planar surface or a feature of size

- Can control orientation and form

An example of a perpendicularity tolerance is shown in Figure 17-2

**Author's Comment**
Perpendicularity tolerances only control deviations of perpendicularity. They cannot locate features or features of size.

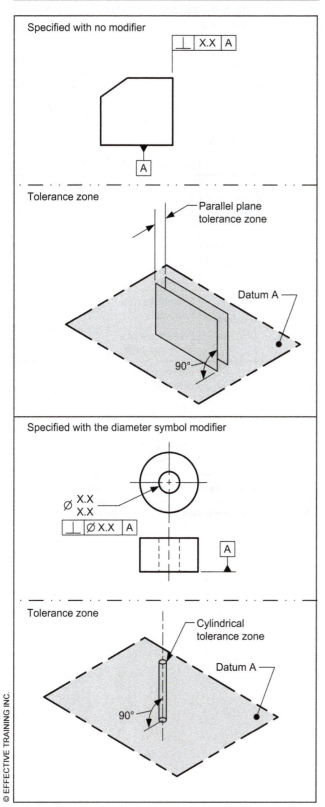

FIGURE 17-2 Common Perpendicularity Tolerance Zones

---

**TECHNOTE 17-1**
**Degrees of Freedom**

Orientation tolerances are constrained only in rotational degrees of freedom relative to their referenced datums; they are not constrained in translational degrees of freedom. Therefore, with orientation tolerances, even in those instances where datum features may constrain all degrees of freedom, the tolerance zone only orients to that datum reference frame.

## Perpendicularity Tolerance Zones

The two common tolerance zones for a perpendicularity tolerance are:

1. The space between two parallel planes

2. The space within a cylinder

The default tolerance zone is two parallel planes. Where the diameter symbol is specified in the feature control frame, the tolerance zone shape is a cylinder. A perpendicularity tolerance zone is always perpendicular to the primary datum referenced in the feature control frame. Where a perpendicularity tolerance is specified, the limits of size must still be met.

Examples of perpendicularity tolerance zones are shown in Figure 17-2.

## Indirect Perpendicularity Tolerances

There are several geometric tolerances that often indirectly affect the perpendicularity deviation of a planar surface or feature of size. Tolerances like position, runout, and profile can also limit perpendicularity deviation. In these applications, the perpendicularity of the toleranced feature is not directly checked. The effect on perpendicularity deviation is a result of the toleranced feature being within the other geometric tolerance zone. If a perpendicularity tolerance and a location tolerance with the same primary datum reference are specified on a feature or feature of size, the perpendicularity tolerance must be a refinement of the location tolerance.

Figure 17-3 shows examples of a profile and a position tolerance that indirectly control perpendicularity. For an example of a runout tolerance that indirectly controls perpendicularity, see Figure 24-12

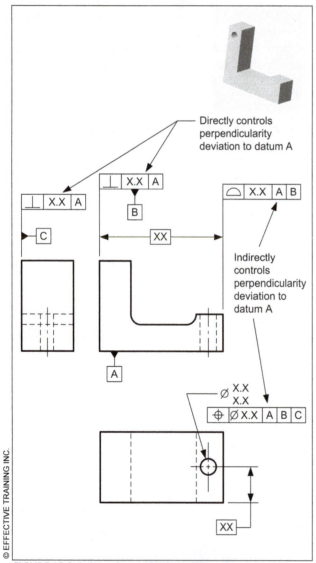

FIGURE 17-3 Indirect Perpendicularity Tolerances

## Modifiers Used With Perpendicularity Tolerances

There are several modifiers that are often used in the tolerance portion of perpendicularity tolerances. The choice of modifier is often related to the functional requirements of the application. The chart in Figure 17-4 shows the common modifiers and examples of functional applications where they could be used.

| Tolerance Modifier | Can Be Applied To | Effect | Functional Application |
|---|---|---|---|
| ⓣ | Feature | Releases the flatness requirement | Orienting of mating surfaces |
| ⓕ * | Feature or feature of size | Releases the restraint requirement | Nonrigid parts with restraint notes |
| ⟨ST⟩ * | Feature or feature of size | Requires statistical process controls | Statistically derived tolerances or tolerances used in statistical tolerance analysis |
| ⌀ | Feature of size | Creates a cylindrical tolerance zone | Holes, external diameters |
| Ⓜ | Feature of size | Permits a bonus tolerance / Permits use of a functional gage | Assembly / Non critical relationships |
| Ⓛ | Feature of size | Permits a bonus tolerance / Requires variable measurement | Minimum distance |
| Ⓟ | Feature of size | Projects the tolerance zone | Fixed fastener assemblies |

\* These modifiers are only introduced in this textbook.

FIGURE 17-4 Modifiers Used with Perpendicularity Tolerances

## TECHNOTE 17-2
### Perpendicularity Tolerance

The following are characteristics of a perpendicularity tolerance:

- In a perpendicularity tolerance application, the toleranced feature must be 90° to the primary datum referenced.
- The default tolerance zone shape is two parallel planes.
- Where the diameter symbol is specified, the tolerance zone shape is a cylinder.
- One or more datums must be specified.
- The tolerance zone is perpendicular to the primary datum specified.
- A perpendicularity tolerance can limit form deviation.
- A perpendicularity tolerance cannot affect location or size.
- A perpendicularity tolerance can only be applied to an individual feature or feature of size.
- A perpendicularity tolerance must be a refinement of other geometric tolerances that affect the orientation of the feature or feature of size.
- Where a perpendicularity tolerance is specified, the limits of size must still be met.

## Specifying Perpendicularity Tolerances

Perpendicularity can be applied to a planar surface or a feature of size. Where a perpendicularity control is applied to a surface, the tolerance zone applies to the surface. Where perpendicularity control is applied to a feature of size, the tolerance zone applies to the axis or center plane of the feature of size. See Figure 17-5.

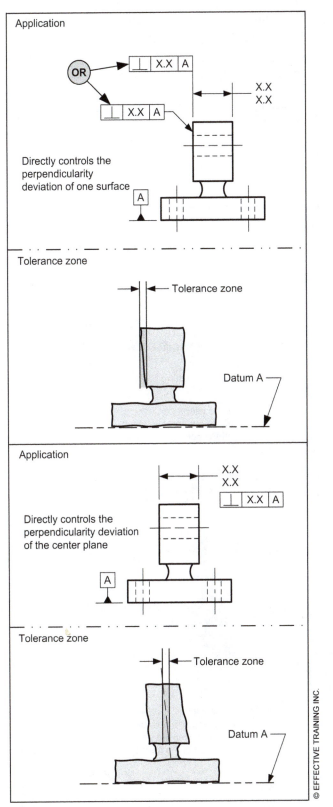

FIGURE 17-5 Specifying Perpendicularity Tolerances on a Surface and a Feature of Size

## Evaluate if a Perpendicularity Tolerance Specification is Standard-Compliant

For a perpendicularity tolerance to be a standard-compliant specification (legal), it must satisfy certain requirements. In order to make it easier to remember what to look for when evaluating the correctness of a geometric tolerance on a drawing, I created a mnemonic.

The requirements are divided into four areas represented by the initials "CARE":

        C - Check for datums

        A - Assess the application

        R - Review the modifiers

        E - Evaluate the tolerance value

A flowchart that identifies the requirements for a standard-compliant perpendicularity tolerance is shown in Figure 17-6.

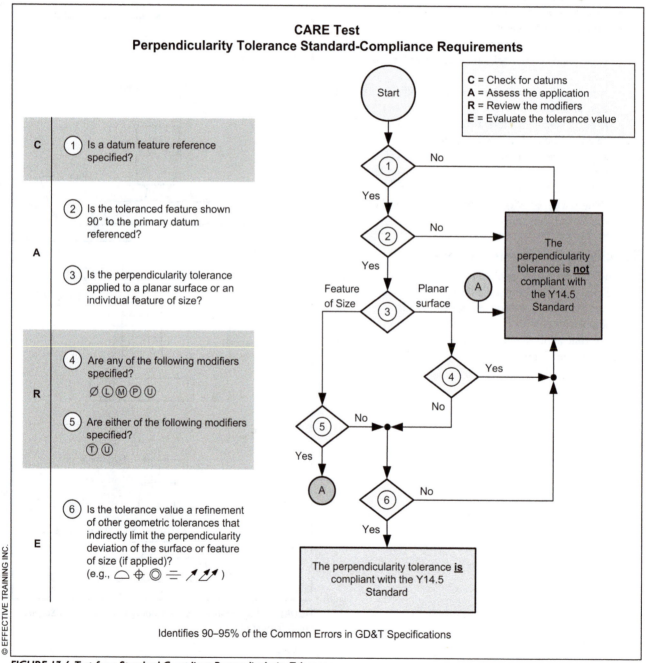

**FIGURE 17-6** *Test for a Standard-Compliant Perpendicularity Tolerance*

## PERPENDICULARITY APPLICATIONS

The following sections describe various perpendicularity applications.

### Real-World Applications of a Perpendicularity Tolerance

There are many applications of a perpendicularity tolerance on a drawing. Four common real-world applications are shown below. These applications limit the perpendicularity deviation of a planar surface or feature of size relative to a datum reference frame for:

- Assembly
- Appearance (customer appeal)
- Establish the relationship between datum features (for improving measurement repeatability)
- Support (guides and stops)

### Perpendicularity Tolerance Applied to a Planar Surface

In certain applications, part function requires the perpendicularity deviation of a surface to be defined relative to a datum.

In the example in Figure 17-7, the perpendicularity deviation of a surface is being defined relative to a datum plane, which is the most common application of a perpendicularity tolerance. Where a perpendicularity tolerance is applied to a planar surface, the following conditions apply:

- The tolerance zone is the space between two parallel planes.
- The tolerance zone is perpendicular to the primary datum plane.
- The tolerance value defines the distance between the tolerance zone planes.
- Since all of the points of the surface must be located within the tolerance zone, the perpendicularity tolerance also limits the flatness of the surface.

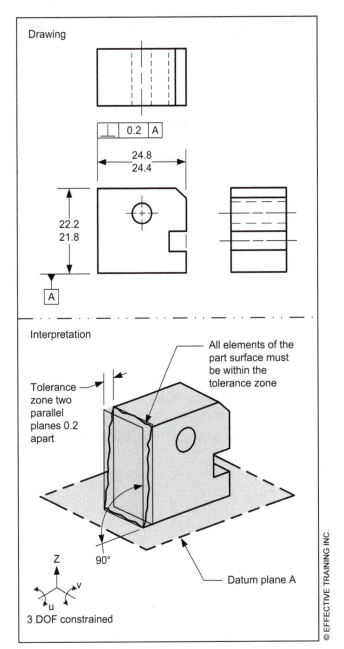

FIGURE 17-7 Perpendicularity Tolerance Applied to a Planar Surface

## Perpendicularity Tolerance With Multiple Datum Feature References

In certain applications, part function requires the perpendicularity deviation of a surface to be defined relative to multiple datums.

Figure 17-8 shows the perpendicularity deviation of a surface being defined relative to two datum planes. Where a perpendicularity tolerance contains two datum references, the tolerance zone is oriented to both datum planes. Where two datum references are used with a perpendicularity tolerance, the tolerance zone is perpendicular to the primary datum reference. The secondary datum reference is used to constrain an additional rotational degree of freedom of the datum reference frame (see Technote 17-2).

Where a perpendicularity tolerance is applied to a planar surface with multiple datum references, the following conditions apply:

- The tolerance zone is the space between two parallel planes.
- The tolerance zone is oriented to both datum planes.
- The tolerance value defines the distance between the tolerance zone planes.
- The perpendicularity tolerance also limits the flatness of the surface.

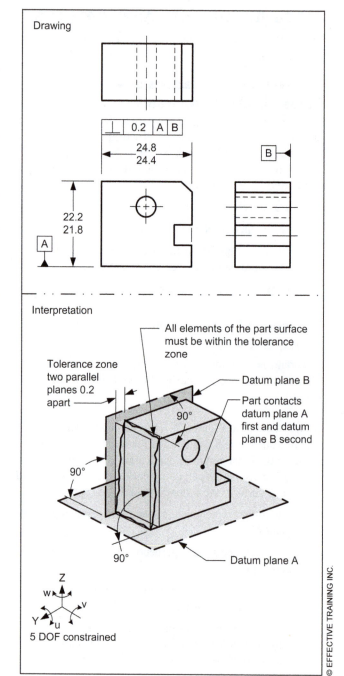

*FIGURE 17-8 Perpendicularity Tolerance With Multiple Datum Feature References Applied to a Planar Surface*

## Perpendicularity Tolerance With the Tangent Plane Modifier Applied to a Planar Surface

In certain applications, part function requires the perpendicularity deviation of the tangent plane of a surface to be defined relative to a datum.

Where a perpendicularity tolerance is intended to apply to the tangent plane of a surface, the tangent plane modifier ⓣ is specified in the tolerance portion of the feature control frame.

The *tangent plane modifier* is a modifier that indicates that a geometric tolerance only applies to the tangent plane of the toleranced surface. Since the tolerance zone only applies to the tangent plane of the toleranced surface, the flatness of the surface is not controlled by the perpendicularity tolerance. Therefore, using the tangent plane modifier is considered a less stringent control.

In the example in Figure 17-9, the perpendicularity deviation of the tangent plane of a surface is being defined relative to a datum plane. Where a perpendicularity tolerance using the tangent plane modifier is applied to a planar surface, the following conditions apply:

- The tolerance zone is the space between two parallel planes.

- The tolerance zone is perpendicular to the datum plane.

- The tolerance value defines the distance between the tolerance zone planes.

- The tangent plane of the surface must be within the tolerance zone.

- The perpendicularity tolerance does not limit the flatness of the surface.

- The tolerance zone applies to the full length and width of the surface.

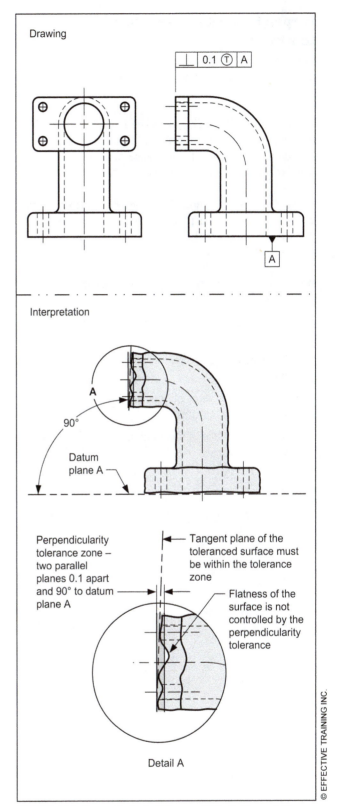

FIGURE 17-9 Perpendicularity Tolerance With the Tangent Plane Modifier

## Perpendicularity Tolerance at RFS Applied to a Width

In certain applications, part function requires the perpendicularity deviation of a width, regardless of its size, to be defined relative to a datum.

Where a perpendicularity tolerance is applied to a width, the feature control frame is placed beneath (or beside) the size dimension. The tolerance zone applies to the center plane of the actual mating envelope of the width.

In the example in Figure 17-10, the allowable orientation deviation of the center plane of the width is defined relative to a datum plane. In this example, the RFS condition applies.

Where perpendicularity is applied to a width at RFS, the following conditions apply:

- The tolerance zone is the space between two parallel planes.

- The tolerance zone is perpendicular to the primary datum plane.

- The tolerance value defines the distance between the tolerance zone planes.

- The center plane of the width must be within the tolerance zone.

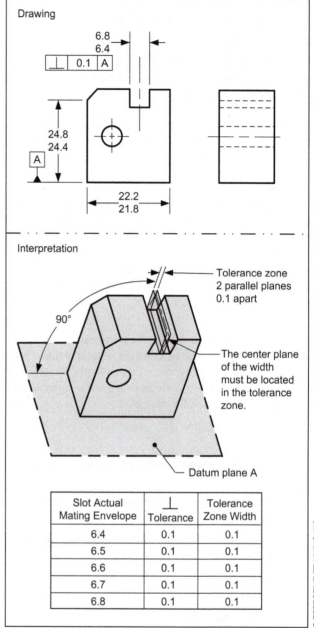

| Slot Actual Mating Envelope | ⊥ Tolerance | Tolerance Zone Width |
|---|---|---|
| 6.4 | 0.1 | 0.1 |
| 6.5 | 0.1 | 0.1 |
| 6.6 | 0.1 | 0.1 |
| 6.7 | 0.1 | 0.1 |
| 6.8 | 0.1 | 0.1 |

**FIGURE 17-10  Perpendicularity Tolerance at RFS Applied to a Width**

## Perpendicularity Tolerance at MMC Applied to a Width

In certain applications, part function requires the perpendicularity deviation of a width (feature of size) to be defined relative to a datum.

Where a perpendicularity tolerance at MMC is used, the function is often assembly and the surface interpretation is often used. In this text, the surface interpretation is illustrated for MMC applications although an axis or center plane interpretation could also be used.

In the example in Figure 17-11, the allowable perpendicularity deviation of the center plane of a width is being defined relative to a datum plane. In this example the MMC modifier is specified in the perpendicularity tolerance. The MMC modifier is often used to ensure the function of assembly.

Where perpendicularity is applied to a width and contains the MMC modifier, the following conditions apply:

- Using the surface interpretation, the tolerance zone is a virtual condition boundary that the part surfaces cannot violate.

- The tolerance zone is perpendicular to the datum plane.

- A boundary check is often used to verify the center plane conformance.

- A bonus tolerance is permissible.

- A fixed gage may be used to verify the perpendicularity requirement.

- The feature of size must also be within its size limits.

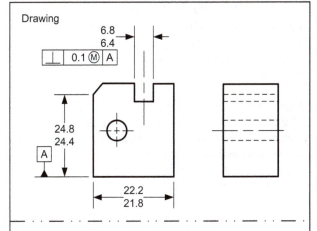

Drawing

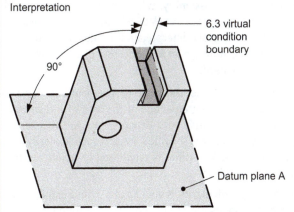

Interpretation

90°

6.3 virtual condition boundary

Datum plane A

| VC (Acceptance Boundary) | Slot Actual Mating Envelope | Total Perpendicularity Tolerance Allowed* |
|---|---|---|
| 6.3 | 6.4  MMC | 0.1 |
| | 6.5 | 0.2 |
| | 6.6 | 0.3 |
| | 6.7 | 0.4 |
| | 6.8  LMC | 0.5 |

\* See Appendix F

© EFFECTIVE TRAINING INC.

**FIGURE 17-11  Perpendicularity Tolerance at MMC Applied to a Width**

### Author's Comment

The Y14.5 standard states that where a geometric tolerance is specified at MMC, there are two interpretations: a boundary and an axis/center plane interpretation. Where there is a conflict between these interpretations, the surface (boundary) interpretation takes precedence. Therefore, in this text, the surface interpretation (VC boundary) is shown. See Appendix F for additional information.

## Perpendicularity Tolerance at MMC Applied to a Cylindrical Feature of Size

In certain applications, part function requires the perpendicularity deviation of a cylindrical feature of size to be defined relative to a datum.

Where a perpendicularity tolerance is applied to a cylindrical feature of size, the feature control frame is placed beneath the size dimension, controlling the orientation of the axis of the actual mating envelope of the cylinder. In most of these cases, the feature control frame contains the diameter symbol to indicate a cylindrical tolerance zone.

Where a perpendicularity tolerance at MMC is used, the function is often assembly. Where the function is assembly, the surface interpretation (VC boundary) is often used. In this text, the surface interpretation is illustrated for MMC applications, although an axis or center plane interpretation could also be used.

In Figure 17-12, the allowable perpendicularity deviation of the actual mating envelope of the axis of a diameter is being defined relative to a datum plane. In this example, the MMC modifier is specified in the perpendicularity tolerance. The MMC modifier is often used to ensure the function of assembly.

Where perpendicularity is applied to a cylindrical feature of size and contains the MMC modifier, the following conditions apply:

- Using the surface interpretation, the tolerance zone is a virtual condition boundary that the part surface cannot violate.

- The tolerance zone is perpendicular to the datum plane.

- A boundary check is often used to verify the axis conformance.

- A bonus tolerance is permissible.

- A fixed gage may be used to verify the perpendicularity requirement.

- The feature of size must also be within its size limits.

***Design Tip***
A common application of a perpendicularity tolerance applied to a feature of size is to control a secondary datum feature relative to the primary datum.

A functional gage may be used to verify the perpendicularity requirement. The gage has a surface to simulate datum plane A. The gage also has a hole that verifies the perpendicularity of the hub diameter. The diameter of the gage hole is equal to the virtual condition of the hub.

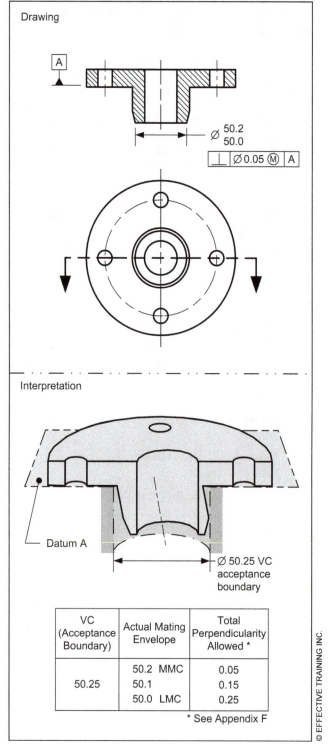

Drawing

Interpretation

| VC (Acceptance Boundary) | Actual Mating Envelope | Total Perpendicularity Allowed * |
|---|---|---|
| 50.25 | 50.2  MMC | 0.05 |
|  | 50.1 | 0.15 |
|  | 50.0  LMC | 0.25 |

\* See Appendix F

**FIGURE 17-12** *Perpendicularity Tolerance at MMC Applied to a Cylindrical Feature of Size*

## The Significant Seven Questions

This section uses the Significant Seven Questions to interpret a perpendicularity tolerance application. The questions are explained in Appendix B. Figure 17-13 contains a perpendicularity tolerance application. The Significant Seven Questions are answered in the interpretation section of the figure. Notice how each answer increases your understanding of the geometric tolerance.

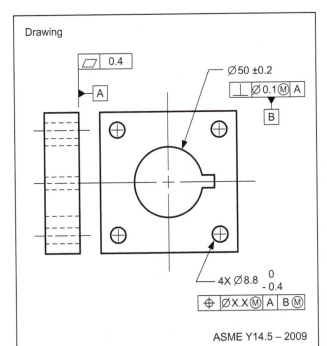

Drawing

⌐ 0.4

Ø 50 ±0.2

⟂ Ø 0.1 Ⓜ A

A

B

4X Ø8.8 ⁰₋₀.₄

⊕ ØX.X Ⓜ A B Ⓜ

ASME Y14.5 – 2009

Interpretation

The Significant Seven Questions

1. Which dimensioning and tolerancing standard applies?
   Answer – ASME Y14.5-2009

2. What does the tolerance apply to?
   Answer – The boundary of the 50 +/- 0.2 diameter

3. Is the specification standard-compliant?
   Answer – Yes

4. What are the shape and size of the tolerance zone?
   Answer – A 49.7 dia virtual condition boundary

5. How much total tolerance is permitted?
   Answer – 0.5 at LMC

6. What are the shape and size of the datum simulators?
   Answer – A plane for datum A and a 49.7 diameter for datum B

7. Which geometry attributes are affected by this tolerance?
   Answer – Orientation, form (straightness)

*FIGURE 17-13 The Significant Seven Questions Applied to a Perpendicularity Tolerance*

# VERIFICATION PRINCIPLES AND METHODS

## Verification Principles for Perpendicularity Tolerances

This section contains a simplified explanation of verification principles and methods that can be used to inspect a perpendicularity tolerance applied to a planar surface or a cylindrical feature of size (at MMC).

When a perpendicularity tolerance is applied to a planar surface, the deviation is verified by finding the distance between two parallel planes that all of the points of the surface are located within. The inspection method must be able to collect a set of points and determine the distance between two planes that contain the set of points. Since location is not a requirement of a perpendicularity tolerance, the inspection method should not be influenced by the location of the toleranced surface.

Verifying a perpendicularity tolerance has two parts:

1. Establishing the relationship specified in the feature control frame between the part and the datum reference frame

2. Verifying the location of the surface, axis, or center plane are within the tolerance zone

The relationship between the part and the datum reference frame may be established by orienting and locating the datum features on the datum planes in the sequence specified in the feature control frame.

When a perpendicularity tolerance (at MMC) is applied to a feature of size, the perpendicularity deviation is often verified by either passing an external feature of size through a virtual condition (a functional gage) or passing a virtual condition (a functional gage) into an internal feature of size.

There are many ways perpendicularity tolerances can be verified. In this section, we will look at two of them.

## Verifying a Perpendicularity Tolerance Applied to a Planar Surface

There are many ways to verify a perpendicularity tolerance. Figure 17-14 shows an example of one method: using a surface plate, a precision square, and a height gage.

A *height gage* is a measuring device used for measuring a vertical distance of a feature from a reference surface. A height gage is used in metrology to verify geometry attributes of parts. The height gage in Figure 17-14 consists of a stand and a dial indicator.

A *precision square* is a gage block that has a precise 90° angle between the bottom surface and one of the sides of the block. The precision square is set on the surface plate. The part is mounted with datum feature A on the precision square and the toleranced surface level with the surface plate. The height gage is moved across the surface plate to move the dial indicator across the part surface. The dial indicator reading must be equal to or less than the perpendicularity tolerance value for the surface to be within specification.

If the surface does not check within its perpendicularity tolerance zone, it does not mean the surface is out of specification. Since only a single datum is referenced in the feature control frame, the part orientation may be adjusted to achieve the minimum dial indicator reading.

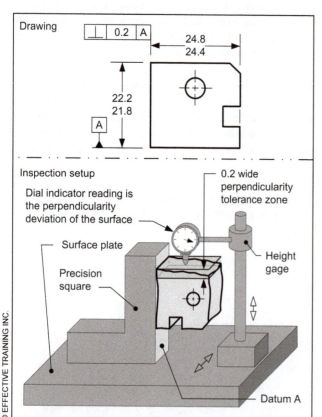

**FIGURE 17-14** *Verifying a Perpendicularity Tolerance Applied to a Planar Surface*

## Verifying a Perpendicularity Tolerance at MMC Applied to a Feature of Size

There are several ways a perpendicularity tolerance (MMC) can be verified. Figure 17-15 shows an example of one method. A common method is to use a functional gage is used. The functional gage verifies that the feature of size fits into its virtual condition.

To verify the perpendicularity tolerance, the part must rest on the datum feature, a minimum of three point contact, and the toleranced feature fit must fit into the virtual condition (acceptance boundary). The size and location of the diameter would be verified separately.

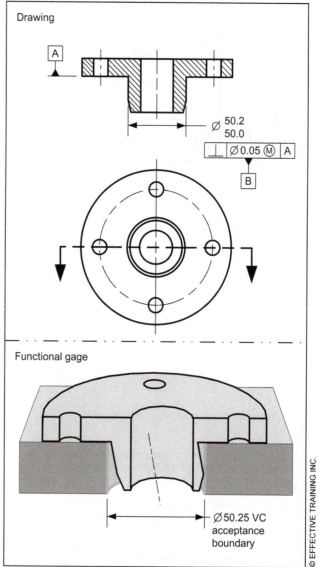

**FIGURE 17-15** *Verifying a Perpendicularity Tolerance at MMC Applied to a Feature of Size*

## SUMMARY

### Key Points

- An implied 90° angle is where center lines and lines depicting features are shown on a 2D orthographic drawing at right angles, and no angle is specified.

- If an implied angle is not controlled with a geometric tolerance, general tolerances apply.

- Using a general tolerance on implied angles has four shortcomings:
  - o The tolerance zone is wedge-shaped. It increases as the surface gets farther from the origin.
  - o There is no datum sequence specified, which means there are at least two ways the part can be measured. Figure 17-1 shows two ways the angle could be verified, using the long side as the implied datum, or using the short side as the implied datum.
  - o The lack of a specified dimension results in the requirement being often overlooked.
  - o Common inspection methods often do not comply with the definition in the standard.

- Perpendicularity is the condition where a surface, axis, or center plane is exactly 90° to a datum plane or axis.

- A perpendicularity tolerance is a geometric tolerance that limits the amount a surface axis, or center plane is permitted to deviate from being perpendicular to a datum.

- The following are requirements of a perpendicularity tolerance.
  - o The default tolerance zone shape is two parallel planes.
  - o Where the diameter symbol is specified, the tolerance zone shape is a cylinder.
  - o One or more datums must be specified.
  - o The tolerance zone is perpendicular to the primary datum specified.
  - o A perpendicularity tolerance can limit form deviation.
  - o A perpendicularity tolerance cannot affect location or size.
  - o A perpendicularity tolerance can only be applied to an individual feature or feature of size.
  - o A perpendicularity tolerance must be a refinement of other geometric tolerances that affect the orientation of the feature or feature of size.
  - o Where a perpendicularity tolerance is specified, the limits of size must still be met.

- Common real-world applications of a perpendicularity tolerance are:
  - o Assembly
  - o Appearance (customer appeal)
  - o Establish the relationship between datum features
  - o Support

- Where a perpendicularity tolerance is applied to a planar surface, it controls the orientation and flatness of the surface.

- Where a perpendicularity tolerance contains the tangent plane modifier, it applies to only the tangent plane of the toleranced surface. It does not control the flatness of the surface.

### Additional Related Topics

*These topics are recommended for further study to improve your understanding of perpendicularity.*

| Topic | Source |
|---|---|
| Modifying note – each line element, each radial element | *ASME Y14.5-2009, Para. 6.4.3* |
| Free state symbol | ETI's *Advanced Concepts of GD&T* textbook |
| MMB calculations | *ASME Y14.5-2009, Para. 4.11.6* |
| Using parallel planes to tolerance an axis | ETI's *Advanced Concepts of GD&T* textbook |

## QUESTIONS AND PROBLEMS

### True and False

*Indicate if each statement is true or false.*

T / F    1. A flatness tolerance applied to a surface is also an indirect perpendicularity control.

T / F    2. A perpendicularity tolerance may use the Ⓣ modifier.

T / F    3. A perpendicularity tolerance may use the Ⓤ modifier.

T / F    4. Two common tolerance zones of a perpendicularity tolerance are two parallel planes or a cylinder.

T / F    5. When verifying a perpendicularity tolerance the part must be oriented on the datum features specified in the feature control frame.

T / F    6. A perpendicularity tolerance may use the Ⓜ modifier.

### Multiple Choice

*Circle the best answer to each statement.*

1. When a perpendicularity tolerance applies to a planar surface:
   A. The tolerance zone is two parallel planes parallel to the primary datum.
   B. The tolerance zone limits the angular variation of a tangent plane.
   C. The tolerance zone limits the variation of high and low points from a basic 90°.
   D. All of the above.

2. When is a perpendicularity tolerance zone cylindrical?
   A. When perpendicularity is applied to a cylindrical feature of size.
   B. When the Ⓜ modifier is placed in the tolerance portion of the feature control frame.
   C. When the ∅ modifier is placed in the tolerance portion of the feature control frame.
   D. All of the above.

3. What is the effect of using the Ⓣ modifier with a perpendicularity tolerance?
   A. Flatness deviations are controlled.
   B. A bonus tolerance is permissible.
   C. A datum feature shift is permissible.
   D. Flatness deviations are not controlled.

4. When a perpendicularity tolerance at RFS is applied to a feature of size:
   A. A virtual condition boundary is created.
   B. The perpendicularity deviation of the axis or center plane of the unrelated actual mating envelope is controlled.
   C. The perpendicularity deviation of the axis or center plane of the related actual mating envelope is controlled.
   D. A bonus tolerance is permitted.

5. Which of the following geometric tolerances could indirectly limit perpendicularity?
   A. ⌒
   B. ▱
   C. ⌐
   D. None of the above

6. A real-world application for a perpendicularity tolerance is:
   A. Seal (lip seal on shaft)
   B. Assembly (clearance holes or fasteners)
   C. Balance (high speed rotation)
   D. Centering (coaxial diameters for rotation or pivoting)

## Application Problems

*The application problems are designed to provide practice on applying the chapter concepts to situations that are similar to on-the-job conditions.*

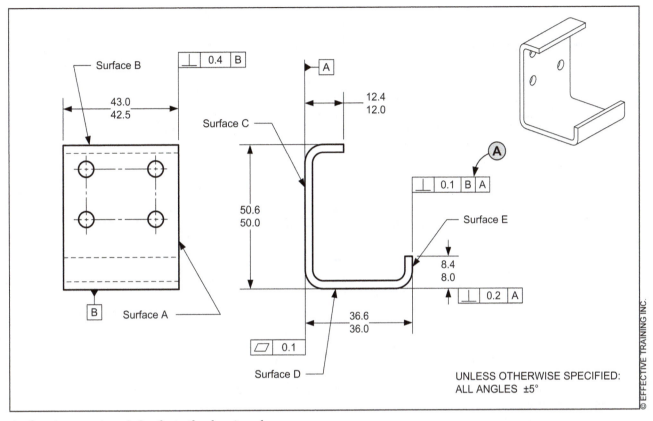

*Application questions 1–5 refer to the drawing above.*

1. What is the shape of the tolerance zone for the perpendicularity tolerance applied to surface D?
   A. Two parallel lines perpendicular to datum A
   B. A tangent plane perpendicular to datum A
   C. Two parallel planes perpendicular to datum A
   D. None of the above

2. For the perpendicularity tolerance applied to surface E, how many degrees of freedom are constrained?
   A. 3         C. 5
   B. 4         D. 6

3. What limits the perpendicularity deviation between surfaces B and C?
   A. Rule #1
   B. The size dimension
   C. The general tolerance for angles
   D. Nothing, it is undefined

4. What limits flatness deviations on the surface labeled "E?"
   A. The perpendicularity tolerance
   B. The size dimension
   C. Rule #1
   D. Flatness variations are not limited

5. Use the Significant Seven Questions to interpret the perpendicularity tolerance labeled "A."

   1) Which dimensioning and tolerancing standard applies?
   _____

   2) What does the tolerance apply to? _____
   _____

   3) Is the specification standard-compliant? _____

   4) What are the shape and size of the tolerance zone?
   _____

   5) How much total tolerance is permitted? _____

   6) What are the shape and size of the datum simulators?
   _____

   7) Which geometry attributes are affected by this tolerance?
   _____

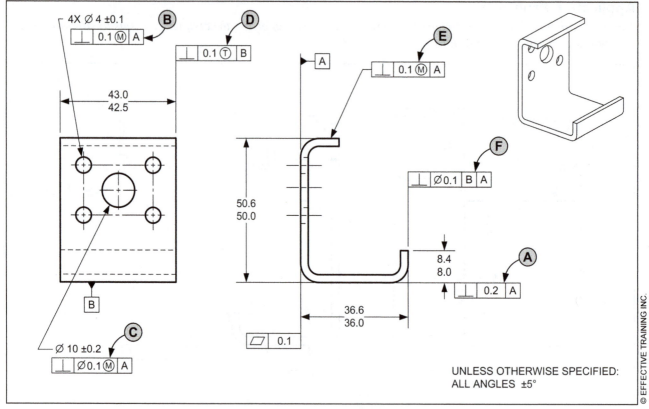

*Questions 6 and 7 refer to the drawing above.*

6. Indicate if each perpendicularity tolerance is standard-compliant. (Hint: use the CARE test.) If a tolerance is not standard-compliant, explain why.

A. ⟂ | 0.2 | A  _____

B. ⟂ | ⌀ 0.1 Ⓜ | A  _____

C. ⟂ | ⌀ 0.1 Ⓜ | A  _____

D. ⟂ | ⌀ 0.1 Ⓣ | B  _____

E. ⟂ | 0.1 Ⓜ | A  _____

F. ⟂ | ⌀ 0.1 | B | A  _____

7. In the space below, drawn and dimension a functional gage for verifying the ⟂ | ⌀0.1Ⓜ | A tolerance labeled "C."

# Parallelism Tolerance

## Goal

Interpret the parallelism tolerance

## Performance Objectives

Upon completing this chapter, you should be able to:

1. Describe what controls the parallelism deviation between two nominally parallel surfaces where no parallelism tolerance is specified (p.232)
2. Describe the terms "parallelism" and "parallelism tolerance" (p.232)
3. Describe two common tolerance zone shapes for a parallelism tolerance (p.232)
4. Recognize when a parallelism tolerance is applied to a planar surface and a feature of size (p.234)
5. List two indirect parallelism tolerances (p.235)
6. Describe the modifiers that may be used in a parallelism tolerance (p.235)
7. Evaluate if a parallelism tolerance is standard-compliant (p.236)
8. Describe real-world applications for a parallelism tolerance (p.237)
9. Interpret a parallelism tolerance applied to a planar surface (p.237)
10. Interpret a parallelism tolerance that uses the tangent plane modifier (p.237)
11. Interpret a parallelism tolerance at RFS applied to a hole (p.238)
12. Interpret a parallelism tolerance at MMC applied to a hole (p.239)
13. Interpret a parallelism tolerance by using the "Significant Seven Questions" (p.240)
14. Understand the verification principles for parallelism (p.241)
15. Describe how a parallelism tolerance applied to a surface is verified (p.241)
16. Describe how to verify a parallelism tolerance at RFS applied to a feature of size (p.242)
17. Describe how to verify a parallelism tolerance at MMC applied to a feature of size (p.242)

## New Terms

- Parallelism
- Parallelism tolerance

## What This Chapter Is About

This is the second of three chapters that explain the orientation tolerances. Orientation tolerances define allowable angular deviations (and sometimes form deviations) on surfaces or features of size. Orientation tolerances cannot control location. In this chapter, we'll look at how to control allowable parallelism deviations on a part.

This chapter explains where to use and how to interpret parallelism tolerances. We will cover the fundamental uses of the parallelism tolerance on surfaces and features of size.

# TERMS AND CONCEPTS

## How Parallelism Deviation is Controlled Where a Parallelism Tolerance is Not Specified

Wherever a planar width (regular feature of size) is shown on a drawing, Rule #1 applies and the parallelism of both surfaces is controlled by the limits of the size dimension. This may be satisfactory in many cases, but there are two shortcomings of using Rule #1 and the size dimension to control parallelism. The first is that the parallelism deviation is limited to the same tolerance as the size dimension. The second is the lack of a datum reference. In Figure 18-1, the part could be inspected using either side as the implied datum. This could result in different measurements by each inspector.

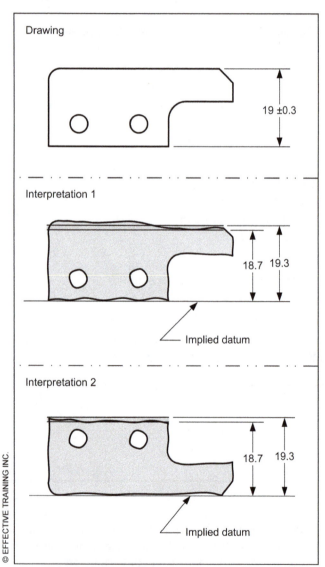

*FIGURE 18-1  Indirect Control of Parallelism Deviation Resulting From Size Dimension and Rule #1*

## Parallelism

*Parallelism* is the condition where a surface, axis, or center plane is equidistant at all points from a datum plane or axis. Since parallelism (the perfect condition) cannot be produced, it is necessary to specify some allowance for parallelism deviation.

## Parallelism Tolerance

A *parallelism tolerance* is a geometric tolerance that limits the amount a surface, axis, or center plane is permitted to deviate from parallelism, relative to a datum reference frame. Since parallelism is an orientation tolerance, it does not control the location of the toleranced feature. Even if the datum reference frame is established from three planar datum features, only the rotational degrees of freedom are constrained, not the translational degrees of freedom. In other words, a parallelism tolerance zone can translate relative to the datums, but not rotate.

A parallelism tolerance:

- Is one of the three direct orientation tolerances
- Must always use a datum reference
- May be applied to a planar surface or a feature of size
- Can control orientation and form

An example of a parallelism tolerance is shown in Figure 18-2.

## Parallelism Tolerance Zones

The two common tolerance zones for a parallelism tolerance are:

1. The space between two parallel planes
2. The space within a cylinder

The default tolerance zone is two parallel planes. Where the diameter symbol is specified in a feature control frame, the tolerance zone shape is a cylinder. A parallelism tolerance zone is parallel to the primary datum feature referenced in the feature control frame. Where a parallelism tolerance is applied to a feature of size, the limits of size must still be met.

Examples of parallelism tolerance zones are shown in Figure 18-2.

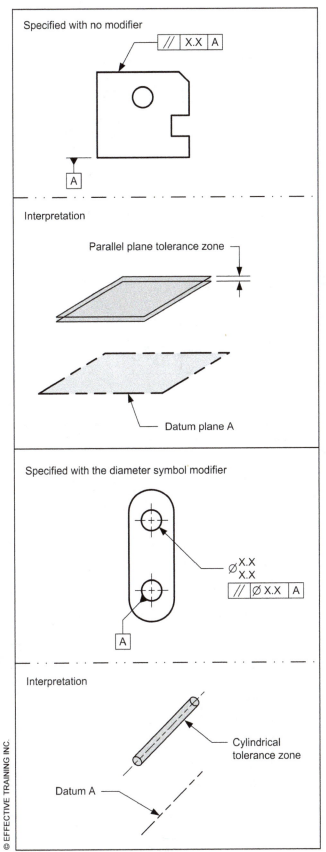

Specified with no modifier

// X.X A

A

Interpretation

Parallel plane tolerance zone

Datum plane A

Specified with the diameter symbol modifier

ø X.X
  X.X

// ø X.X A

A

Interpretation

Cylindrical tolerance zone

Datum A

**FIGURE 18-2  Parallelism Tolerance Zones**

***Author's Comment***
Parallelism tolerances only control deviations of parallelism. They cannot locate features or features of size.

**TECHNOTE 18-1**
**Degrees of Freedom**

An orientation tolerance can only constrain rotational degrees of freedom relative to the datum referenced. Even where the datum features may constrain all degrees of freedom, the tolerance zone only orients to the datum reference frame.

***Author's Comment***
It is not common, but a parallelism tolerance may have two additional tolerance zone shapes invoked by adding these modifying notes to the feature control frame: "EACH ELEMENT" and EACH RADIAL ELEMENT." The use of these notes is beyond the scope of this text and is covered in ETI's *Advanced Concepts of GD&T* textbook.

## Specifying Parallelism Tolerances on a Planar Surface and a Feature of Size

A parallelism tolerance may be applied to a planar surface or a feature of size. Where parallelism is applied to a surface, the tolerance zone is located by the surface. Where it applies to a feature of size, the tolerance zone applies to the axis or center plane of the feature of size. The axis or center plane must be within the tolerance zone.

Where a parallelism tolerance is applied to a planar surface, the feature control frame is attached to an extension line from the surface, or a leader line directed to the surface. Where a parallelism tolerance is applied to a feature of size, the feature control frame is placed below or beside the size dimension.

Figure 18-3 shows examples of a parallelism tolerance applied to a surface and a feature of size. Note the location of the tolerance zone in each application.

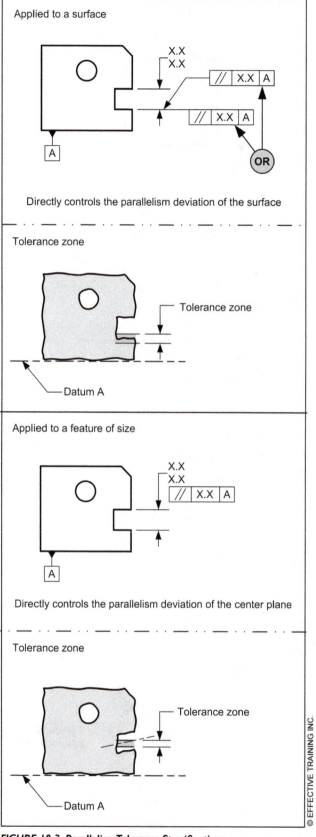

**FIGURE 18-3  Parallelism Tolerance Specifications**

© EFFECTIVE TRAINING INC.

## Indirect Parallelism Tolerances

There are several geometric tolerances that indirectly affect the parallelism deviation of a planar surface or feature of size. Location tolerances, such as position and profile of a surface, can indirectly limit parallelism deviation. A size dimension or a dim origin dimension can also limit parallelism deviation.

In these applications, the parallelism of the toleranced surface or feature of size is not directly inspected. The effect on parallelism deviation is a result of the toleranced feature being within the location tolerance zone.

If a parallelism tolerance and a location tolerance with the same primary datum reference frame are specified on a feature or feature of size, the parallelism tolerance must be a refinement of the location tolerance.

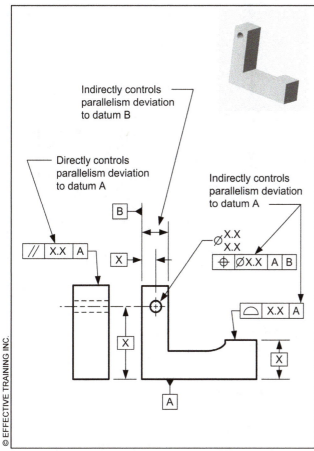

**FIGURE 18-4 Indirect Control of Parallelism Deviation**

## Modifiers Used With Parallelism Tolerances

There are several modifiers used in the tolerance portion of a parallelism tolerance feature control frame. The choice of modifier is often related to the functional requirements of the application.

Figure 18-5 shows the common modifiers and examples of functional applications where they could be used.

| Tolerance Modifier | Can Be Applied To | Effect | Functional Application |
|---|---|---|---|
| Ⓣ | Feature | Releases the flatness requirement | Orienting of mating surfaces |
| Ⓕ * | Feature or feature of size | Releases the restraint requirement | Nonrigid parts with restraint notes |
| ⟨ST⟩ * | Feature or feature of size | Requires statistical process controls | Statistically derived tolerances or tolerances used in statistical tolerance analysis |
| ⌀ | Feature of size | Creates a cylindrical tolerance zone | Holes, external diameters |
| Ⓜ | Feature of size | Permits a bonus tolerance / Permits use of a functional gage | Assembly / Non critical relationships |
| Ⓛ * | Feature of size | Permits a bonus tolerance / Requires variable measurement | Minimum distance |
| Ⓟ | Feature of size | Projects the tolerance zone | Fixed fastener assemblies |

\* These modifiers are only introduced in this textbook.

**FIGURE 18-5 Common Modifiers Used With Parallelism Tolerances**

---

### TECHNOTE 18-2
### Parallelism

The following are characteristics of a parallelism tolerance:

- The toleranced feature must be nominally parallel relative to the primary datum referenced.
- The default tolerance zone shape is two parallel planes.
- Where the diameter symbol is specified, the tolerance zone shape is a cylinder.
- One or more datums must be referenced.
- The tolerance zone is parallel to the primary datum specified.
- A parallelism tolerance can limit form deviation.
- A parallelism tolerance cannot affect location or size.
- A parallelism tolerance can only be applied to an individual feature or feature of size.
- A parallelism tolerance must be a refinement of other geometric tolerances that affect the orientation of the feature or feature of size.
- Where a parallelism tolerance is specified, the limits of size must still be met.

## Evaluate if a Parallelism Tolerance Specification is Standard-Compliant

For a parallelism tolerance to be a standard-compliant specification (legal), it must satisfy certain requirements. In order to make it easier to remember what to look for when evaluating the correctness of a geometric tolerance on a drawing, I created a mnemonic.

The requirements are divided into four areas represented by the initials "CARE":

     C - Check for datums

     A - Assess the application

     R - Review the modifiers

     E - Evaluate the tolerance

A flowchart that identifies the requirements for a standard-compliant parallelism tolerance is shown in Figure 18-6.

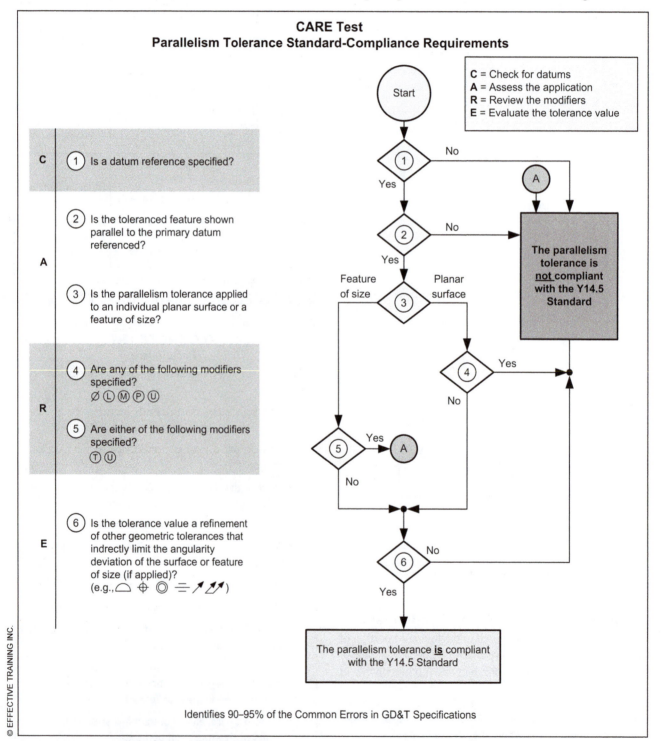

FIGURE 18-6 *Test for a Standard-Compliant Parallelism Tolerance*

# PARALLELISM APPLICATIONS

The following sections describe various parallelism tolerance applications.

### Real-World Applications

This section lists three common real-world applications of a parallelism tolerance. These applications limit the parallelism deviation of a planar surface or feature of size relative to a datum reference frame for:

- Support (distribute load and reduce wear)
- Assembly (maintain a uniform gap)
- Performance (linear motion)

## Parallelism Tolerance Applied to a Planar Surface

In certain applications, the function of a part requires defining the parallelism deviation of a surface relative to a datum. In Figure 18-7, a parallelism tolerance is applied to a planar surface relative to a planar datum. This is the most common application of parallelism. Where a parallelism tolerance is applied to a planar surface, the following conditions apply:

- The tolerance zone is the space between two parallel planes.
- The tolerance zone is parallel to the primary datum plane.
- The tolerance value defines the distance between the tolerance zone planes.
- The parallelism tolerance also limits the flatness of the surface.
- The tolerance zone applies to the full extent of the surface.

## Parallelism With the Tangent Plane Modifier Applied to a Planar Surface

In some applications, the function of the part requires controlling the parallelism deviation of the high points of a surface. The low points do not affect the function of the part.

Where a parallelism tolerance is intended to apply to the high points of the surface only, the tangent plane modifier is specified in the tolerance portion of the feature control frame. The tolerance zone applies to a plane tangent to the high points of the surface. Because the parallelism tolerance does not apply to the low points, the flatness of the surface is not controlled. Therefore, using the tangent plane modifier is considered a less stringent requirement.

In Figure 18-8, a parallelism tolerance with the tangent plane modifier is applied to a planar surface relative to an opposed planar surface. In these types of applications, the following conditions apply:

- The tolerance zone is the space between two parallel planes.
- The tolerance zone is parallel to the primary datum plane.
- The tolerance value defines the distance between the tolerance zone planes.
- The tangent plane of the surface must be within the tolerance zone.
- The parallelism tolerance does not limit the flatness of the surface.
- The tolerance zone applies to the full extent of the surface.

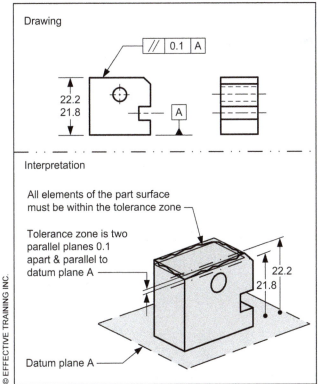

*FIGURE 18-7  Parallelism Tolerance Applied to a Planar Surface*

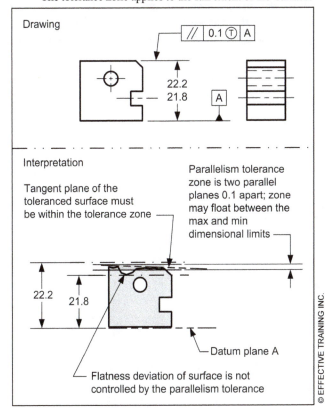

*FIGURE 18-8  Parallelism Tolerance With the Tangent Plane Modifier*

## Parallelism Tolerance at RFS Applied to a Hole

In certain applications, part function requires defining the parallelism deviation of a hole or other cylindrical feature of size (regardless of its size) relative to a datum.

Where a parallelism tolerance is applied to a hole, the feature control frame is placed beneath the size dimension. The tolerance zone applies to the axis of the hole.

In the example in Figure 18-9, the allowable orientation deviation of the axis of the hole is defined relative to a datum plane. In this example, the RFS condition applies.

Where a parallelism tolerance is applied to a hole at RFS, the following conditions apply:

- A diameter symbol may be used in the feature control frame to indicate that the tolerance zone is the space within a cylinder.

- The tolerance zone is parallel to the primary datum plane.

- The axis of the hole must be within the tolerance zone.

**Author's Comment**
Where a parallelism tolerance is applied to a cylindrical feature of size, the feature control frame may or may not contain a diameter symbol modifier. If the datum reference is a single planar datum, the diameter symbol is optional.
If the datum reference is an axis or includes a secondary datum, the diameter symbol does have an effect. In these cases, the diameter symbol is usually specified.

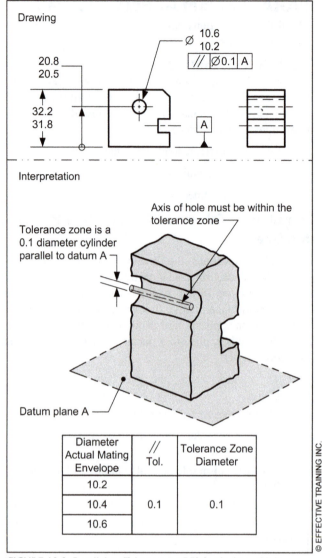

| Diameter Actual Mating Envelope | //<br>Tol. | Tolerance Zone Diameter |
|---|---|---|
| 10.2 | | |
| 10.4 | 0.1 | 0.1 |
| 10.6 | | |

**FIGURE 18-9 Parallelism Tolerance at RFS Applied to a Hole**

## Parallelism Tolerance at MMC Applied to a Hole

To indicate that the parallelism tolerance applies to a hole or other cylindrical feature of size, the feature control frame is placed below or beside the size dimension. In many cases, when applied to a hole or other cylindrical feature of size, a diameter symbol is specified preceding the tolerance value to indicate a cylinder shaped tolerance zone.

Where a parallelism tolerance at MMC is used, the function is often assembly, so the virtual condition is the important characteristic. In this text, the surface interpretation (VC boundary) is illustrated for MMC applications, although an axis or center plane interpretation could also be used.

In Figure 18-10, a parallelism tolerance is applied to a hole at MMC. When the MMC modifier is specified, the tolerance zone is a virtual condition around which the surface of the hole must fit. The MMC modifier is often used to ensure assembly.

Where parallelism is applied to a hole or other cylindrical feature of size at MMC, the following conditions apply:

- The tolerance zone is a virtual condition.

- The tolerance zone is parallel to the datum features referenced.

- The virtual condition size is equal to the MMC minus the parallelism tolerance (for a hole).

- The surface of the hole must fit around its virtual condition.

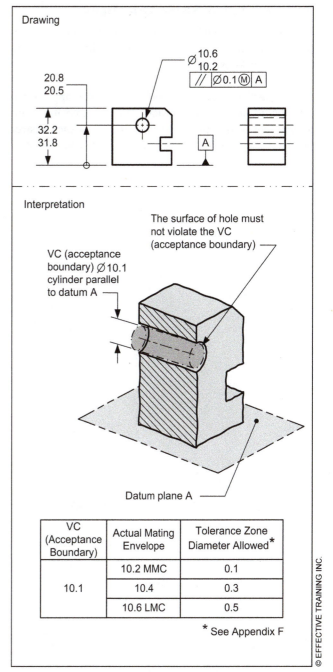

Drawing

Interpretation

The surface of hole must not violate the VC (acceptance boundary)

VC (acceptance boundary) Ø10.1 cylinder parallel to datum A

Datum plane A

| VC (Acceptance Boundary) | Actual Mating Envelope | Tolerance Zone Diameter Allowed* |
|---|---|---|
| 10.1 | 10.2 MMC | 0.1 |
| | 10.4 | 0.3 |
| | 10.6 LMC | 0.5 |

\* See Appendix F

© EFFECTIVE TRAINING INC.

**FIGURE 18-10  Parallelism Tolerance at MMC Applied to a Hole**

## The Significant Seven Questions

This section uses the Significant Seven Questions to interpret a parallelism tolerance application. The questions are explained in Appendix B.

Figure 18-11 contains a parallelism tolerance application. The Significant Seven Questions are answered in the interpretation section of the figure. Notice how each answer increases your understanding of the geometric tolerance.

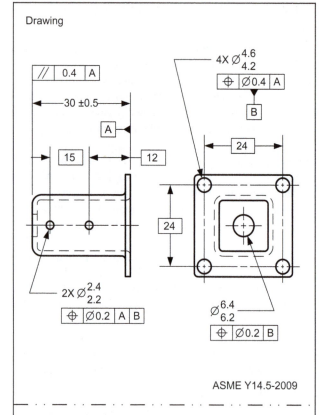

Drawing

ASME Y14.5-2009

Interpretation

The Significant Seven Questions

1. Which dimensioning and tolerancing standard applies?
   Answer – ASME Y14.5-2009

2. What does the tolerance apply to?
   Answer – A planar surface

3. Is the specification standard-compliant?
   Answer – Yes

4. What are the shape and size of the tolerance zone?
   Answer – The space between two parallel planes 0.4 apart and parallel to datum A

5. How much total tolerance is permitted?
   Answer – 0.4

6. What are the shape and size of the datum simulators?
   Answer – A planar surface for datum A

7. Which geometry attributes are affected by this tolerance?
   Answer – Orientation, form (flatness)

© EFFECTIVE TRAINING INC.

*FIGURE 18-11 The Significant Seven Questions Applied to a Parallelism Tolerance*

# VERIFICATION METHODS AND PRINCIPLES

## Verification Principles for Parallelism Tolerances

This section contains a simplified explanation of verification principles and methods that can be used to inspect a parallelism tolerance applied to a planar surface, a cylindrical feature of size (RFS), and a cylindrical feature of size (at MMC).

When a parallelism tolerance is applied to a planar surface, the deviation is verified by finding the distance between two planes that are parallel to the datum within which all points of the surface are located. The inspection method must be able to collect a set of points and determine the distance between the two planes that contains the set of points. Since location is not a requirement of a parallelism tolerance, the inspection method should not be influenced by the location of the toleranced surface.

Verifying a parallelism tolerance has two parts:

1. Establishing the relationship specified in the feature control frame between the part and the datum reference frame.
2. Verifying the location of the surface, axis, or center plane are within the tolerance zone.

The relationship between the part and the datum reference frame may be established by orienting and locating the datum features on the datum planes in the sequence specified in the feature control frame.

When a parallelism tolerance (at MMC) is applied to a feature of size, the parallelism deviation is often verified by either passing an external feature of size through a virtual condition (a functional gage) or passing a virtual condition (a functional gage) into an internal feature of size.

When a parallelism tolerance (MMC) is applied to a feature of size, the deviation is often verified by passing the feature of size through a virtual condition.

There are many ways parallelism tolerances can be verified. In this section we will look at three of them.

## Verifying a Parallelism Tolerance Applied to a Planar Surface

There are many ways to verify a parallelism tolerance. Figure 18-12 shows an example of one way to inspect a parallelism tolerance applied to a planar surface. In this method, a surface plate is used to simulate the datum plane, and a dial indicator is used to measure the amount of parallelism deviation.

The part datum feature is rested on the surface plate. The toleranced surface is parallel to the surface plate. The dial indicator is zeroed on the surface and moved randomly across the part surface. If the dial indicator reading is smaller than the parallelism tolerance value, the surface is within the parallelism tolerance zone.

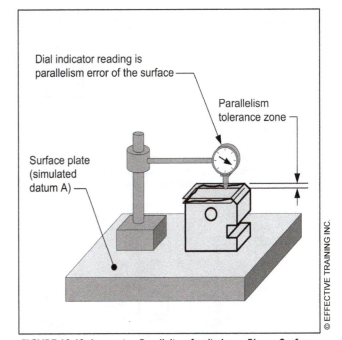

**FIGURE 18-12 Inspecting Parallelism Applied to a Planar Surface**

## Verifying a Parallelism Tolerance at RFS Applied to a Feature of Size

There are many ways to verify a parallelism tolerance. Figure 18-13 shows an example of one method: using a surface plate, a gage pin, and a height gage.

The surface plate simulates the primary datum. The part is rested on the surface plate. A "best fit" gage pin is placed into the hole to establish the unrelated actual mating envelope of the hole. The height gage is used to measure a point on each end of the pin to establish the parallelism deviation of the hole relative to the datum.

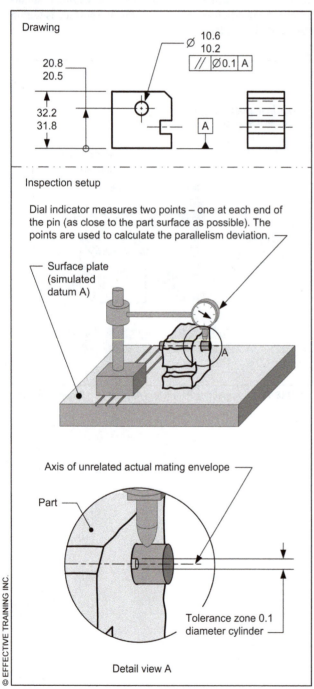

FIGURE 18-13 Verifying a Parallelism Tolerance at RFS Applied to a Hole

## Verifying a Parallelism Tolerance at MMC Applied to a Feature of Size

A functional gage may be used to verify the parallelism requirement for the part in Figure 18-14. The gage has a surface to simulate the datum plane and a fixed-size gage pin that verifies the parallelism of the hole. Because the parallelism tolerance does not control the location of the hole, the pin must be able to move up and down to allow for the hole location deviation. The size of the gage pin is equal to the virtual condition size of the hole (10.1 mm diameter).

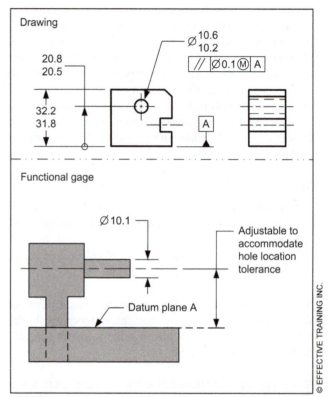

FIGURE 18-14 Verifying a Parallelism Tolerance at MMC Applied to a Hole

## SUMMARY

### Key Points

- Parallelism is the condition where a surface, axis, or center plane is equidistant at all points from a datum plane or datum axis.

- A parallelism tolerance is a geometric tolerance that limits the amount a surface, axis, or center plane is permitted to deviate from parallelism relative to a datum(s).

- Location tolerances, such as position and profile can also limit parallelism.

- Wherever a width regular feature of size is shown on a drawing, Rule #1 applies, and the parallelism of both surfaces is controlled by the limits of the size dimension.

- Using a size dimension to control parallelism deviation has two shortcomings:
  o The parallelism deviation is limited to the same tolerance as the size dimension
  o The lack of a datum reference

- Three common real-world applications of a parallelism tolerance on a drawing:
  o Support (distribute load and reduce wear)
  o Assembly (maintain a uniform gap)
  o Performance (linear motion)

- Requirements of a parallelism tolerance:
  o The default tolerance zone shape is two parallel planes.
  o Where the diameter symbol is specified, the tolerance zone shape is a cylinder.
  o One or more datums must be specified.
  o The tolerance feature must be parallel to the primary datum feature.
  o A parallelism tolerance can limit form deviation.
  o A parallelism tolerance cannot affect location or size.
  o A parallelism tolerance can only be applied to an individual feature or feature of size.
  o A parallelism tolerance must be a refinement of other geometric controls that affect the orientation of the feature or feature of size.
  o Where applied to a feature of size, the limits of size must still be met.

- Where a parallelism tolerance is applied to a planar surface, it controls the orientation and flatness of the surface.

## Additional Related Topics

*These topics are recommended for further study to improve your understanding of parallelism.*

| Topic | Source |
|---|---|
| Modifying note - each line element | *ASME Y14.5-2009, Para. 6.4.3* |
| Free state symbol | ETI's *Advanced Concepts of GD&T* textbook |
| Using angularity to control parallelism | *ASME Y14.5-2009, Para. 6.6* |
| Using parallelism tolerances with multiple datum references | ETI's *Advanced Concepts of GD&T* textbook |
| Using parallelism to tolerance an axis | ETI's *Advanced Concepts of GD&T* textbook |

## QUESTIONS AND PROBLEMS

**Website Bonus Materials**
Additional questions are available at our website. To access bonus materials for this textbook, please visit:
www.etinews.com        www.etinews.com/textbookbonus

### True and False

*Indicate if each statement is true or false.*

T / F    1.  A parallelism tolerance always requires a datum reference.

T / F    2.  A parallelism tolerance may use the ⓣ modifier.

T / F    3.  A parallelism tolerance may use the ⓤ modifier.

T / F    4.  When no parallelism tolerance is specified, the parallelism deviation of the sides of a planar feature of size are undefined.

T / F    5.  A parallelism tolerance may not be applied to a center plane.

T / F    6.  The applied to a planar surface tolerance zone for a parallelism tolerance is two parallel lines.

### Multiple Choice

*Circle the best answer to each statement.*

1.  Which statement describes parallelism?
    A.  All points of a surface perpendicular to the axis are equidistant from that axis
    B.  A surface is equidistant at all points from a datum plane or axis
    C.  A surface of revolution having all points equidistant from a common axis
    D.  None of the above

2.  A real-world application of a parallelism tolerance is:
    A.  Assembly (maintain a uniform gap)
    B.  Seal (lip seal on a shaft)
    C.  Balance (high speed rotating parts)
    D.  None of the above

3.  Where a parallelism tolerance is applied to a surface, _____ of the surface must be located within the tolerance zone.
    A.  The highest points
    B.  The tangent plane
    C.  All points
    D.  The three high points

4.  What is the effect of specifying the ⓣ modifier in a parallelism tolerance?
    A.  The flatness of the toleranced surface is controlled
    B.  The flatness of the toleranced surface is not controlled
    C.  The tolerance zone becomes a tangent plane
    D.  The tangent plane of the toleranced surface may be outside the tolerance zone

5.  A _____ tolerance can be an indirect parallelism control.
    A.  Profile of a surface
    B.  Flatness
    C.  Perpendicularity
    D.  None of the above

6.  When verifying a parallelism tolerance, the relationship between the part and _____ must be established.
    A.  Datum reference frame
    B.  Tolerance zone
    C.  Implied datums
    D.  Location of the toleranced surface

## Application Problems

*The application problems are designed to provide practice on applying the chapter concepts to situations that are similar to on-the-job conditions.*

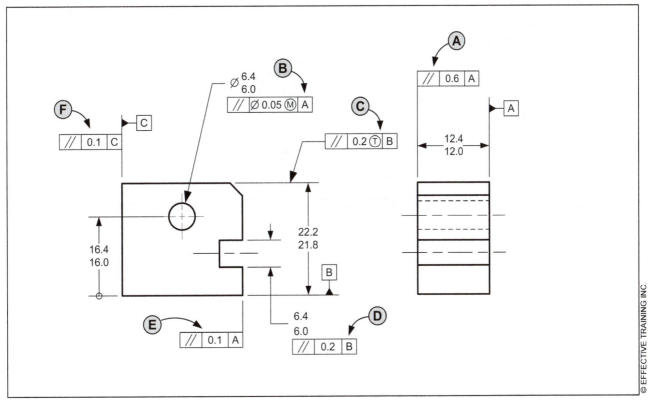

*Questions 1 and 2 refer to the drawing above.*

1.  For each parallelism tolerance labeled above, indicate if it is a standard-compliant specification. If a tolerance is not standard-compliant, explain why not. (Hint: use the CARE test.)

    (A) _____

    (B) _____

    (C) _____

    (D) _____

    (E) _____

    (F) _____

2.  In the space below, draw and dimension a functional gage for verifying the parallelism tolerance labeled "B."

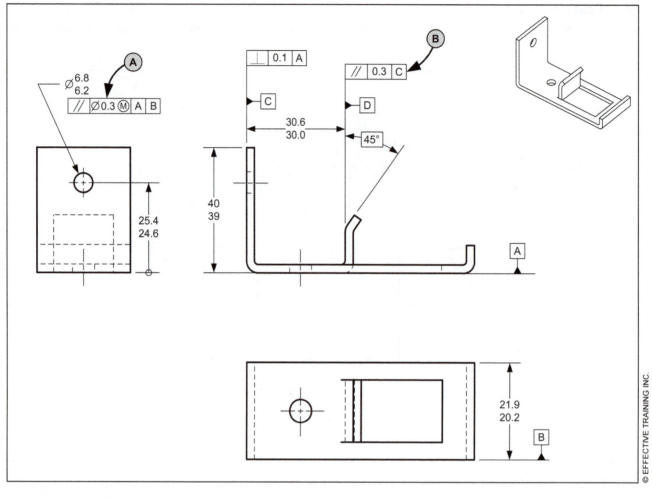

*Questions 3 through 5 refer to the drawing above.*

3. Use the Significant Seven Questions to interpret the parallelism tolerance labeled "A."

   1) Which dimensioning and tolerancing standard applies?

   _____

   2) What does the tolerance apply to?_____

   _____

   3) Is the specification standard-compliant?_____

   4) What are the shape and size of the tolerance zone?

   _____

   5) How much total tolerance is permitted?_____

   6) What are the shape and size of the datum simulators?

   _____

   7) Which geometry attributes are affected by this tolerance?

   _____

4. Describe how to verify the parallelism tolerance labeled "A."

   _____

   _____

   _____

5. Describe how to verify the parallelism tolerance labeled "B."

   _____

   _____

   _____

6. Describe how to verify a parallelism tolerance applied to a feature of size at RFS.

   _____

   _____

# Angularity Tolerance

## Goal

Interpret the angularity tolerance

## Performance Objectives

Upon completing this chapter, you should be able to:

1. Describe the terms "angularity" and "angularity tolerance" (p.248)
2. Describe two common tolerance zone shapes for an angularity tolerance (p.248)
3. Recognize when an angularity tolerance is applied to a planar surface or a feature of size (p.248)
4. List two indirect angularity tolerances (p.249)
5. Describe the alternative practice for using an angularity tolerance to control orientation (p.250)
6. List which modifiers should be used in an angularity tolerance (p.250)
7. Evaluate if an angularity tolerance specification is standard-compliant (p.251)
8. Describe real-world applications for an angularity tolerance (p.252)
9. Interpret an angularity tolerance applied to a planar surface (p.252)
10. Explain the effect of using multiple datum feature references with an angularity tolerance (p.253)
11. Interpret an angularity tolerance that uses the tangent plane modifier (p.254)
12. Interpret an angularity tolerance at RFS applied to a hole (cylindrical feature of size) (p.254)
13. Interpret an angularity tolerance at MMC applied to a hole (cylindrical feature of size) (p.255)
14. Interpret an angularity tolerance by using the "Significant Seven Questions" (p.256)
15. Understand verification principles for angularity (p.257)
16. Describe how an angularity tolerance applied to a surface is verified (p.257)
17. Describe a functional gage for verifying an angularity tolerance at MMC (p.258)

## New Terms

- Angularity
- Angularity tolerance
- Sine bar

## What This Chapter Is About

This is the third of three chapters that explain the orientation tolerances. Orientation tolerances define allowable angular deviations (and sometimes form deviations) on planar surfaces or features of size. Orientation tolerances cannot control location. In this chapter, we'll look at how to control allowable angular deviations on a part.

This chapter explains where to use and how to interpret angularity tolerances. We will cover the fundamental uses of the angularity tolerance on surfaces and features of size.

# TERMS AND CONCEPTS

## Angularity

*Angularity* is the condition where a surface, axis, or center plane is at the specified angle to a datum plane or datum axis. For a part surface, axis, or center plane to achieve the condition known as angularity is to say the surface, axis, or center plane is to be at exactly the specified angle. Since perfect angularity cannot be manufactured, it is necessary to specify allowances for angular relationships. This is often done with an angularity tolerance.

## Angularity Tolerance

An *angularity tolerance* is a geometric tolerance that limits the amount a surface, axis, or center plane is permitted to deviate from a basic angle relative to a datum reference frame. Since angularity is an orientation tolerance, it does not control the location of the toleranced feature. Even if the datum reference frame is established from three planar datum features, only the rotational degrees of freedom are constrained, not the translational degrees of freedom. In other words, the tolerance zone is free to translate relative to the datum features, but not to rotate.

An angularity tolerance:

- Is one of the three direct orientation tolerances

- Must always use a datum reference

- May be applied to a planar surface or a feature of size

- Can control orientation and form

An example of an angularity tolerance is shown in Figure 19-2.

## Angularity Tolerance Zones

The two common tolerance zones for an angularity tolerance are:

1. The space between two parallel planes

2. The space within a cylinder

The default tolerance zone is two parallel planes. Where the diameter symbol is specified in the feature control frame, the tolerance zone shape is a cylinder. An angularity tolerance zone is oriented by the specified basic dimensions relative to the primary datum referenced in the feature control frame. Where an angularity tolerance is applied to a feature of size, the limits of size must still be met. Examples of angularity tolerance zones are shown in Figure 19-1.

## Specifying Angularity Tolerances on a Planar Surface and a Feature of Size

An angularity tolerance may be applied to a planar surface or a feature of size. Where angularity is applied to a surface, the tolerance zone is located at the surface. Where it applies to a feature of size, the tolerance zone applies to the axis or center plane of the feature of size. The axis or center plane must be within the tolerance zone.

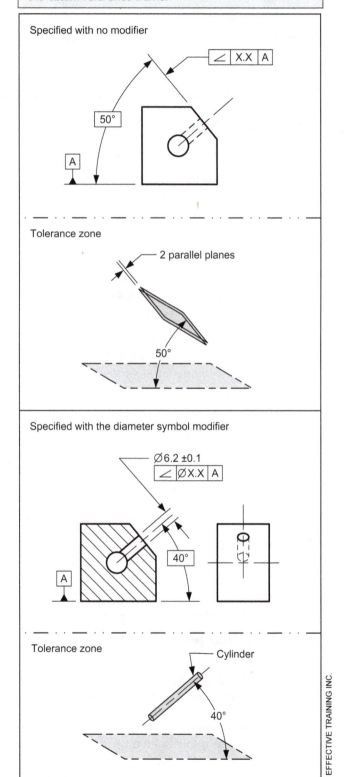

*FIGURE 19-1 Common Tolerance Zones for Angularity*

© EFFECTIVE TRAINING INC.

Where an angularity tolerance is applied to a planar surface, the feature control frame is attached to an extension line from the surface or a leader line directed to the surface. Where an angularity tolerance is applied to a feature of size, the feature control frame is placed below or beside the size dimension. Figure 19-2 shows examples of an angularity tolerance applied to a surface and a feature of size. Note the location of the tolerance zone in each application.

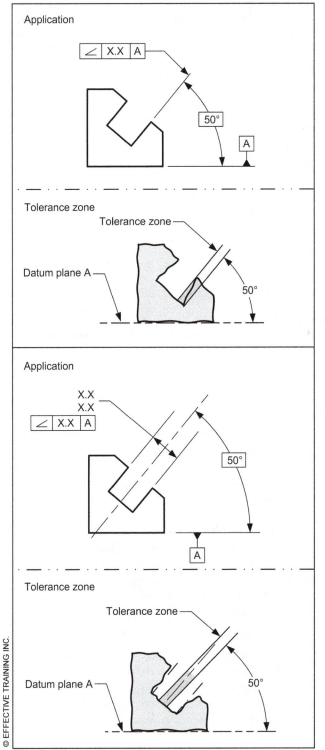

**FIGURE 19-2 Angularity Tolerance Specifications**

## Indirect Angularity Tolerances

There are several location tolerances that indirectly affect the angularity of a planar surface or feature of size, such as position and profile. In these applications, the angularity of the toleranced feature is not directly inspected. The effect on the angularity deviation is a result of the toleranced feature being within the location tolerance zone. If an angularity and location tolerance with the same datum reference frame are specified on a planar surface or feature of size, the angularity tolerance must be less than the location tolerance.

***Author's Comment***
The Y14.5-2009 standard allows the angularity tolerance to be used in place of perpendicularity or parallelism.

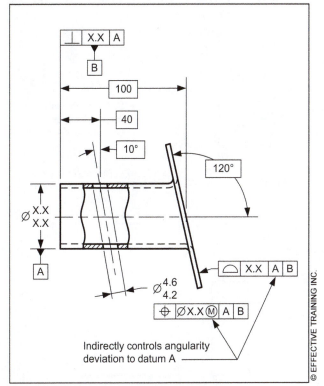

**FIGURE 19-3 Angularity and Location Tolerances**

***Author's Comment***
It is not common, but an angularity tolerance may have two additional tolerance zone shapes. They are invoked by adding these modifying notes to the feature control frame: "EACH ELEMENT" and EACH RADIAL ELEMENT." The use of these notes is beyond the scope of this text and are covered in the *Advanced Concepts of GD&T* textbook.

## Alternative Practice for Applying Angularity Tolerances

As an alternative practice, an angularity tolerance may be used to control perpendicular and parallel relationships. In these cases, an implied basic angle exists between the toleranced surface and the datum(s).

### Author's Comment

Using an angularity tolerance in place of a perpendicularity or parallelism tolerance places a greater burden on the reader of the drawing. Although this practice is shown in the standard, I recommend using the conventional method of specifying perpendicularity and parallelism tolerances.

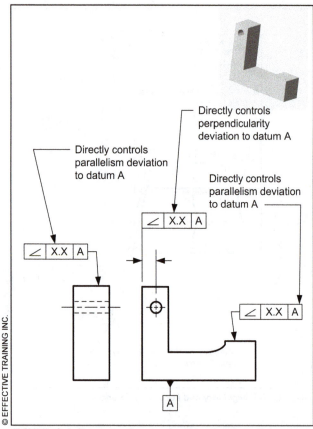

© EFFECTIVE TRAINING INC.

Directly controls perpendicularity deviation to datum A

Directly controls parallelism deviation to datum A

Directly controls parallelism deviation to datum A

∠ X.X A

∠ X.X A

∠ X.X A

A

**FIGURE 19-4** *Alternative Practice for Using Angularity Tolerances to Control Parallelism and Perpendicularity*

## Modifiers Used With Angularity Tolerances

There are several modifiers that are often used in the tolerance portion of a feature control frame. The choice of modifier is often related to the functional requirements of the application.

Figure 19-5 shows common modifiers and examples of functional applications where they could be used.

| Tolerance Modifier | Can Be Applied To | Effect | Functional Application |
|---|---|---|---|
| ⓣ | Feature | Releases the flatness requirement | Orienting of mating surfaces |
| Ⓕ * | Feature or feature of size | Releases the restraint requirement | Nonrigid parts with restraint notes |
| ⟨ST⟩ * | Feature or feature of size | Requires statistical process controls | Statistically derived tolerances or tolerances used in statistical tolerance analysis |
| ⌀ | Feature of size | Creates a cylindrical tolerance zone | Holes, external diameters |
| Ⓜ | Feature of size | Permits a bonus tolerance / Permits use of a functional gage | Assembly / Non critical relationships |
| Ⓛ * | Feature of size | Permits a bonus tolerance / Requires variable measurement | Minimum distance |
| Ⓟ | Feature of size | Projects the tolerance zone | Fixed fastener assemblies |

* These modifiers are only introduced in this textbook.

© EFFECTIVE TRAINING INC.

**FIGURE 19-5** *Common Modifiers Used With Angularity Tolerances*

### TECHNOTE 19-2
### Angularity Tolerance

The following are characteristics of an angularity tolerance:

- The toleranced feature must be dimensioned with a basic angle relative to the primary datum feature referenced.
- The default tolerance zone shape is two parallel planes.
- Where the diameter symbol is specified, the tolerance zone shape is a cylinder.
- One or more datum feature references must be specified.
- The tolerance zone is oriented by the basic angle to the primary datum specified.
- An angularity tolerance can limit form deviation.
- An angularity tolerance cannot affect location or size.
- An angularity tolerance can only be applied to an individual feature or feature of size.
- The tolerance value must be a refinement of other geometric tolerances that affect the orientation of the feature or feature of size.

## Evaluate if an Angularity Tolerance Specification is Standard-Compliant

For an angularity tolerance to be a standard-compliant specification (legal), it must satisfy certain requirements. In order to make it easier to remember what to look for when evaluating the correctness of a geometric tolerance on a drawing, I created a mnemonic.

The requirements are divided into four areas represented by the initials "CARE":

C - Check for datums

A - Assess the application

R - Review the modifiers

E - Evaluate the tolerance

A flowchart that identifies the requirements for a standard-compliant angularity tolerance is shown in Figure 19-6.

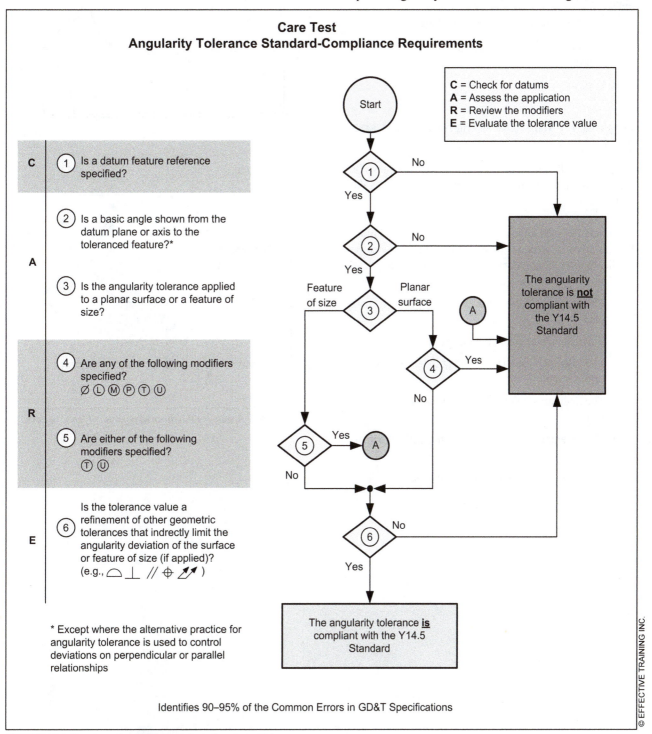

**Care Test**
**Angularity Tolerance Standard-Compliance Requirements**

C = Check for datums
A = Assess the application
R = Review the modifiers
E = Evaluate the tolerance value

**C**
1. Is a datum feature reference specified?

**A**
2. Is a basic angle shown from the datum plane or axis to the toleranced feature?*

3. Is the angularity tolerance applied to a planar surface or a feature of size?

**R**
4. Are any of the following modifiers specified?
⌀ Ⓛ Ⓜ Ⓟ Ⓣ Ⓤ

5. Are either of the following modifiers specified?
Ⓣ Ⓤ

**E**
6. Is the tolerance value a refinement of other geometric tolerances that indirectly limit the angularity deviation of the surface or feature of size (if applied)?
(e.g., ⌒ ⊥ // ⊕ ✎ )

* Except where the alternative practice for angularity tolerance is used to control deviations on perpendicular or parallel relationships

The angularity tolerance is **not** compliant with the Y14.5 Standard

The angularity tolerance **is** compliant with the Y14.5 Standard

Identifies 90–95% of the Common Errors in GD&T Specifications

© EFFECTIVE TRAINING INC.

*FIGURE 19-6 Test for a Standard-Compliant Angularity Tolerance*

# ANGULARITY APPLICATIONS

The following sections describe various angularity tolerance applications.

## Real-World Applications of an Angularity Tolerance

This section lists three real-world applications of an angularity tolerance. These applications limit the angularity deviation of a planar surface or feature of size relative to a datum reference frame for:

- Providing clearance (between adjacent parts)
- Maintaining lead-in angles (engagement between mating parts)
- Support

## Angularity Tolerance Applied to a Planar Surface

In certain applications, the function of a part requires limiting the angular deviations of a surface relative to a datum plane.

In Figure 19-7, an angularity tolerance is applied to a surface relative to a datum plane. This is the most common application of an angularity tolerance. In this application, the following conditions apply:

- The tolerance zone is the space between two parallel planes
- The distance between the planes is 0.3 mm
- The tolerance zone is oriented at the basic angle to the primary datum
- The angularity tolerance limits the flatness of the surface
- The tolerance zone applies to the full extent of the surface
- The surface locates the tolerance zone

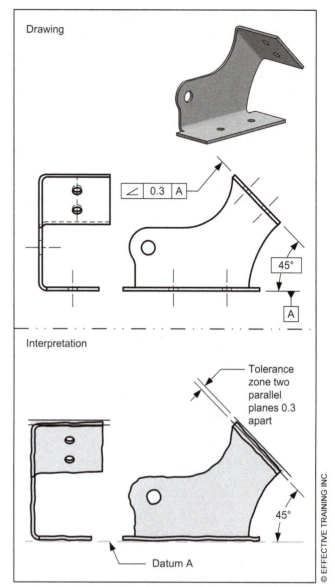

**FIGURE 19-7  Angularity Tolerance Applied to a Planar Surface**

© EFFECTIVE TRAINING INC.

## Angularity Tolerance With Multiple Datum Feature References

In certain applications, the function of a part requires limiting the angular deviations of a surface relative to multiple datum features.

In Figure 19-8, an angularity tolerance is applied to a surface relative to two datum planes. In this application, the following conditions apply:

- The tolerance zone is the space between two parallel planes

- The distance between the planes is 0.3mm

- The tolerance zone is at a basic 45° to datum plane A and 90° to datum plane B

- The angularity tolerance limits the flatness of the surface

- The tolerance zone applies to the full extent of the surface

- The surface locates the tolerance zone

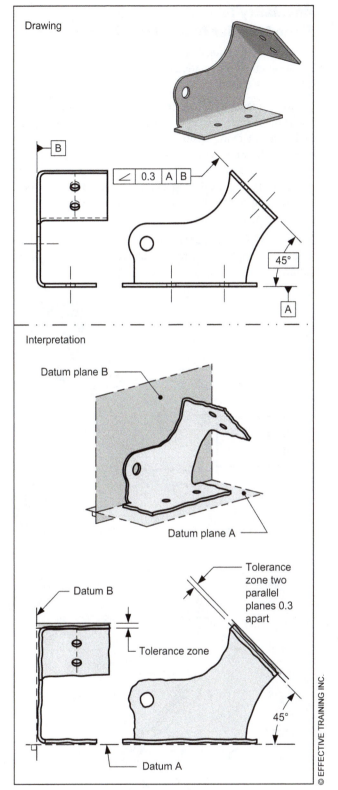

FIGURE 19-8 Angularity Applied to a Planar Surface With Two Datum Feature References

© EFFECTIVE TRAINING INC.

## Angularity Tolerance With the Tangent Plane Modifier Applied to a Planar Surface

In certain applications, the function of a part requires defining the angularity deviation of the tangent plane of a surface relative to a datum.

Where an angularity tolerance is intended to apply to the high points of the surface only, the tangent plane modifier is specified in the tolerance portion of the feature control frame. The tolerance zone applies to a plane tangent to the high points of the surface. Because the angularity tolerance does not apply to the low points, the flatness of the surface is not controlled. Therefore, using the tangent plane modifier is considered a less stringent requirement.

In Figure 19-9, an angularity tolerance with the tangent plane modifier is applied to a planar surface relative to a planar datum.

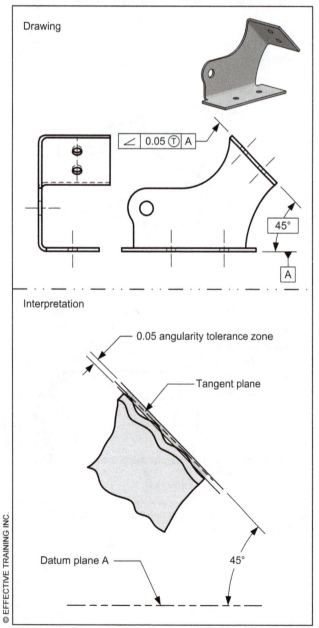

**FIGURE 19-9 Angularity Tolerance With the Tangent Plane Modifier**

Where the tangent plane modifier is specified, the following conditions apply:

- The tolerance zone is the space between two parallel planes.
- The tolerance zone is oriented relative to the primary datum feature referenced.
- The tolerance value defines the distance between the tolerance zone planes.
- The tangent plane of the surface must lie within the tolerance zone.
- The angularity tolerance does not limit the flatness of the surface.
- The tolerance zone applies to the full extent of the surface.
- The high points of the surface locates the tolerance zone.

## Angularity Tolerance at RFS Applied to a Cylindrical Feature of Size

In certain applications, the function of a part requires defining the angularity deviation of a cylindrical feature of size relative to a datum.

To indicate the angularity tolerance applies to the axis of a feature of size, the feature control frame is placed below or beside the size dimension. Where an angularity tolerance applies to a hole, it controls the orientation of the axis. In most cases when applied to a cylindrical feature of size, a diameter symbol precedes the tolerance value to indicate a cylinder-shaped tolerance zone.

In Figure 19-10, an angularity tolerance is applied to a hole at RFS (per Rule #2). Where the RFS condition applies, the tolerance zone is a fixed size. The axis of the toleranced feature must be within the tolerance zone. The following conditions apply:

- The tolerance zone is the space within a fixed-size cylinder.
- The angularity tolerance value defines the size of the tolerance zone cylinder.
- The axis of the hole must be within the tolerance zone.
- The tolerance zone is oriented to datum A with a basic dimension in one direction and to datum an implied 90° in the other direction.
- The tolerance zone is oriented parallel to datum B.

Drawing

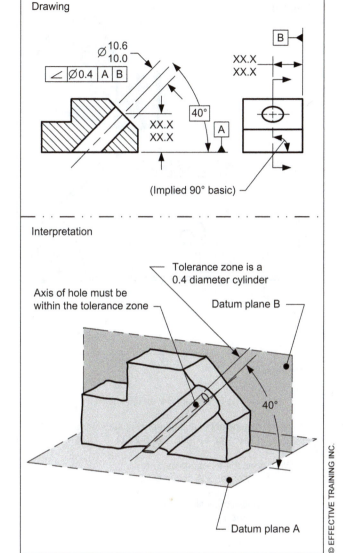

(Implied 90° basic)

Interpretation

Tolerance zone is a
0.4 diameter cylinder

Axis of hole must be
within the tolerance zone

Datum plane B

40°

Datum plane A

© EFFECTIVE TRAINING INC.

**FIGURE 19-10  Angularity Tolerance at RFS Applied to a Hole**

## Angularity at MMC Applied to a Cylindrical Feature of Size

Where an angularity tolerance applies to a cylindrical feature of size, it controls the orientation of the axis, and, in most cases, a diameter symbol precedes the tolerance value to indicate a cylinder shaped tolerance zone.

Where an angularity tolerance at MMC is used, the function is often assembly. and the virtual condition is the important characteristic. In this text, the surface interpretation (VC boundary) is illustrated for MMC applications, although an axis or center plane interpretation could also be used.

In Figure 19-11, an angularity tolerance is applied to a hole at MMC. When the MMC modifier is specified, the tolerance zone is a virtual condition that the surface of the hole must fit around. The MMC modifier is often used to ensure assembly.

Where angularity is applied to a cylindrical feature of size at MMC, the following conditions apply:

- The tolerance zone is a virtual condition.
- The tolerance zone is oriented at the basic angles to the datum reference frame.
- The virtual condition size is equal to the MMC minus the angularity tolerance (for a hole).
- The surface of the hole must fit around its virtual condition boundary.
- A fixed-size gage element may be used to verify the angularity requirement.
- A bonus tolerance is permissible.

When calculating the bonus tolerance, the size of the related actual mating envelope is used.

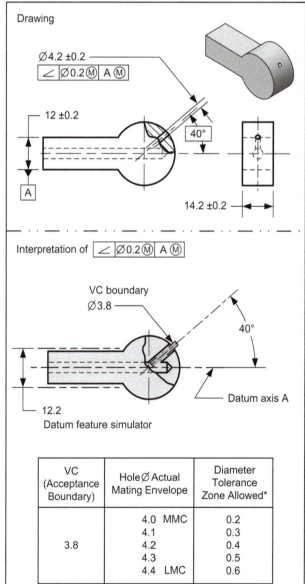

Drawing

Ø4.2 ±0.2

∠ Ø0.2Ⓜ A Ⓜ

12 ±0.2

40°

A

14.2 ±0.2

Interpretation of  ∠ Ø0.2Ⓜ A Ⓜ

VC boundary
Ø3.8

40°

12.2
Datum feature simulator

Datum axis A

| VC (Acceptance Boundary) | Hole Ø Actual Mating Envelope | Diameter Tolerance Zone Allowed* |
|---|---|---|
| 3.8 | 4.0   MMC | 0.2 |
| | 4.1 | 0.3 |
| | 4.2 | 0.4 |
| | 4.3 | 0.5 |
| | 4.4   LMC | 0.6 |

\* See Appendix F

© EFFECTIVE TRAINING INC.

**FIGURE 19-11  Angularity Tolerance at MMC Applied to a Hole**

## The Significant Seven Questions

This section uses the Significant seven questions to interpret an angularity tolerance application. The significant seven questions are explained in Appendix B.

Figure 19-12 contains an angularity application. The Significant Seven Questions are answered in the interpretation of the figure. Notice how each answer increases your understanding of the geometric tolerance.

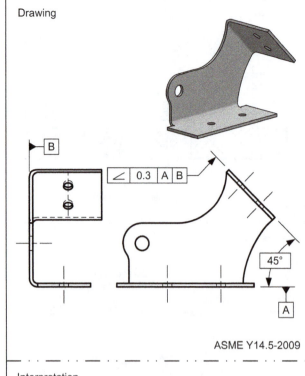

Drawing

ASME Y14.5-2009

Interpretation

The Significant Seven Questions

1. Which dimensioning and tolerancing standard applies?
   Answer – ASME Y14.5-2009

2. What does the tolerance apply to?
   Answer – A planar surface

3. Is the specification standard-compliant?
   Answer – Yes

4. What are the shape and size of the tolerance zone?
   Answer – The space between two parallel planes 0.3 apart at a 45° angle to datum A

5. How much total tolerance is permitted?
   Answer – 0.3

6. What are the shape and size of the datum simulators?
   Answer – A planar surface for datum A

7. Which geometry attributes are affected by this tolerance?
   Answer – Orientation, form (flatness)

© EFFECTIVE TRAINING INC.

**FIGURE 19-12 The Significant Seven Questions Applied to an Angularity Tolerance**

# VERIFICATION PRINCIPLES AND METHODS

## Verification Principles for Angularity Tolerances

This section contains a simplified explanation of verification principles and methods that can be used to inspect an angularity tolerance applied to a planar surface or a feature of size (at MMC). Verifying an angularity tolerance has two parts:

1. Establishing the relationship specified in the feature control frame between the part and the datum reference frame

2. Verifying the location of the surface, axis, or center plane is within the tolerance zone or VC (acceptance boundary)

The relationship between the part and the datum reference frame may be established by orienting and locating the datum features on the datum planes in the sequence in the feature control frame.

When an angularity tolerance is applied to a planar surface, the angularity deviation of a surface is verified by finding the distance between two parallel planes, at a specified angle, that all of the points of the surface are located between. The inspection method must be able to collect a set of points from the surface. Then, determine the distance between two planes, at a specified angle, that contain the set of points.

When an angularity tolerance (at MMC) is applied to a feature of size, the angularity deviation is often verified by either passing an external feature of size through a virtual condition (a functional gage) or passing a virtual condition at the basic angle (a functional gage) into an internal feature of size.

## Verifying Angularity Tolerances Applied to a Surface

There are many ways to verify an angularity tolerance. Figure 19-13 shows one way to inspect an angularity tolerance applied to a planar surface using a surface plate, a sine plate, a dial (or digital) indicator mounted on an indicator stand and gage blocks.

A *sine bar* is a precision inspection tool used to orient parts at any angle, usually between zero and sixty degrees. It is so named because the sine of the angle determines the height of a stack of gage blocks used to set the angle. It consists of two hardened precision ground plates with two precision ground rolls (pins) attached to the top plate that are set a fixed distance apart.

The sine bar is set at the basic angle and the part is mounted on it. This establishes the datum plane and sets the tolerance zone parallel to the surface plate. The indicator is then zeroed on the surface and moved around. The full indicator movement (difference between highest and lowest points) must be equal to or less than the angularity tolerance value.

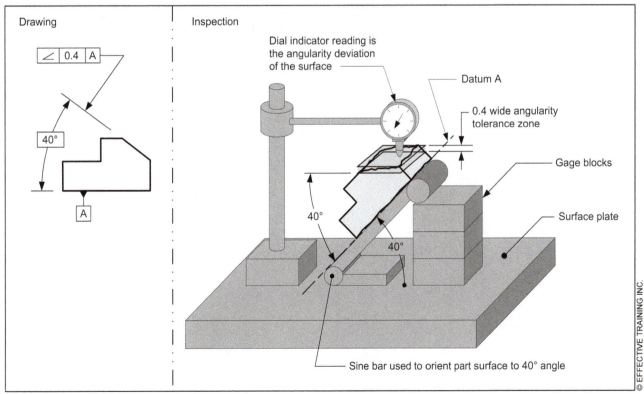

FIGURE 19-13 Inspecting Angularity Tolerance Applied to a Planar Surface

## Functional Gage for Verifying an Angularity Tolerance at MMC

A functional gage may be used to verify the angularity requirement. The gage has a datum feature simulator of two parallel planes 12.2 apart, simulating datum A. The virtual condition is simulated with a gage pin. The gage cylinder represents the ⌀3.8 virtual condition boundary at the basic angle from datum A. The angularity virtual condition is not located. The gage has a large clearance hole with a thumbscrew that allows the guide plate and gage pin to be relocated as necessary to verify the angularity requirement.

The angularity tolerance on the drawing in Figure 19-14 specifies the tolerance value of 0.2 at MMC. The clearance between the gage pin and the hole permits angular variations in excess of 0.2 through the bonus tolerance. The angularity datum feature reference of A uses the MMB modifier. This fixes the gage boundary at 12.2. The clearance between the datum feature simulator and the produced size of the datum feature permits a datum feature shift.

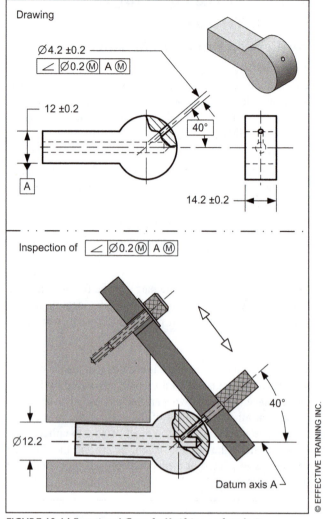

**FIGURE 19-14 Functional Gage for Verifying an Angularity Tolerance at MMC**

## SUMMARY

### Key Points

- Angularity is the condition where a surface, axis, or center plane is at the specified angle to a datum plane or datum axis.

- An angularity tolerance is a geometric tolerance that limits the amount a surface axis, or center plane is permitted to deviate from a basic angle relative to a datum.

- Requirements of an angularity tolerance:

  o The default tolerance zone shape is two parallel planes.
  o Where the diameter symbol is specified, the tolerance zone shape is a cylinder.
  o One or more datums must be specified.
  o The tolerance zone must be oriented with a basic angle to the primary datum specified.
  o An angularity tolerance can limit form deviation.
  o An angularity tolerance cannot affect location or size.
  o An angularity tolerance can only be applied to an individual feature or feature of size.
  o An angularity tolerance must be a refinement of other geometric tolerances that affect the orientation of the feature or feature of size.
  o Where an angularity tolerance is specified, the limits of size must still be met.

- Where an angularity tolerance is applied to a planar surface, it controls the orientation and flatness of the surface.

- Where an angularity tolerance contains the tangent plane modifier, it applies only to the tangent plane of the toleranced surface. It does not control the flatness of the surface.

- A dial indicator is an instrument used to accurately measure small linear distances and is frequently used to check the variation during the inspection process of a part.

### Additional Related Topics

*These topics are recommended for further study to improve your understanding of angularity.*

| Topic | Source |
|---|---|
| Modifying note - each line element | *ASME Y14.5-2009, Para. 6.4.3* |
| Free state symbol | ETI's *Advanced Concepts of GD&T* textbook |
| Using angularity to control perpendicularity and parallelism | *ASME Y14.5-2009, Para. 6.6* |

## QUESTIONS AND PROBLEMS

### True and False

*Indicate if each statement is true or false.*

T / F  1. When an angularity tolerance is applied to a planar surface, the tolerance zone is two parallel planes.

T / F  2. A flatness tolerance may be an indirect angularity tolerance.

T / F  3. An angularity tolerance may use the ⓣ modifier.

T / F  4. An angularity tolerance requires a single datum feature reference and a basic angle dimension to that datum.

T / F  5. An angularity tolerance may be applied to a 90° angle.

T / F  6. An angularity tolerance may affect orientation, form, and location.

## Multiple Choice

*Circle the best answer to each statement.*

1. Which geometric tolerance may provide an indirect angularity tolerance?
   A.  ⊕
   B.  ◎
   C.  ◻
   D.  ∠

2. When an angularity tolerance is applied to a hole RFS:
   A.  Variations in circularity are limited.
   B.  The axis must be within the tolerance zone.
   C.  The tolerance zone is oriented by being tangent to the surface of the hole.
   D.  All of the above

3. When an angularity tolerance is applied to a hole MMC:
   A.  A virtual condition exists.
   B.  The tolerance zone is oriented at the basic angle(s) to the datum reference frame.
   C.  A bonus tolerance is permissible as the size departs from the MMC condition.
   D.  All of the above

4. A functional gage may be used to verify an angularity tolerance applied:
   A.  To a planar surface
   B.  With the Ⓣ modifier
   C.  To a planar feature of size RFS
   D.  With the Ⓜ modifier

5. A real-world application for an angularity tolerance is:
   A.  Seal (lip seal on a shaft)
   B.  Assembly (of mating parts)
   C.  Providing clearance (between adjacent parts)
   D.  All of the above

## Application Problems

*The application problems are designed to provide practice on applying the chapter concepts to situations that are similar to on-the-job conditions.*

*Questions 1–8 refer to the drawing on the next page.*

1. The angularity tolerance labeled "D" requires that the surface must be within a 0.3mm wide tolerance zone that is:
   A.  45° relative to datum A only
   B.  45° relative to datum A and 90° relative to datum B
   C.  ±5° relative to datum A
   D.  ±5° relative to datum A and 90° relative to datum B

2. Describe the tolerance zone for the angularity tolerance labeled "D." _____

_____

3. Describe the effect of the Ⓣ modifier in the angularity tolerance labeled "D." _____

_____

4. Which geometric tolerance could be used in place of the angularity tolerance labeled "A" that would result in the same requirement?
   A.  Perpendicularity
   B.  Parallelism
   C.  Cylindricity
   D.  None of the above

5. Is the angularity tolerance labeled "A" a standard-compliant specification? _____ If no, explain.

_____

_____

6. Is the angularity tolerance labeled "B" a standard-compliant specification? _____ If no, explain.

_____

_____

7. Is the angularity tolerance labeled "C" a standard-compliant specification? _____ If no, explain.

_____

_____

8. Is the angularity tolerance labeled "D" a standard-compliant specification? _____ If no, explain.

_____

_____

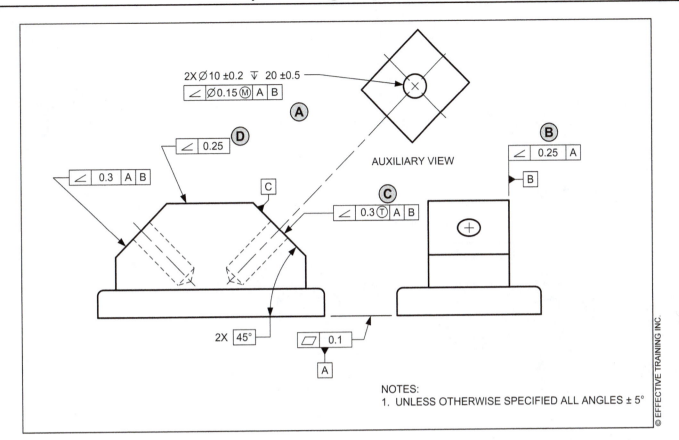

2X Ø10 ±0.2 ▽ 20 ±0.5

∠ | Ø0.15Ⓜ | A | B

(A)

(D)

∠ | 0.25

AUXILIARY VIEW

(B)

∠ | 0.25 | A

B

∠ | 0.3 | A | B

C

(C)

∠ | 0.3Ⓣ | A | B

2X | 45°

▱ | 0.1

A

NOTES:
1. UNLESS OTHERWISE SPECIFIED ALL ANGLES ± 5°

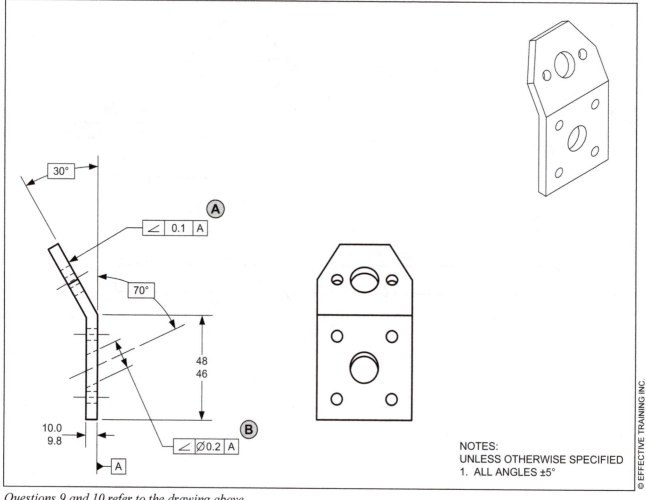

NOTES:
UNLESS OTHERWISE SPECIFIED
1.  ALL ANGLES ±5°

*Questions 9 and 10 refer to the drawing above.*

9.  Use the Significant Seven Questions to interpret the parallelism tolerance labeled "A."

1) Which dimensioning and tolerancing standard applies?

_____

2) What does the tolerance apply to?_____

_____

3) Is the specification standard-compliant?_____

4) What are the shape and size of the tolerance zone?

_____

5) How much total tolerance is permitted?_____

6) What are the shape and size of the datum simulators?

_____

7) Which geometry attributes are affected by this tolerance?

_____


10. Use the Significant Seven Questions to interpret the parallelism tolerance labeled "B."

1) Which dimensioning and tolerancing standard applies?

_____

2) What does the tolerance apply to?_____

_____

3) Is the specification standard-compliant?_____

4) What are the shape and size of the tolerance zone?

_____

5) How much total tolerance is permitted?_____

6) What are the shape and size of the datum simulators?

_____

7) Which geometry attributes are affected by this tolerance?

_____

# Position Tolerance - Introduction

## Goal

Interpret the position tolerance

---

## Performance Objectives

Upon completing this chapter, you should be able to:

1. Describe the terms, "true position" and "position tolerance" (p.264)
2. Describe the geometry attributes that a position tolerance can affect (p.264)
3. Describe two common tolerance zone shapes for a position tolerance (p.264)
4. List the requirements of a position tolerance (p.264)
5. Describe the modifiers that can be used with a position tolerance (p.265)
6. Explain when each material condition modifier (MMC, LMC, or RFS) should be used in a position tolerance (p.266)
7. Explain the surface interpretation for a position tolerance (p.268)
8. Explain the axis/center plane interpretation for a position tolerance (p.270)
9. List six advantages of using a position tolerance (p.270)
10. Describe four types of feature of size relationships commonly toleranced with a position tolerance (p.271)
11. Evaluate if a position tolerance specification is standard-compliant (p.271)

---

## New Terms

- Axis/center plane interpretation
- Position tolerance
- Surface interpretation
- True position

---

## What This Chapter Is About

The next four chapters cover the position tolerance. A position tolerance defines the orientation and location of features of size. In this chapter, we'll look at the basic concepts involved with position tolerances. We'll explain where to use and how to interpret position tolerances and cover their fundamental concepts.

The position tolerance is among the most commonly used geometric tolerances. Learning about the fundamental principles of position tolerances will prepare you to be able to interpret position applications on drawings.

# TERMS AND CONCEPTS

## True Position

*True position* is the theoretically exact location of a feature of size as established by basic dimensions. Since true position (the perfect condition) cannot be produced, it is necessary to specify some allowance for position deviation.

## Position Tolerance

A *position tolerance* is a geometric tolerance that limits the amount the center point, axis, or center plane of a feature of size is permitted to deviate from true position.

A position tolerance:

- Is one of the three location tolerances

- May or may not use a datum feature reference

- Can only be applied to features of size

- Directly controls location, and may indirectly control orientation and form

An example of a position tolerance is shown in Figure 20-2.

## Geometry Attributes Affected by Position Tolerances

A position tolerance directly controls the geometry attribute of location, and indirectly controls the geometry attributes of orientation, and form. The chart in Figure 6-4 (page 66) shows the geometric tolerances and the geometry attributes that they can control.

## Position Tolerance Zones

There are two common tolerance zones shapes for a position tolerance:

1. The space between two parallel planes

2. The space within a cylinder

The default tolerance zone is two parallel planes. Where the diameter symbol is specified in the feature control frame, the tolerance zone shape is a cylinder. A position tolerance zone is located and oriented by the specified or implied basic dimensions to the datum features referenced in the feature control frame. A position tolerance applied to a feature of size does not affect its size. The limits of size must still be met. Examples of position tolerance zones are shown in Figure 20-1.

A position tolerance may be used in single segment feature control frame, a multiple single segment feature control frame or in a composite feature control frame. An example of a single segment feature control frame is shown in figure 20-2. Examples of multiple single-segment and composite feature control frames are shown in Chapter 22.

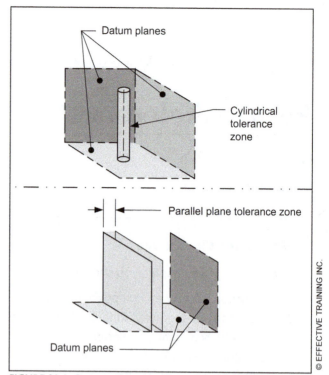

**FIGURE 20-1** *Common Tolerance Zones for a Position Tolerance*

© EFFECTIVE TRAINING INC.

## Position Tolerance Requirements

There are four requirements to correctly specify a position tolerance. They are described below and illustrated in Figure 20-2:

- A position tolerance can only be applied to a feature of size or a pattern of features of size.

- Basic dimensions must be used to define the true position of the toleranced feature(s) of size.

- Datum references must be specified in the feature control frame, except when applied to coaxial features of size. (Does not apply to lower segments of composite position tolerances.)

- RFS/RMB is the default condition per Rule #2

A position tolerance is always applied to a feature of size or a pattern of features of size. The feature control frame is placed below or beside the size dimension. See Figure 20-2 (1).

Basic dimensions are required to locate and orient the feature(s) of size relative to the datum features referenced in the feature control frame. This includes basic dimensions that define the spacing between features of size. See Figure 20-2 (2).

In most cases, datum feature references are required in the feature control frame. See Figure 20-2 (3). Two exceptions are a coaxial features of size application and the lower segment of a composite feature control frame. A coaxial features of size example is shown in Figure 20-2 (4) and is explained in Chapter 21. Composite position tolerancing is discussed in ETI's *Advanced Concepts of GD&T (2009)* textbook.

Rule #2 (see Chapter 7) is the default condition for the tolerance value and size datum features MMC, MMB, or LMC, LMB. Where functional requirements allow, the MMC or LMC modifiers should be specified. See the balloons labeled "4" in Figure 20-2.

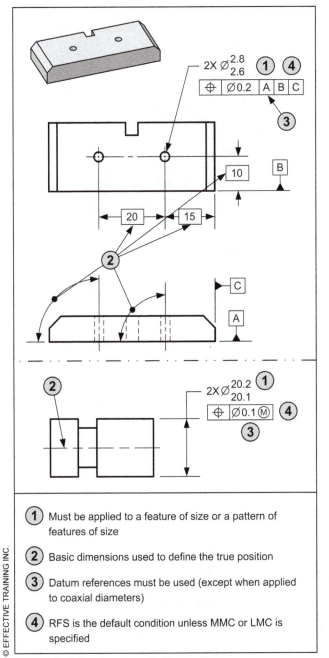

1. Must be applied to a feature of size or a pattern of features of size

2. Basic dimensions used to define the true position

3. Datum references must be used (except when applied to coaxial diameters)

4. RFS is the default condition unless MMC or LMC is specified

*FIGURE 20-2 Requirements of a Position Tolerance*

## Modifiers Used With Position Tolerances

There are several modifiers that are often used in the tolerance portion of a position tolerance feature control frame. The choice of modifier is often related to the functional requirements of the application.

The chart in Figure 20-3 shows four of the common modifiers with examples of functional applications where they could be used.

| Tolerance Modifier | Functional Application | Effect |
|---|---|---|
| ⌀ | Cylindrical features of size | Creates a cylindrical tolerance zone |
| Ⓕ * | Nonrigid parts with restraint notes | Releases the restraint requirement |
| Ⓟ | Bolted joint assemblies<br><br>Fixed fastener applications | Projects the tolerance zone |
| ⟨ST⟩ * | Where the tolerance is based on statistical analysis | Requires statistical process controls |

\* These modifiers are only introduced in this textbook.

*FIGURE 20-3 Guide for Selecting Modifiers in a Position Tolerance Feature Control Frame*

## Material Condition Modifiers Used With Position Tolerance

The chart in Figure 20-4 shows the three material condition modifiers, examples of functional applications where they could be used, and their relative cost to produce.

| Tolerance Modifier | Functional Application | Effect | Relative Cost to Produce and Verify |
|---|---|---|---|
| Ⓜ | Assembly<br><br>Not critical relationships | Permits a bonus tolerance<br><br>Results in a boundary tolerance zone<br><br>Functional gaging | Lowest |
| Ⓛ | Minimum distance | Permits a bonus tolerance<br><br>Results in a boundary tolerance zone<br><br>Variable measurement | Greater than MMC, less than RFS |
| None (RFS per Rule #2) | Alignment<br><br>Centering | RFS applies per Rule #2<br><br>Results in an axis tolerance zone<br><br>Variable measurement | Higher than MMC and LMC |

*FIGURE 20-4 Guide for Selecting Material Condition Modifiers in a Position Tolerance Feature Control Frame*

## Position Tolerance Zone Interpretations

The tolerance zone for a position tolerance can be interpreted as an axis (or center plane) tolerance zone or a virtual condition boundary tolerance zone. Where a position tolerance applies at RFS, it is interpreted as an axis/center plane tolerance zone. Where a position tolerance applies at MMC (or LMC), it is typically interpreted as a boundary tolerance zone.

---

### TECHNOTE 20-2
#### Position Tolerance Zones

As a convention in this text, a boundary tolerance zone is used in position tolerance at MMC applications and an axis (or center plane) tolerance zone is used in position tolerance at RFS applications.

---

## Surface Interpretation (VC Boundary)

Where a position tolerance applies at MMC, the *surface interpretation* applies: the tolerance zone is a virtual condition (acceptance boundary), located at true position, which the surface(s) of the tolerance features(s) of size must not violate. To illustrate, Figure 20-5 shows the resulting boundary zones from a position tolerance specified at MMC applied to a pattern of holes.

In this application, the position tolerance specified in the feature control frame applies when the holes are at their MMC size. The virtual condition boundaries are located at the true position of each hole and perpendicular to datum plane A. The holes must be within their size limits, and located and oriented so that no surface element lies inside the virtual condition. Therefore, the position tolerance is both a location and orientation tolerance. The maximum amount the holes can deviate from perpendicular to datum plane A is limited by the virtual condition.

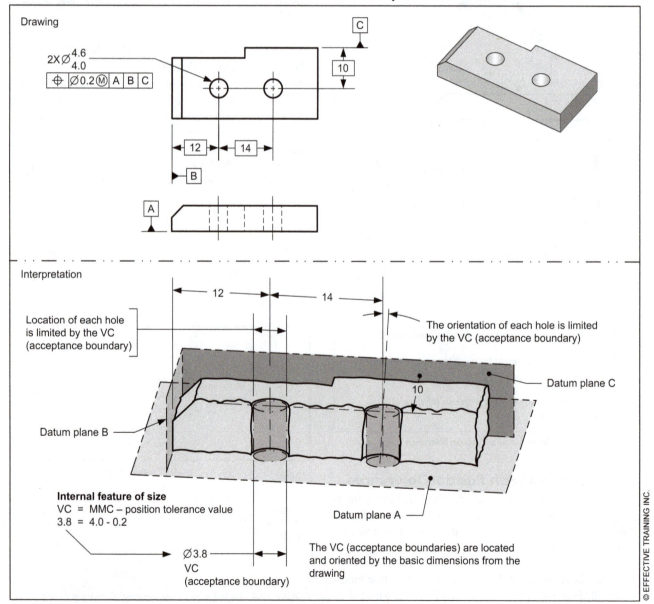

**FIGURE 20-5 Surface Interpretation for Position Tolerance (MMC) Applied to a Pattern of Holes**

The size of the virtual condition is equal to the MMC size, minus the position tolerance. If a functional gage is used to verify the location of the holes, the gage pins will be made to the virtual condition size and perpendicular to the primary datum feature simulator on the gage. The holes must be located and oriented so that they can fit around their respective gage pins and allow a minimum three-point contact with the primary datum feature simulator, two-point contact with the secondary datum feature simulator, and one-point contact with the tertiary datum feature simulator.

Where a position tolerance is specified at MMC and applied to a width type feature of size, the surface interpretation also applies. In Figure 20-6, the virtual condition for the tab width is equal to its MMC size plus the position tolerance. The offset (location) of the tab from datum center plane B and its perpendicularity deviation from datum plane A is limited by its virtual condition.

---

**TECHNOTE 20-3**
**Surface Interpretation Over**
**Axis/Center Plane Interpretation**

This text uses the surface interpretation for geometric tolerances applied at MMC. See Appendix F for the rationale.

---

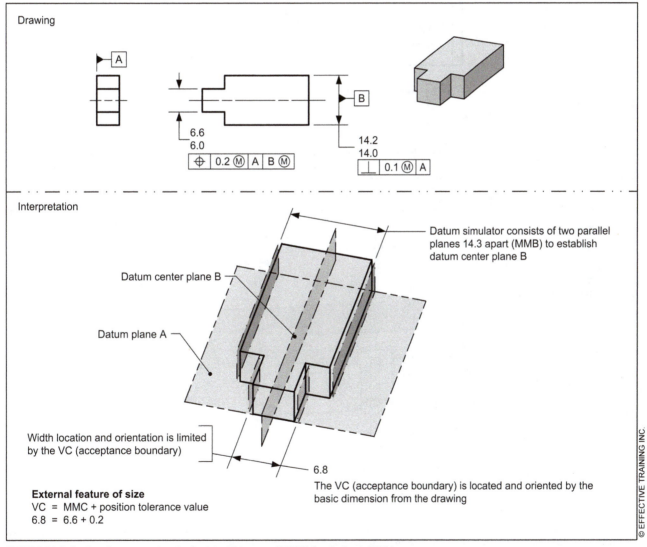

**FIGURE 20-6  Surface Interpretation for Position Tolerance (MMC) Applied to a Width**

## Axis/Center Plane Interpretation

Where a position tolerance applies at RFS, the axis/center plane interpretation is used. In the *axis/center plane interpretation*, the axis (or center plane) of the feature of size must be within the tolerance zone. In Chapter 5 (page 55), an axis is defined as the axis of the unrelated actual mating envelope. (For a refresher on unrelated actual mating envelope, see Figure 5-15 on page 54.)

In an RFS application, the position tolerance zone is a constant value, regardless of the size of the toleranced feature of size. For example, in Figure 20-7, the tolerance zone diameters are 0.2, and the axes must be located inside the tolerance zone. The tolerance zones are centered on the true position of the holes and perpendicular to datum plane A.

Therefore, the position tolerance is both a location and orientation tolerance. The maximum the axis of the holes can deviate from perpendicular to datum plane A is limited by the position tolerance zone.

---

**TECHNOTE 20-4**
**Position Tolerance RFS**

Where a position tolerance is applied at RFS:

- The axis interpretation is used.
- The tolerance zone is a constant value (fixed-size) zone.
- The axis of the toleranced feature of size must be located within the tolerance zone.
- The orientation is controlled relative to the primary datum feature referenced.

---

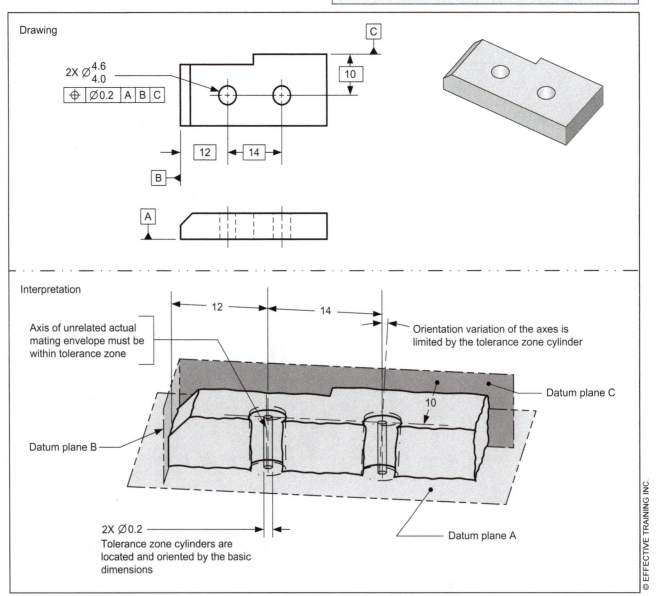

**FIGURE 20-7 Axis Interpretation for Position Tolerance (RFS) Applied to a Pattern of Holes**

Where a position tolerance is specified at RFS and applied to a width, the center plane interpretation applies. In Figure 20-8, the tolerance zone for the tab width is two parallel planes 0.2 apart centered about datum center plane B and perpendicular to datum plane A. The tolerance zone limits how much the center plane of the tab can be offset from datum center plane B and how much it can deviate from perpendicular to datum plane A.

---

### TECHNOTE 20-5
### Position Tolerance Summary

- A position tolerance can be applied to a feature of size or a pattern of features of size.
- Basic dimensions are used to define the true position of features of size.
- The default tolerance zone for a position tolerance is two parallel planes.
- Where a diameter symbol is indicated in a position tolerance, a cylindrical tolerance zone applies.
- In a position tolerance, one or more datum features must be referenced, except where applied to coaxial features of size.
- A position tolerance zone is located at true position.
- A position tolerance zone is oriented to the primary datum.
- Per Rule #2, RFS is the default; MMC or LMC must be specified.
- Where a position tolerance applies at RFS, the axis interpretation applies.
- Where a position tolerance applies at MMC (or LMC), the surface interpretation applies.
- A position tolerance zone can limit orientation.
- The size limits of a feature of size must still be met.

---

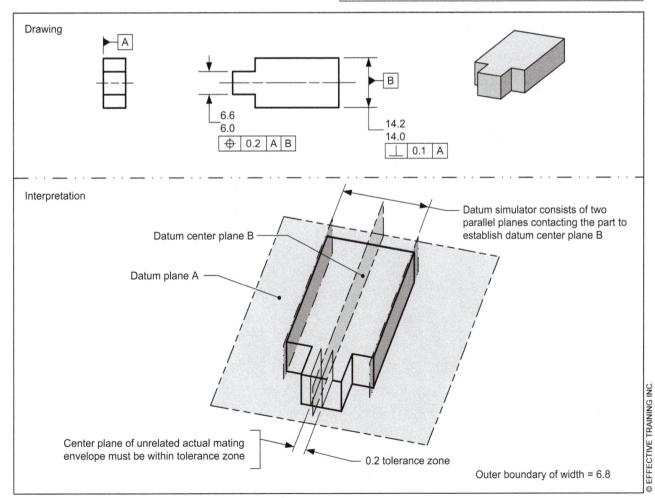

Drawing

6.6
6.0
⊕ | 0.2 | A | B

14.2
14.0
⊥ | 0.1 | A

Interpretation

Datum center plane B

Datum plane A

Datum simulator consists of two parallel planes contacting the part to establish datum center plane B

Center plane of unrelated actual mating envelope must be within tolerance zone

0.2 tolerance zone

Outer boundary of width = 6.8

© EFFECTIVE TRAINING INC.

*FIGURE 20-8  Center Plane Interpretation for a Position Tolerance at RFS Applied to a Width*

## Advantages of Using a Position Tolerance

In comparison to coordinate tolerancing, position tolerances offer many advantages. Six important advantages of using position tolerances are listed below:

1. Provides 57% larger tolerance zones

2. Permits bonus and datum feature shift

3. Prevents tolerance accumulation

4. Permits the use of a functional gage

5. Protects the part function

6. Lowers manufacturing and inspection costs

These advantages will be obvious as position tolerances are explained in the next several chapters. However, it is good to be aware of the advantages as you learn about the position tolerance, so I will highlight them.

The use of cylindrical tolerance zones is common with position tolerances. The advantage of cylindrical tolerance zones is shown in Figure 20-9. The 57% additional tolerance of the location of a feature of size that results from a cylindrical tolerance zone is a significant advantage for manufacturing.

Additional tolerances (e.g., bonus and datum shift) can easily add 100-200% or more additional tolerance (in certain cases) without affecting product function.

The use of basic dimensions to establish the true position of the toleranced features of size eliminates tolerance accumulation. See Figure 20-10.

Where the MMC modifier is specified, the surface interpretation allows the use of functional gages and protects the part function.

The additional tolerances for manufacturing, as well as the use of functional gages for inspection, result in lower manufacturing and inspection costs.

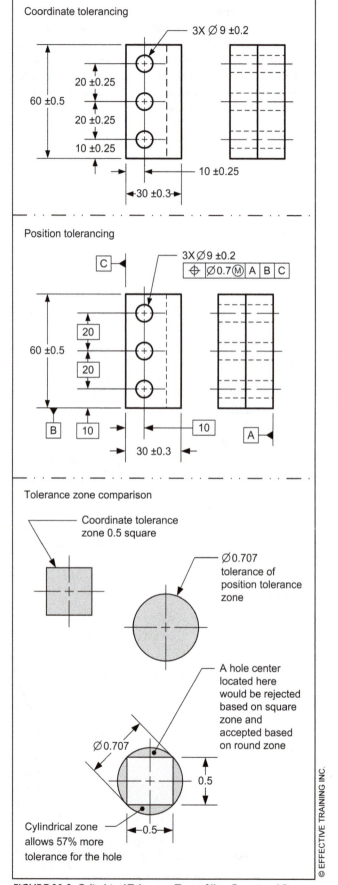

FIGURE 20-9  *Cylindrical Tolerance Zones Allow Functional Parts to Be Used*

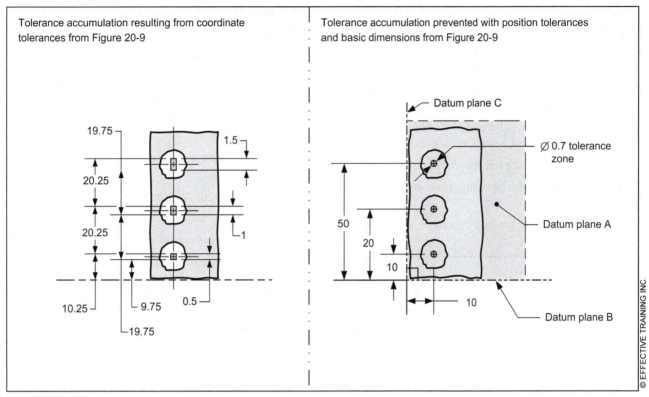

Tolerance accumulation resulting from coordinate tolerances from Figure 20-9

Tolerance accumulation prevented with position tolerances and basic dimensions from Figure 20-9

*FIGURE 20-10  Use of Basic Dimensions Prevents Unwanted Tolerance Accumulation*

© EFFECTIVE TRAINING INC.

## Feature of Size Relationships Controlled by Position

Position tolerances are commonly used to control four types of feature of size relationships:

1. The distance (spacing) between features of size, such as holes, bosses, tabs, slots, etc.

2. The location of a pattern or individual feature of size, such as holes, bosses, tabs, slots, etc.

3. The coaxiality between features of size.

4. The symmetrical relationship between features of size.

An example of each type of relationship is shown in Figure 20-11.

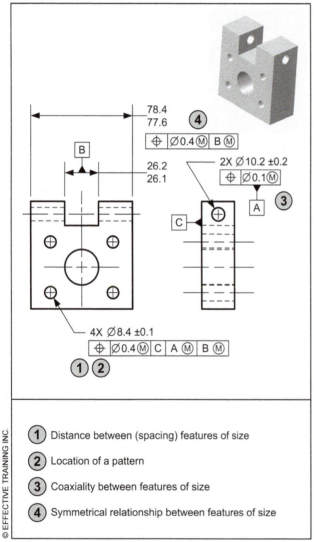

FIGURE 20-11  Examples of Using Position Tolerance to Control Feature of Size Relationships

## Evaluate if a Single-Segment Position Tolerance Specification is Standard-Compliant

For a position tolerance to be a standard-compliant specification (legal), it must satisfy certain requirements. In order to make it easier to remember what to look for when evaluating the correctness of a geometric tolerance on a drawing, I created a mnemonic.

The requirements are divided into four areas represented by the initials "CARE":

C - Check for datums

A - Assess the application

R - Review the modifiers

E - Evaluate the tolerance

A flowchart that identifies the requirements for a standard-compliant single-segment position tolerance is shown in Figure 20-12.

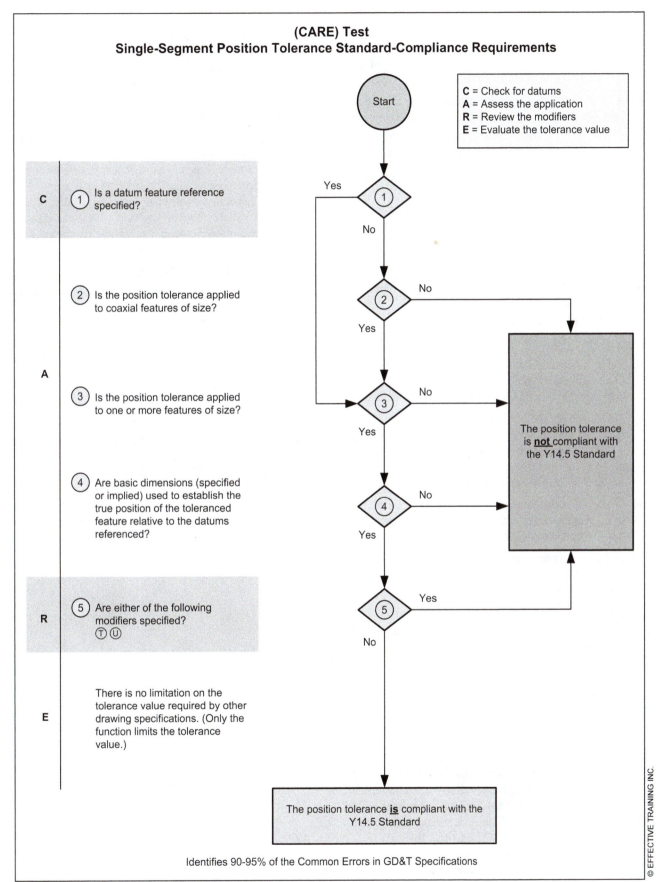

**(CARE) Test**
**Single-Segment Position Tolerance Standard-Compliance Requirements**

Start

C = Check for datums
A = Assess the application
R = Review the modifiers
E = Evaluate the tolerance value

**C**

① Is a datum feature reference specified?

② Is the position tolerance applied to coaxial features of size?

**A**

③ Is the position tolerance applied to one or more features of size?

④ Are basic dimensions (specified or implied) used to establish the true position of the toleranced feature relative to the datums referenced?

**R**

⑤ Are either of the following modifiers specified?
Ⓣ Ⓤ

**E**

There is no limitation on the tolerance value required by other drawing specifications. (Only the function limits the tolerance value.)

① Yes → No

② No → Yes

③ No → Yes

④ No → Yes

⑤ Yes → No

The position tolerance is **not** compliant with the Y14.5 Standard

The position tolerance **is** compliant with the Y14.5 Standard

Identifies 90-95% of the Common Errors in GD&T Specifications

© EFFECTIVE TRAINING INC.

*FIGURE 20-12 Test for Standard-Compliant Position Tolerance*

# SUMMARY

## Key Points

- True position is the theoretically exact location of a feature of size as established by basic dimensions.

- A position tolerance is a geometric tolerance that defines the allowable deviation of a feature of size from true position.

- Position tolerances may also limit orientation deviation.

- Requirements of a position tolerance. A position tolerance must…
  - o Be applied to a feature of size or pattern of features of size
  - o Have the location and spacing of the feature(s) of size defined by basic dimensions
  - o Have datums specified (unless it is applied to coaxial features of size or a lower segment of a composite tolerance)
  - o Have modifiers specified per Rule #2
  - o RFS/RMB is the default condition per Rule #2

- Basic dimensions defining true positions may be specified or implied. It is common to have basic implied zero offset and basic implied 90° or 180° dimensions.

- The choice of modifier in a position tolerance is often related to the functional requirements of the application:
  - o MMC for assembly or non-critical relationships
  - o LMC for minimum distance
  - o RFS for alignment or centering

- Where the surface interpretation is used, the surface(s) of the feature of size must not violate the virtual condition (acceptance boundary).

- In the axis/center plane interpretation, the axis (or center plane) of the feature of size must be within the tolerance zone.

- A summary of position tolerance information:
  - o Position applies to a feature of size or a pattern of features of size.
  - o Basic dimensions are used to define the true position of the feature of size.
  - o The default tolerance zone shape is two parallel planes.
  - o Where a diameter symbol is indicated, a cylinder-shaped tolerance zone applies.
  - o One or more datums must be referenced (except where applied to coaxial features of size).
  - o The tolerance zone is centered at the true position.
  - o The tolerance zone is oriented to the primary datum.
  - o RFS is the default; MMC or LMC must be specified.
  - o Where applied at RFS, the axis interpretation applies.
  - o Where applied at MMC (or LMC), the surface interpretation applies.
  - o Position can limit orientation.
  - o The size limits of a feature of size must still be met.
  - o The tolerance zone applies to the full extent of the feature of size.

- Six advantages of using a position tolerance:
  - o Provides 57% larger tolerance zones
  - o Permits bonus and datum feature shift
  - o Prevents tolerance accumulation
  - o Permits the use of a functional gage
  - o Protects the part function
  - o Lowers manufacturing and inspection costs

- Position tolerances are commonly used to control four types of part relationships:
  - o The distance (spacing) between features of size within a pattern, such as holes, bosses, tabs, slots, etc.
  - o The location of a pattern or individual feature of size
  - o The coaxiality between cylindrical features of size
  - o The symmetry between planar width features of size

## Additional Related Topics

*These topics are recommended for further study to improve your understanding of position.*

| Topic | Source |
|---|---|
| Simultaneous requirements | *ASME Y14.5-2009, Para. 7.5.4* |
| Position with profile | *ASME Y14.5-2009, Para. 8.8* |
| Conical tolerance zones | *ASME Y14.5-2009, Para. 7.4.3* |
| Position of threaded features | *ASME Y14.5-2009, Paras. 2.9 & 7.4.1* |

## QUESTIONS AND PROBLEMS

***Website Bonus Materials***
Additional questions are available at our website. To access bonus materials for this textbook, please visit:
www.etinews.com     www.etinews.com/textbookbonus

## True and False

*Indicate if each statement is true or false.*

T / F     1. True position is a geometric tolerance that locates a feature of size.

T / F     2. A position tolerance must always be supplemented with an orientation tolerance.

T / F     3. A position tolerance may be used to locate surfaces.

T / F     4. The axis interpretation is only used when a position tolerance is applied at MMC.

T / F     5. A round tolerance zone (across the corners of a square tolerance zone) provides more tolerance.

T / F     6. The advantages of position tolerances includes lower manufacturing costs.

## Multiple Choice

*Circle the best answer to each statement.*

1. One requirement of a position tolerance is:
   A. The diameter modifier must be specified.
   B. True position must be specified.
   C. A datum feature reference must be specified.
   D. All of the above

2. A position tolerance be applied to:
   A. A planar surface
   B. A feature of size
   C. A line element
   D. All of the above

3. A position tolerance may also be an indirect _____ tolerance.
   A. Location
   B. Orientation
   C. Cylindricity
   D. Size

4. How does the use of a position tolerance benefit manufacturing?
   A. Cylindrical tolerance zones are 57% larger
   B. Permits a bonus tolerance
   C. Permits a datum feature shift
   D. All of the above

5. What type of relationships is a position tolerance used to control?
   A. Distance between features of size
   B. Coaxial relationships
   C. Symmetrical relationships
   D. All of the above

## Application Problems

*The application problems are designed to provide practice on applying the chapter concepts to situations that are similar to on-the-job conditions.*

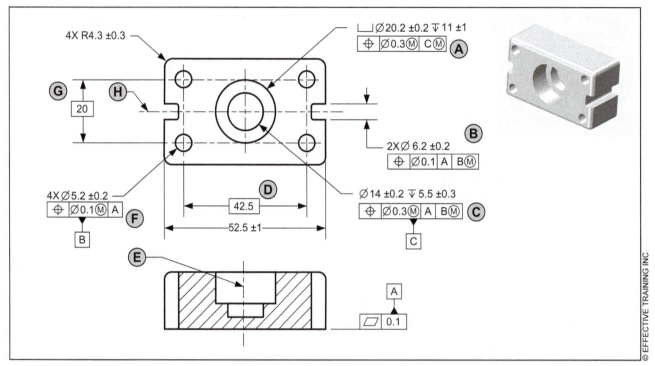

*Application questions 1-5 refer to the drawing above.*

1. Which feature control frames should be interpreted using a surface (VC boundary) interpretation?
   A. A, B, and C
   B. B, C, and F
   C. A, C, and F
   D. None of the above

2. Which feature control frame(s) are to be interpreted using an axis/center plane interpretation?
   A. B
   B. A and F
   C. B and C
   D. F

3. Describe the true position of the hole with position tolerance labeled "A."
   A. Perpendicular to datum A and 14.25 from datum B
   B. Perpendicular to datum A and 10 from datum B
   C. Parallel and coaxial to datum C
   D. Both A and B

4. What is the permissible bonus tolerance for the position tolerance labeled "A?"
   A. 0.2
   B. 0.3
   C. 0.4
   D. 0.5

5. Indicate the type of part relationship being controlled with each position tolerance.

| | Spacing | Location | Coaxiality | Symmetrical Relationship |
|---|---|---|---|---|
| A | | | | |
| B | | | | |
| C | | | | |
| F | | | | |

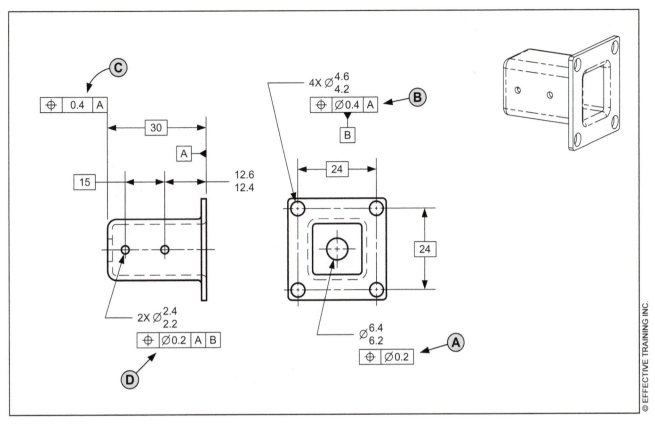

*Question 6 refers to the drawing above.*

6. For each tolerance of position listed, indicate if it is a legal (standard-compliant) specification. If a tolerance is not standard-compliant, explain why.

(A) _____

(B) _____

(C) _____

(D) _____

# Position Tolerance - RFS and MMC

## Goal

Interpret position tolerance applications RFS and MMC

## Performance Objectives

Upon completing this chapter, you should be able to:

1. List five conditions that apply where a position tolerance is indicated at RFS (p.280)
2. List five conditions that apply where a position tolerance is indicated at MMC (p.280)
3. Describe real-world applications for position tolerances (p.280)
4. Interpret a position tolerance at RFS applied to a hole (p.280)
5. Interpret a position tolerance at RFS applied to a slot (p.281)
6. Interpret a position tolerance at RFS applied to coaxial features of size (p.282)
7. Interpret a position tolerance at RFS using the "Significant Seven Questions" (p.282)
8. Interpret a position tolerance at RFS applied to a pattern of holes (p.283)
9. Interpret a position tolerance at MMC applied to a hole (p.284)
10. Interpret a position tolerance at MMC using the "Significant Seven Questions" (p.284)
11. Interpret a position tolerance at MMC applied to coaxial features of size (p.285)
12. Interpret a position tolerance at MMC applied to a pattern of holes (p.286)
13. Understand the verification principles for position tolerances (p.287)
14. Describe how to inspect a position tolerance (RFS) (p.288)
15. Describe how to inspect a position tolerance at MMC (p.289)

## New Term

Coordinate measuring machine (CMM)

## What This Chapter Is About

This is the second of four chapters about position tolerances. In this chapter, we'll look at the basic RFS and MMC applications of position tolerances.

This chapter explains how to interpret position tolerances and covers their fundamental uses and interpretation. Position tolerances are most commonly applied with the MMC modifier. Understanding position tolerances at MMC and RFS will help you make important decisions that affect the function and cost of products.

# TERMS AND CONCEPTS

## Position Tolerance (RFS)

In certain cases, the function of a part may require locating a feature of size at RFS. An example is shown in Figure 21-1. Where a position tolerance is applied at RFS, five conditions apply:

- No bonus tolerance is permitted.
- The axis/center plane interpretation applies.
- The tolerance zone is located and oriented by the basic dimensions.
- Variable measurement is required.
- The deviations of orientation and location of the toleranced feature of size are controlled.

## Position Tolerance (MMC)

In certain cases, the function of a part may require that a feature of size be located at MMC. An example is shown in Figure 21-2. Where a position tolerance is applied at MMC, five conditions apply:

- Bonus tolerance is permitted.
- The surface interpretation applies.
- The boundary zone is located and oriented by the basic dimensions.
- A functional gage may be used.
- The deviations of orientation and location of the toleranced feature of size are controlled.

# POSITION TOLERANCE APPLICATIONS

The following sections describe interpretation of various position tolerance applications.

## Real-World Applications of a Position Tolerance

There are many common real-world applications of a position tolerance on a drawing. These applications limit the location deviation of a pattern or a single feature of size relative to a datum reference frame for:

- Assembly (clearance holes for fasteners)
- Alignment (linear motion or mechanisms)
- Centering (coaxial features of size for rotating or pivoting)
- Locating non-critical features of size (lightening holes)
- Controlling minimum distances (wall thickness, machine stock on a casting)

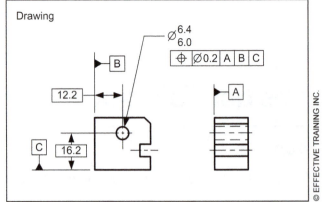

**FIGURE 21-1 Position Tolerance (RFS) with No Material Condition Modifier Applied at RFS Per Rule #2**

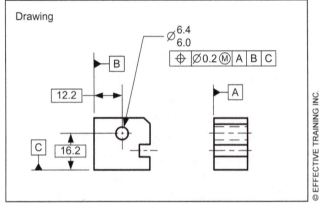

**FIGURE 21-2 Position Tolerance with MMC Modifier Specified**

## Position Tolerance RFS Applied to a Hole

In certain cases, the function of a part may require a feature of size be located and oriented regardless of feature size relative to the outside edges of a part. In Figure 21-3, a tolerance of position at RFS is applied to the hole size dimension. In this application, the following conditions apply:

- The tolerance zone is a fixed-size cylinder of $\emptyset 0.2$.
- The tolerance zone is perpendicular to datum plane A.
- The tolerance zone is located at the true position of the hole.
- The axis of the unrelated actual mating envelope must lie within the zone.
- The tolerance zone limits the location, and orientation of the hole.
- Bonus tolerance is not allowed.
- Rule #1 applies to the hole.
- The hole must be within its size limits.
- The worst-case boundary is $\emptyset 5.8 = (6.0 - 0.2)$.

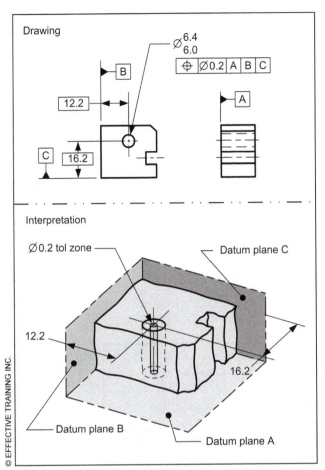

FIGURE 21-3 Position Tolerance (RFS) Used to Define Hole Location and Orientation

## Position Tolerance RFS Applied to a Slot

In certain cases, the function of a part may require a planar feature of size to be located and oriented regardless of feature size relative to the outside edges of a part.

In Figure 21-4, a tolerance of position at RFS is applied to the slot size dimension. In this application, the following conditions apply:

- The tolerance zone is the space between two parallel planes 0.1 apart.
- The tolerance zone is perpendicular to datum plane A.
- The tolerance zone is located at the true position of the slot.
- The center plane (or the actual mating envelope) of the slot must lie within the zone.
- The tolerance zone limits the location and orientation of the slot.
- Bonus tolerance is not allowed.
- Rule #1 applies to the slot.
- The slot must be within its size limits.
- The worst-case boundary is 5.0 = (5.1 - 0.1).

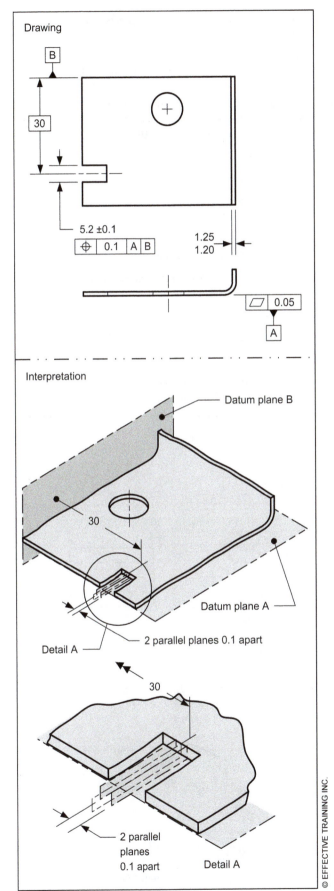

FIGURE 21-4 Position Tolerance (RFS) Used to Define Slot Location and Orientation

## Position Tolerance RFS Applied to Coaxial Features of Size

In certain cases, the function of a part may require that the coaxial relationship between two features of size be controlled regardless of feature size. In Figure 21-5, a tolerance of position at RFS is applied to the size dimension of one of the features of size. The other feature of size is the datum feature. In this application, the following conditions apply:

- The tolerance zone is a fixed-size cylinder of ∅0.4.
- The tolerance zone is centered about the datum axis.
- The datum axis is the axis of the datum feature simulator expanding in the internal feature of size.
- The axis of the external feature of size must be within the tolerance zone.
- The tolerance zone limits location and parallelism between the holes.
- Bonus tolerance is not allowed.
- Rule #1 applies to both features of size.
- Both features of size must be within their size limits.
- The worst-case boundary of the outside feature of size is ∅17.0 = (16.6 + 0.4).

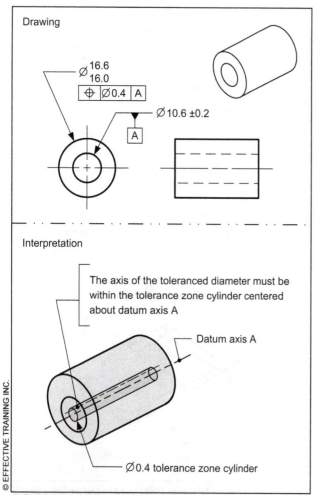

FIGURE 21-5 Position Tolerance (RFS) Used to Define Location and Orientation of Coaxial Features of Size

## The Significant Seven Questions

This section uses the Significant Seven Questions to interpret a position tolerance at RFS application. The questions are explained in Appendix B.

Figure 21-6 contains a position tolerance application. In the interpretation section of the figure the significant seven questions are answered. Notice how each answer increases your understanding of the geometric tolerance.

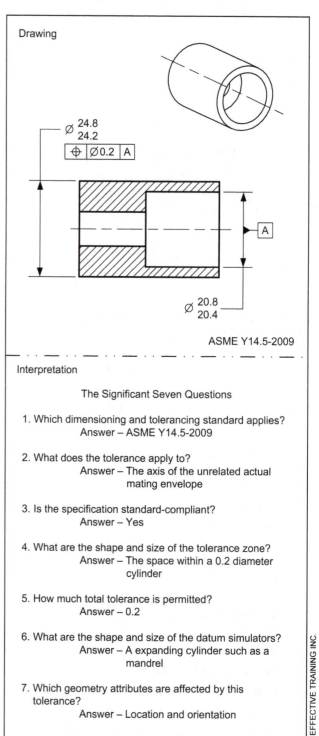

FIGURE 21-6 The Significant Seven Questions Applied to a Position Tolerance

## Position Tolerance RFS Applied to a Pattern of Holes

In certain cases, the function of a part may require a pattern of holes be located and oriented at regardless of feature size relative to the outside edges of a part. In Figure 21-7, a tolerance of position at RFS is applied to the size dimension of the pattern of holes. In this application, the following conditions apply:

- The tolerance zones are four fixed-size cylinders of $\varnothing 0.3$.
- The tolerance zones are perpendicular to datum plane A.
- The tolerance zones are located at the true position of each hole.
- The axis (of the actual mating envelope) of each hole must be within the zone.
- The tolerance zone limits the location, orientation, and spacing between the holes.
- Bonus tolerance is not allowed.
- Rule #1 applies to each hole.
- Each hole must be within its size limits.
- The worst-case boundary is $\varnothing 7.7 = (8.0 - 0.3)$.

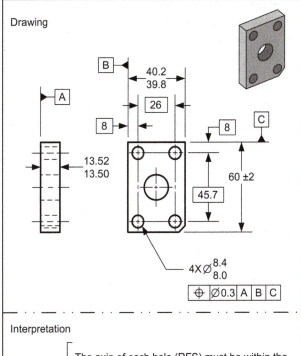

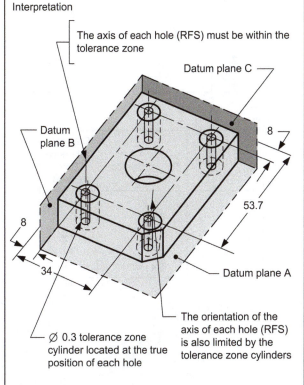

**FIGURE 21-7  Position Tolerance (RFS) Used to Define Location and Orientation of a Pattern of Holes**

## Position Tolerance MMC Applied to a Hole

In certain cases, the function of a part may allow more location and orientation tolerance for a feature of size when it is not at its MMC size limit. In Figure 21-3, a tolerance of position at MMC is applied to the hole size dimension. In this application, the following conditions apply:

- The tolerance zone is a virtual condition cylinder of ∅5.6 perpendicular to datum A.

- The virtual condition is located at the true position of the hole.

- The virtual condition limits the location and orientation of the hole.

- The hole must be within its size limits, and Rule #1 applies.

- The maximum permissible bonus is 0.4.

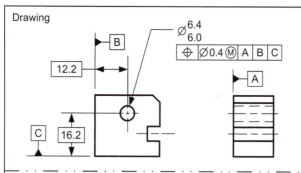

Drawing

Interpretation

| VC (Acceptance Boundary) | Hole Actual Mating Envelope | Diameter Tolerance Zone Allowed* |
|---|---|---|
| 5.6 | 6.0 MMC | 0.4 |
| | 6.1 | 0.5 |
| | 6.2 | 0.6 |
| | 6.3 | 0.7 |
| | 6.4 LMC | 0.8 |

*See Appendix F

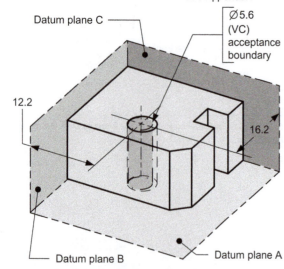

**FIGURE 21-8  Position Tolerance (MMC) Used to Define Hole Location and Orientation**

## The Significant Seven Questions

This section uses the Significant Seven Questions to interpret a position tolerance MMC application. The questions are explained in Appendix B. Figure 21-9 contains a position tolerance at MMC applied to a hole pattern application. The significant seven questions are answered in the interpretation section of the figure. Notice how each answer increases your understanding of the geometric tolerance.

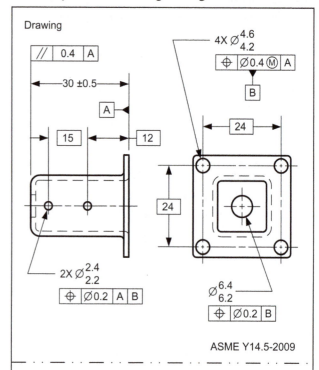

Drawing

ASME Y14.5-2009

Interpretation

The Significant Seven Questions

1. Which dimensioning and tolerancing standard applies?
   Answer – ASME Y14.5-2009

2. What does the tolerance apply to?
   Answer – A pattern of 4 holes

3. Is the specification standard-compliant?
   Answer – Yes

4. What are the shape and size of the tolerance zone?
   Answer – Four 3.8 diameter VC acceptance boundaries perpendicular to datum A and spaced apart by the basic dimensions shown

5. How much total tolerance is permitted?
   Answer – 0.8 when the holes are at LMC

6. What are the shape and size of the datum simulators?
   Answer – A planar surface for datum A

7. Which geometry attributes are affected by this tolerance?
   Answer – Orientation relative to datum A, spacing between the holes

**FIGURE 21-9 The Significant Seven Questions Applied to a Position Tolerance**

## Position Tolerance MMC Applied to Coaxial Features of Size

In certain cases, the function of a part may allow more tolerance between coaxial diameters when they are not at their MMC size limits. There are three common part configurations involving coaxial features of size:

- Opposed features of size (i.e., one external and one internal feature of size)

- Non-opposed features of size of the same size (i.e., two external features of size)

- Non-opposed features of size of different size (i.e., two external features of size)

In Figure 21-10, the function of the part allows more tolerance for coaxiality when the features of size are not at MMC. A position tolerance at MMC is applied to the I.D. size dimension. The O.D. is referenced as the primary datum at MMB.

In this application, the following conditions apply:

- The tolerance zone is a virtual condition cylinder of $\emptyset15.7 = (16.2 - 0.5)$.

- The tolerance zone is centered about the datum axis.

- The datum axis is the axis of the O.D. MMB ($\emptyset20.6$).

- The surface of the I.D. must not violate the virtual condition.

- The virtual condition limits location and orientation.

- Maximum bonus is $0.2 = (16.4 - 16.2)$.

- Maximum datum feature shift is $0.1 = (20.1 - 20.0)$.

- Rule #1 applies to both features of size (individually).

- Both features of size must be within their size limits.

Figure 12-11 shows a part with three coaxial non-opposed features of size of the same size. The function of the part requires the three features of size to be coaxial to each other when they are at MMC. More tolerance is allowed when they are not at MMC. A position tolerance at MMC is applied to the size dimension of the features of size diameters. A datum feature reference is not required because the three features of size are an implied self-datum since they locate and orient the part in the assembly. In this application, the following conditions apply:

- The tolerance zone is a virtual condition cylinder of $\emptyset20.12 = (20.02 + 0.1)$.

- The surface elements must not violate the virtual condition.

- The virtual condition limits location and orientation relative to each other.

- Maximum bonus is $0.04 = (20.02 - 19.98)$.

- Maximum datum feature shift is zero. (Because they are self-datums and their size was used to calculate bonus, it cannot be used again for datum feature shift.)

- Rule #1 applies to all three features of size (individually).

- All three features of size must be within their size limits.

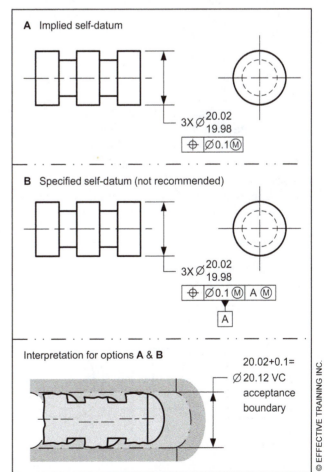

FIGURE 21-11 Position Tolerance (MMC) Used to Define Location and Orientation of Non-Opposed Coaxial Features of Size

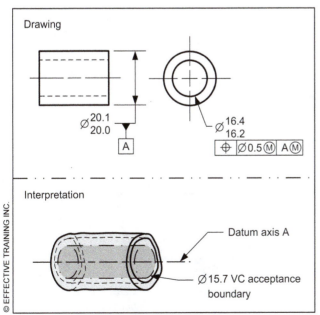

FIGURE 21-10 Position Tolerance (MMC) Used to Define Location and Orientation of Opposed Coaxial Features of Size

Figure 12-12 shows a part with two coaxial non-opposed features of size of different size. The function of the part requires the two features of size to be coaxial to each other when they are at MMC so they can fit into two coaxial cylinder bores in the mating part. More tolerance is allowed when they are not at MMC. A position tolerance at MMC is applied to the size dimension of each feature of size. The features of size are referenced as co-primary datums at MMC because they both locate and orient the part in the assembly. In this application the following conditions apply:

- The tolerance zone is two coaxial virtual condition boundaries.

- The virtual condition for datum A is $\emptyset24.65 = (24.6 + 0.05)$.

- The virtual condition for datum B is $\emptyset34.65 = (34.6 + 0.05)$.

- The surface elements of both features of size must not violate their virtual condition boundaries.

- The virtual condition boundaries limit location and orientation of both features of size relative to each other.

- Maximum bonus for datum A is $0.2 = (24.6 - 24.4)$.

- Maximum bonus for datum B is $0.2 = (34.6 - 34.4)$.

- Maximum datum feature shift is zero. (They are self-datums, and their size was used to calculate bonus; it cannot be used again for datum shift.)

- Rule #1 applies to both features of size (individually).

- Both features of size must be within their size limits.

---

### TECHNOTE 21-1
### Position Tolerance Applied to
### Coaxial Features of Size

When a position tolerance is used on coaxial features of size without a datum feature reference, it controls the relationship between the coaxial features of size but does not establish a relationship to any datums.

---

## Position Tolerance MMC Applied to a Pattern of Holes

In certain cases, the function of a part may allow more location and orientation tolerance for a pattern of holes, when they are not at MMC, relative to the outside edges of a part. In Figure 21-13, a tolerance of position at MMC is applied to the size dimension of the pattern of holes. In this application, the following conditions apply:

- The tolerance zones are four virtual condition cylinders of $\emptyset7.7 = (8.0 - 0.3)$.

- The virtual conditions are perpendicular to datum plane A.

- The virtual conditions are located at the true position of each hole.

- The surface of each hole must not violate its virtual condition.

- The virtual condition limits the location, orientation, and spacing between the hole.

- The maximum bonus is $0.4 = (8.4 - 8.0)$.

- Datum feature shift is not allowed.

- Rule #1 applies to each hole.

- Each hole must be within its size limit.

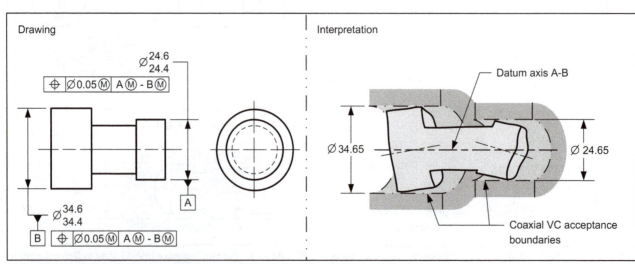

FIGURE 21-12  *Position Tolerance at MMC/MMB Applied to Non-Opposed Coaxial Features of Size*

Drawing

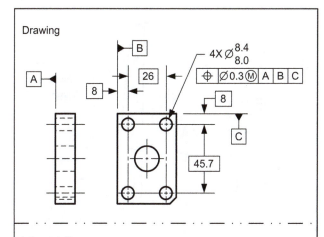

Interpretation

| VC (Acceptance Boundary) | Hole Actual Mating Envelope | Diameter Tolerance Zone Allowed* |
|---|---|---|
| | 8.0 MMC | 0.3 |
| | 8.1 | 0.4 |
| 7.7 | 8.2 | 0.5 |
| | 8.3 | 0.6 |
| | 8.4 LMC | 0.7 |

*See Appendix F

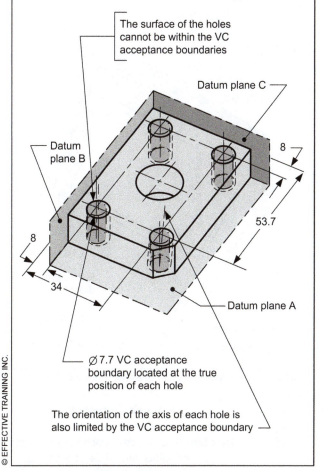

The surface of the holes cannot be within the VC acceptance boundaries

Ø 7.7 VC acceptance boundary located at the true position of each hole

The orientation of the axis of each hole is also limited by the VC acceptance boundary

# VERIFICATION PRINCIPLES AND METHODS

## Verification Principles for Position Tolerances

This section contains a simplified explanation of verification principles and methods that can be used to inspect a position tolerance applied to features of size at RFS and at MMC.

Verifying a position tolerance applied to a feature of size at RFS has three parts:

1. Establishing the relationship specified in the feature control frame between the part and the datum reference frame

2. Establishing the axis or center plane of the toleranced feature of size

3. Verifying that the axis or center plane of the toleranced feature of size is within the tolerance zone

Verifying a position tolerance applied to a feature of size at MMC involves three parts:

1. Establishing the relationship specified in the feature control frame between the part and the datum reference frame

2. Establishing the VC (acceptance boundary)

3. Verifying that the toleranced feature of size does not violate the VC (acceptance boundary)

FIGURE 21-13 Position Tolerance (MMC) Used to Define Location and Orientation of a Pattern of Holes

## How to Verify a Position Tolerance (RFS)

When verifying a position tolerance RFS, a measurement system that provides a numerical value for a measurement is typically used. One such device is a coordinate measuring machine. Coordinate measurement machines come in many different styles and sizes. A picture of a bridge style coordinate measuring machine is shown in Figure 21-14.

A *coordinate measuring machine (CMM)* is a precision three-axis machine used to collect and analyze surface points from a part to verify the geometric requirements (size, location, orientation, and form) of part features. A CMM has four main components:

1. The machine (table and movable frame)

2. A measuring probe

3. A computer control system

4. Software

A standard touch probe CMM uses scales and transducers to determine the precise X, Y, and Z coordinates of the probe, relative to the machine origin. When the probe touches the surface of a workpiece, a sensor is triggered so the computer can calculate the precise X, Y, and Z coordinates at the point of contact. After collecting a sufficient number of surface points, the data set is then analyzed by the measurement software. The software uses various algorithms to analyze the surface point data set to construct features to determine the size, location, orientation, and form of a part feature.

Now let's look at a simple explanation of how to inspect the position tolerance of a hole using a CMM. Figure 21-15 shows the part from Figure 21-1 located in a CMM fixture. The fixture is placed on the CMM surface plate (table). The datum feature simulator surfaces are probed following the 3-2-1 rule, and the CMM measurement software creates the datum reference frame.

Next, the part is placed on the fixture and a best-fit gage pin is pushed into the hole to simulate its actual mating envelope. The pin is probed near its top and close to where it protrudes from the hole to derive an axis. The measurement software projects the axis to datum plane A. The top surface around the hole is also probed to establish a plane that intersects the axis. The measurement software then compares the two intersecting points of the axis to the true position of the hole and reports the axis deviation from true position. See Figure 21-15.

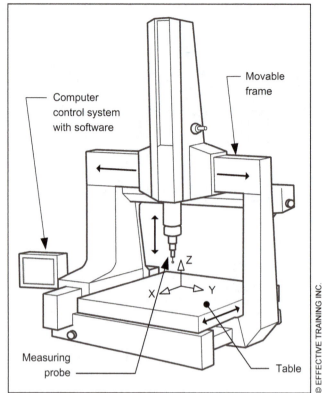

FIGURE 21-14  Coordinate Measuring Machine (CMM)

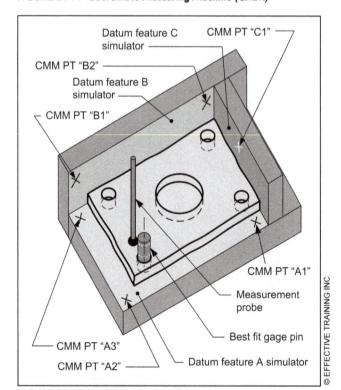

FIGURE 21-15  Verifying a Position Tolerance Using a CMM

**Author's Comment**

If a fixture is not used and the part datum features are directly probed to establish the datum reference frame, a sufficient number of points should be collected to minimize the measurement error caused by the form and orientation deviations of the part datum features.

**Author's Comment**

If a best-fit gage pin is not used and the part feature is directly probed, a sufficient number of points should be collected to minimize the measurement error caused by the form deviations of the part feature.

## How to Verify a Position Tolerance (MMC)

When a position tolerance at MMC is verified, a functional gage is often used. (See Chapter 8.) In Figure 21-16, the part from Figure 21-13 is shown on the functional gage used to verify the position tolerance of the holes. When the part is located against the datum feature simulators, the gage pins made to the virtual condition size of the holes must fit into the holes. If the pins fit, the part passes. If the pins do not fit, the part fails. A functional gage only provides attribute (pass or fail) data.

The design and tolerancing of functional gages is covered in the standard, "Dimensioning and Tolerancing Principles for Gages and Fixtures ASME Y14.43." This standard covers:

- Gaging policies
- Gage design
- Plug gages
- Functional gages
- Gage tolerances
- Gage usage
- CMM fixtures

There are many benefits of using a functional gage:

- The gage represents the virtual condition.
- Parts can be verified quickly.
- A functional gage is economical to produce.
- No special skills are required to read the gage or interpret results.
- In some cases, a functional gage can check several part characteristics simultaneously (i.e., location and orientation of several features of size).

Although a functional gage provides several benefits for verifying position tolerances at MMC, their use is not mandatory. A position tolerance at MMC may also be verified with variable measurement system such as a CMM.

### Author's Comment
In this text, we describe the functional gage pins as being the virtual condition size with no discussion on gage tolerances. Gage tolerancing is an extensive topic and is not required to understand dimensioning and tolerancing of parts.

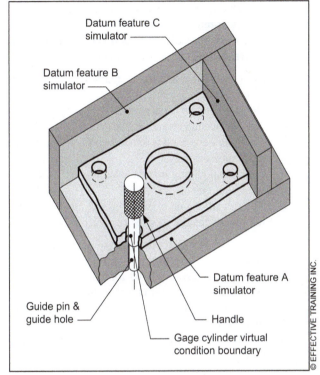

© EFFECTIVE TRAINING INC.

**FIGURE 21-16** *Verifying a Position Tolerance (MMC) With a Functional Gage*

## SUMMARY

### Key Points

- Where a position tolerance is applied at RFS, five conditions apply:
  - o No bonus tolerance is permitted.
  - o The axis/center plane interpretation applies.
  - o The tolerance zone is located and oriented by the basic dimensions.
  - o Variable measurement is used to verify the position requirement.
  - o The deviations of orientation and location of the toleranced feature of size are controlled.

- Where a position tolerance is applied at MMC, five conditions apply:
  - o Bonus tolerance is permitted.
  - o The surface interpretation applies.
  - o The boundary zone is located and oriented by the basic dimensions.
  - o Functional gaging may be used to verify the position requirement.
  - o The deviations of orientation and location of the toleranced feature of size are controlled.

- Common real-world applications of a position tolerance are:
  - o Assembly (clearance holes for fasteners)
  - o Alignment (linear motion or mechanisms)
  - o Centering (coaxial features of size for rotating or pivoting)
  - o Locating non-critical features of size (lightening holes)

- When a position tolerance is used on coaxial features of size without a datum feature reference, it controls the relationship between the coaxial features of size and establishes an implied self-datum.

- A coordinate measuring machine (CMM) is a precision three axis machine used to collect and analyze surface points from a part to verify the geometric requirements (size, location, orientation, and form) of part features.

### Additional Related Topics

*These topics are recommended for further study to improve your understanding of position RFS and MMC.*

| Topic | Source |
|---|---|
| Simultaneous requirements | *ASME Y14.5-2009, Para. 7.5.4* |
| Position tolerances at LMC | *ASME Y14.5-2009, Para. 7.3.5* |
| Functional gages | ETI's *Functional Gaging and Measurement* textbook |

## QUESTIONS AND PROBLEMS

**Website Bonus Materials**
Additional questions are available at our website. To access bonus materials for this textbook, please visit:
www.etinews.com/textbookbonus

### True and False

*Indicate if each statement is true or false.*

T / F    1. When a position tolerance at RFS is specified, the axis/center plane interpretation applies.

T / F    2. When a position tolerance at MMC is specified, the axis/center plane interpretation applies.

T / F    3. A position tolerance can be applied to a pattern of holes.

T / F    4. When a position tolerance at MMC is applied to a hole, Rule #1 is overridden.

T / F    5. When a position tolerance at RFS is applied to a hole, Rule #1 is overridden.

T / F    6. A position tolerance at RFS is often verified with a functional gage.

T / F    7. When a position tolerance at MMC is applied to coaxial features of size, a datum feature reference is not required.

T / F    8. The verification principles are the same for verifying a position tolerance applied at MMC or RFS.

### Multiple Choice

*Circle the best answer to each statement.*

1. Which condition(s) apply where a position tolerance is applied at RFS?
   A. A functional gage may be used to verify the position requirement.
   B. The axis/center plane interpretation is used.
   C. Orientation deviations are not controlled.
   D. All of the above

2. Which condition(s) apply where a position tolerance is applied at MMC?
   A. A bonus tolerance is permitted.
   B. The acceptance boundary is located and oriented by basic dimensions.
   C. A functional gage may be used to verify the position tolerance.
   D. All of the above

3. A position tolerance at RFS may be used to limit locational deviations of:
   A. A pattern of holes
   B. Coaxial features of size
   C. A slot
   D. All of the above

4. How does the bonus tolerance concept apply where a position tolerance at MMC is applied to a hole?
   A. The hole must be at LMC to receive a bonus tolerance.
   B. The hole receives additional size tolerance based on how well it is located.
   C. The hole receives additional location tolerance based on the produced hole size.
   D. The hole receives less size tolerance in exchange for more location tolerance.

5. When a position tolerance is applied to coaxial features of size, which of the following is required?
   A. A datum feature reference
   B. The position tolerance must be applied at RFS
   C. The position tolerance must be applied at MMC
   D. None of the above

## Application Problems

*The application problems are designed to provide practice applying the chapter concepts to situations that are similar to on-the-job conditions.*

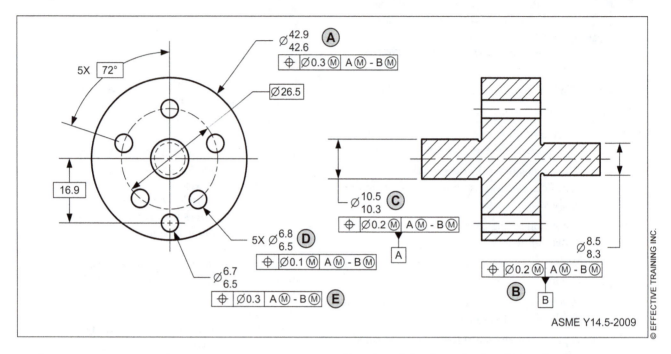

ASME Y14.5-2009

*Application questions 1-3 refer to the drawing above.*

1. The permissible bonus tolerance for the position tolerance labeled "D" is:
   A. 0
   B. 0.1
   C. 0.3
   D. 0.4

2. The position tolerance labeled "D" controls:
   A. The spacing between 5-hole pattern
   B. The coaxiality of the 5-hole pattern to the datum axis
   C. The parallelism of the 5-hole pattern to the datum axis
   D. All of the above

3. How do datum feature references at MMB affect the position tolerance labeled "D"?
   A. It permits a datum feature shift.
   B. It permits a bonus tolerance.
   C. It supports the bonus tolerance by specifying a fixed-size datum feature simulator.
   D. The MMB specification has no effect and could be removed.

*Questions 4-6 refer to the drawing on the next page.*

4. Use the Significant Seven Questions to interpret the position tolerance labeled "A."

   1) Which dimensioning and tolerancing standard applies?
   _____

   2) What does the tolerance apply to? _____
   _____

   3) Is the specification standard-compliant?_____

   4) What are the shape and size of the tolerance zone?
   _____

   5) How much total tolerance is permitted?_____

   6) What are the shape and size of the datum simulators?
   _____

   7) Which geometry attributes are affected by this tolerance?
   _____

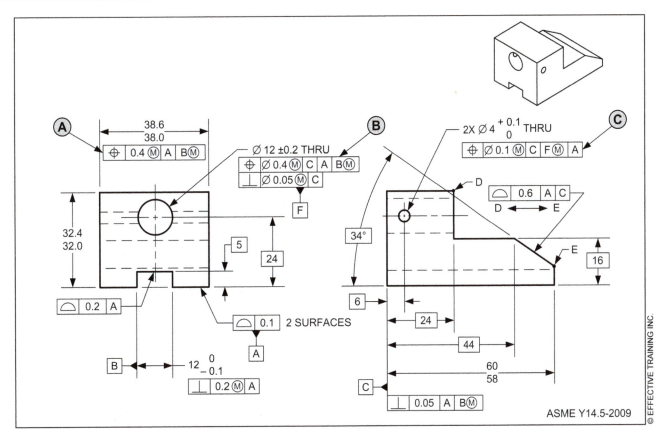

ASME Y14.5-2009

© EFFECTIVE TRAINING INC.

5. Use the Significant Seven Questions to interpret the position tolerance labeled "B."

1) Which dimensioning and tolerancing standard applies?

_____

2) What does the tolerance apply to? _____

_____

3) Is the specification standard-compliant?_____

4) What are the shape and size of the tolerance zone?

_____

5) How much total tolerance is permitted?_____

6) What are the shape and size of the datum simulators?

_____

7) Which geometry attributes are affected by this tolerance?

_____

6. Use the Significant Seven Questions to interpret the position tolerance labeled "C."

1) Which dimensioning and tolerancing standard applies?

_____

2) What does the tolerance apply to?_____

_____

3) Is the specification standard-compliant?_____

4) What are the shape and size of the tolerance zone?

_____

5) How much total tolerance is permitted?_____

6) What are the shape and size of the datum simulators?

_____

7) Which geometry attributes are affected by this tolerance?

_____

# Position Tolerance - Special Applications

22

## Goal

Interpret special applications of position tolerances

## Performance Objectives

Upon completing this chapter, you should be able to:

1. Describe the term "projected tolerance zone" (p.296)
2. Describe a real-world application of the projected tolerance zone modifier (p.296)
3. Interpret a position tolerance using the projected tolerance zone modifier (p.297)
4. Describe how a position tolerance with the projected tolerance zone modifier can be verified (p.297)
5. Describe what a multiple single-segment position tolerance is (p.298)
6. Describe a real-world application of a multiple single-segment position tolerance (p.298)
7. Interpret a multiple single-segment position tolerance (p.298)
8. Describe what a composite position tolerance is (p.298)
9. Describe a real-world application of a composite position tolerance (p.300)
10. Interpret a composite position tolerance (p.300)
11. Interpret a position tolerance applied to a hole (MMC) at an angle to the datums (p.301)
12. Interpret a position tolerance used as a bidirectional tolerance (p.302)
13. Interpret a position tolerance applied to an elongated hole (p.303)
14. Interpret a position tolerance used to tolerance a symmetrical relationship (p.304)
15. Interpret a position tolerance using the LMC modifier (p.305)
16. Interpret a position tolerance used to tolerance the spacing and orientation of a hole pattern (p.306)
17. Explain what zero tolerance at MMC tolerancing is (p.307)
18. List three benefits of zero tolerance at MMC dimensioning (p.308)
19. Describe a real-world application for using zero tolerance at MMC (p.309)
20. Interpret a zero tolerance at MMC position application (p.310)

## New Terms

- Bidirectional position tolerance
- BOUNDARY
- Composite position tolerance
- Multiple single-segment position tolerance
- Projected tolerance zone
- Tolerance analysis chart
- Zero tolerance at MMC

## What This Chapter Is About

This is the third of four chapters about position tolerances. In this chapter, we'll look at position tolerances in a selection of applications. This chapter highlights the flexibility of the position tolerance and explains how to interpret position tolerance applications that are common in industry.

# TERMS AND CONCEPTS, APPLICATIONS, AND VERIFICATION PRINCIPLES

This chapter covers a selection of position tolerance concepts and applications. To make the applications easier to learn, the application and verification information is combined with the terms and concepts.

## Projected Tolerance Zone

The fundamental rules in Y14.5 specify that all tolerances apply to the full length, width, and depth of a feature of size. In the case of a hole, the tolerance zone applies for the full depth of the hole. In certain cases, the function of a part requires that a tolerance zone exist outside the hole. In these applications, a projected tolerance zone is used.

A *projected tolerance zone* is a tolerance zone that is projected (moved) outside of the feature of size. The height that the tolerance zone is projected must be specified. A projected tolerance zone exists wherever the projected tolerance zone modifier (see page 64) is specified in a feature control frame. Examples of a tolerance zone that applies inside the hole and a tolerance zone that is projected outside the hole are shown in Figure 22-1.

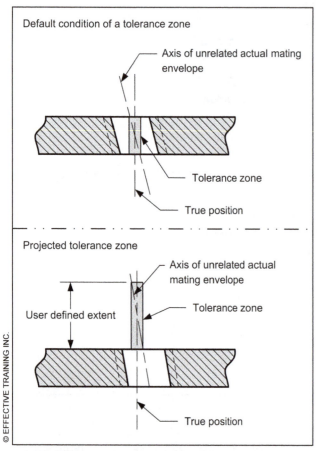

FIGURE 22-1 Tolerance Zone Examples (Projected and Not Projected)

## Real-World Applications of a Projected Tolerance Zone

A projected tolerance zone should be considered for use where the orientation deviation inside a position tolerance zone could result in interference with the mating part. One real world application is in bolted joint assemblies where the thickness of the part with the clearance hole is greater than the depth of the threaded hole.

In Figure 22-2A, without a projected tolerance zone, the orientation deviation of the fastener may result in an interference condition with the mating part. In Figure 22-2B, a projected tolerance zone is used restricting the orientation of the fastener (but not the location) to ensure assembly with the mating part.

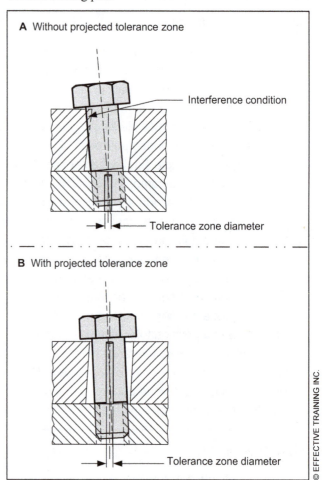

FIGURE 22-2 Impact of Projected Tolerance Zone on the Orientation of a Fastener

The use of a projected tolerance zone permits the use of the largest position tolerance while eliminating the risk of orientation deviation creating an interference condition. The height of a projected tolerance zone should be equal to the maximum height of the mating part. The axis of the hole must reside in the projected tolerance zone. Since the height of the projected zone is longer than the depth of the hole, the orientation deviation of the hole is more restricted than the location deviation for the hole.

***Design Tip***
A rule of thumb in bolted joint applications: wherever a hole has a fastener (or dowel pin) extended from the part surface, and the length of the extended portion is greater than the depth of the fastener inside the part, a projected tolerance zone modifier should be specified.

## Projected Tolerance Zone Interpretation

Figure 22-3 shows a position tolerance application using a projected tolerance zone. Note the use of the projected tolerance zone modifier and the specification of the projected tolerance zone height in the feature control frame. The height of the projected tolerance zone should be equal to the maximum thickness of the mating part. The projected tolerance zone is located at the true position of the hole.

***Design Tip***
A rule of thumb in bolted joint applications: wherever the height of the clearance hole is less than the depth of the threaded hole, a projected tolerance zone modifier is not recommended.

## How to Verify a Projected Tolerance Zone

A projected tolerance zone is easily verified using a gage pin threaded into the hole. The length of the pin protruding from the hole is equal to the projected height. The axis of the pin represents the axis of the unrelated actual mating envelope of the fastener passing through the mating part. The portion of the pin protruding out of the hole can be measured using a CMM. If the axis of the pin lies within the projected tolerance zone, the part passes; if not, the part fails.

A functional gage can also be used to verify the projected tolerance zone. The functional gage is designed with receiver holes that must fit around the gage pins that protrude from the threaded holes while resting on the primary datum feature.

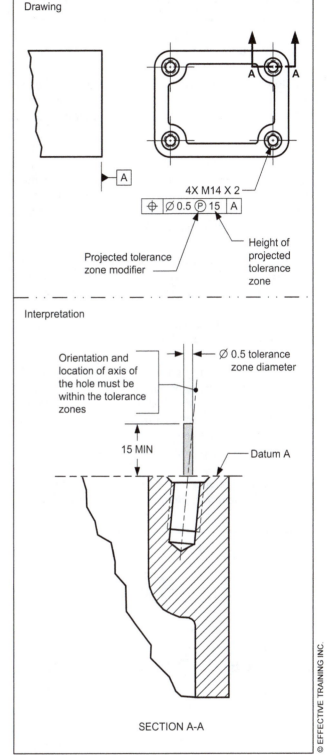

**FIGURE 22-3** *Projected Tolerance Zone Application*

© EFFECTIVE TRAINING INC.

## Multiple Single-Segment Position Tolerance

A *multiple single-segment position tolerance* is two or more single-segment position tolerances applied to the same feature of size or pattern of features of size.

A multiple single-segment position tolerance is used to define the location, orientation, and spacing of a pattern of features of size. The requirements for the location of the pattern and are different from those for the spacing and orientation of the pattern. The tolerance is specified by stacking two or more individual feature control frames with their left sides aligned, as shown in Figure 22-4. A multiple single-segment feature control frame typically contains two or three segments, but it may contain as many segments as needed to convey the design requirements.

Each segment of a multiple single segment position tolerance creates different tolerance requirements for location, spacing and / or orientation of a pattern. The uppermost segment is the pattern-locating tolerance. It specifies a location tolerance for the pattern of features of size relative to the datums references specified.

Each lower segment may create an additional location, orientation or spacing requirement on the pattern relative to the datums references specified. Each lower segment can control location or orientation of a pattern. All of the datums references of the lower segments cannot be identical to the datum feature references in the upper segment.

If datum feature references are not specified in a lower segment, the segment only controls spacing or alignment between the features of size in the pattern.

## Real-World Application of a Multiple Single-Segment Position Tolerance

A multiple single-segment position tolerance is often used where the function of the part requires a tighter tolerance for the spacing and orientation of a pattern of features of size than it does for the location of the pattern. Applications may include mounting holes for:

1. Cover and spacer plates

2. Holes for switch, gauge, and light fixture mounting brackets

3. Mounting holes for alternators, compressors, oil pumps, filters, etc.

## Multiple Single-Segment Position Tolerance Interpretation

In Figure 22-5, the upper position tolerance controls the location of the hole pattern. The lower position tolerance controls the spacing between the holes and their perpendicularity to datum plane A. Each position tolerance is a separate requirement that must be inspected separately.

**Design Tip**

In multiple single-segment position tolerancing, the lower feature control frame may have one, two, or three datums referenced, and may have different datums, but it cannot repeat the same datum reference frame from the upper feature control frame.

## Composite Position Tolerance

A pattern of features of size often requires several levels of control for location, spacing, alignment, and / or orientation. These multiple levels of control may be created with multiple single-segment position tolerances or with composite position tolerances. This section introduces how to specify these multiple levels of controls using composite position tolerances.

Composite position tolerances look similar to multiple single-segment position tolerances, but they have different interpretations and requirements. Composite position tolerances are a complex topic and are only introduced in this text.

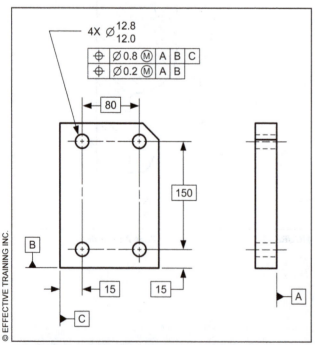

*FIGURE 22-4 Multiple Single-Segment Position Tolerance*

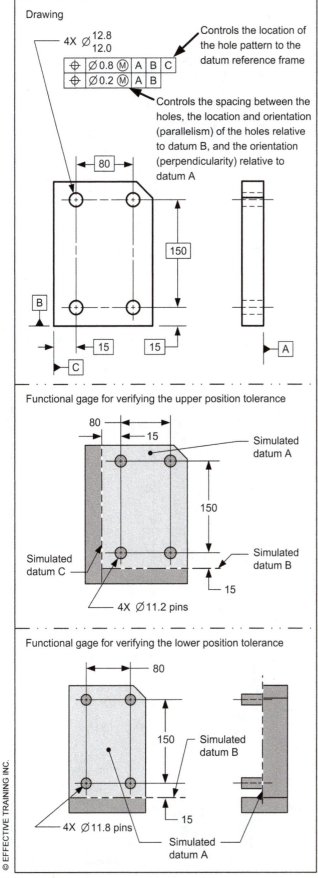

Drawing

4X Ø 12.8 / 12.0

| ⊕ | Ø0.8 Ⓜ | A | B | C |
| ⊕ | Ø0.2 Ⓜ | A | B |

Controls the location of the hole pattern to the datum reference frame

Controls the spacing between the holes, the location and orientation (parallelism) of the holes relative to datum B, and the orientation (perpendicularity) relative to datum A

80

150

B

15   15

C

A

Functional gage for verifying the upper position tolerance

80

15

Simulated datum A

150

Simulated datum C

Simulated datum B

15

4X Ø11.2 pins

Functional gage for verifying the lower position tolerance

80

150   Simulated datum B

15

4X Ø11.8 pins

Simulated datum A

**FIGURE 22-5** *Multiple Single-Segment Position Tolerance Interpretation*

A *composite position tolerance* is a feature control frame that contains multiple segments with a single entry of a position tolerance symbol that is applicable to all horizontal segments of the feature control frame. See Figure 22-6. Each horizontal segment has rules for how to specify the tolerance value, modifiers, and datum feature references. A composite position feature control frame typically contains two or three segments, but it may contain as many segments as needed to convey the design requirements.

A composite position tolerance can only be applied to patterns of features of size. Each segment creates different tolerance requirements for location, spacing and / or orientation of the pattern. The uppermost segment is the pattern-locating tolerance. It specifies a location tolerance for the pattern of features of size relative to the datums references specified. Depending upon the datums referenced, the uppermost segment can constrain translation and rotational degrees of freedom.

Each lower segment creates an orientation or spacing requirement on the pattern relative to the datums references specified. The lower segments cannot control location, they can only constrain rotational degrees of freedom. Therefore, the lower segments can only control orientation, alignment, and spacing of a pattern,

If datum references are not specified in a lower segment, the segment only controls spacing or alignment between the features of size in the pattern.

A major difference between multiple single-segment and composite position tolerances is shown below:

- In a multiple single-segment positions tolerance, each segment can constrain translational and rotational degrees of freedom (i.e., control location).

- In a composite position tolerance, only the uppermost segment can constrain translational and rotational degrees of freedom. The lower segments can only control rotational degrees of freedom (i.e., orientation only).

**Author's Comment**
This text only introduces the topic of composite position tolerances. To learn about the requirements and limitations, see ETI's *Advanced Concepts of GD&T (2009)* textbook.

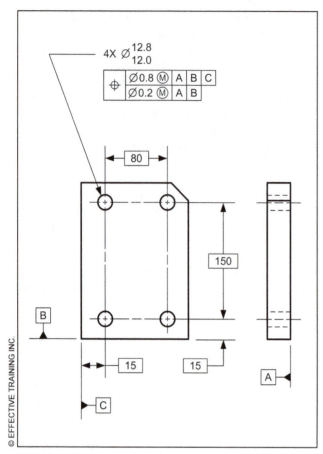

*FIGURE 22-6 Composite Position Tolerance*

## Real-World Applications of a Composite Position Tolerance

A composite position tolerance is often used on a pattern of features of size where the function of the part allows a larger location tolerance for the pattern but requires a tighter tolerance for the spacing and orientation (usually to several datums). A composite position tolerance is also used to control alignment between coaxial features of size to a tighter tolerance than their location or orientation. Applications may include:

1. Name plate mounting holes to control parallelism between the edge of the name plate and the edge of the mating part.

2. Alignment between bearing bores for camshafts, crankshafts, bushing bores for linear actuators, etc.

3. Alignment between pivot holes, hinge pin holes, etc.

## Composite Position Tolerance Interpretation

In Figure 22-7, the uppermost segment of the composite position tolerance controls the location of the hole pattern. The lower segment of the position tolerance controls the spacing between the holes and their orientation to datums A and B. Notice that when verifying the lower segment, only the rotational degrees of freedom are constrained relative to datums A and B. Simply put, the basic dimensions relative to the datums do not apply to the lower segment.

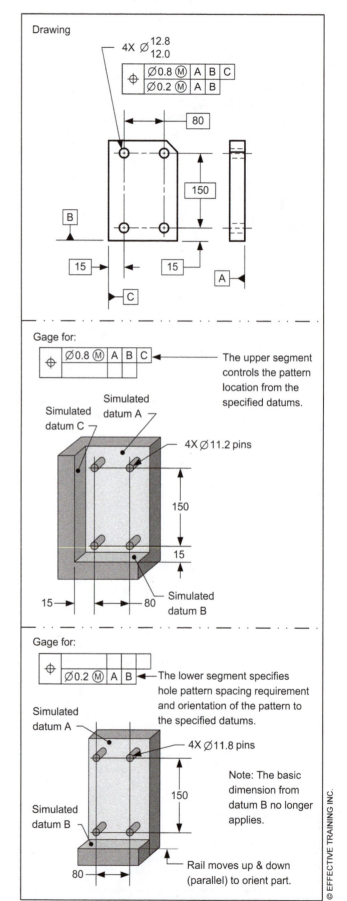

*FIGURE 22-7 Composite Position Tolerance Interpretation*

## Position Tolerance (MMC) Applied to a Pattern of Holes at Any Orientation

In certain cases, the function of a part may require that a pattern of holes that are not parallel or perpendicular to the datums be located and oriented relative to a datum reference frame. Figure 22-8 shows an example.

In this example, the following conditions apply:

- The tolerance zones are four virtual condition cylinders of $\emptyset 5.6 = (6.0 - 0.4)$.

- The virtual conditions are at a basic 30° to datum plane B.

- The virtual conditions are located at the true position of each hole.

- The surface of each hole must not violate its virtual condition.

- The virtual condition limits the location, orientation, and spacing between the holes.

- Maximum bonus is $0.2 = (6.2 - 6.0)$.

- Datum feature shift is not allowed (datum A referenced RMB).

- Rule #1 applies to each hole.

- Each hole must be within its size limits.

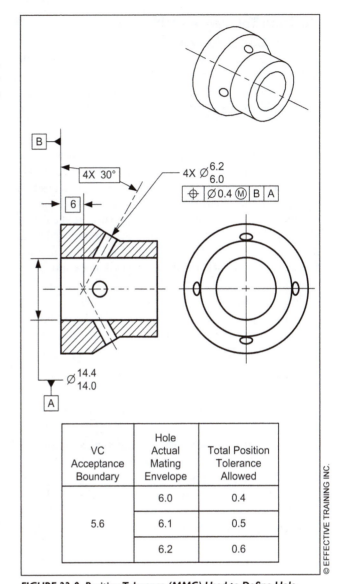

| VC Acceptance Boundary | Hole Actual Mating Envelope | Total Position Tolerance Allowed |
|---|---|---|
| 5.6 | 6.0 | 0.4 |
| | 6.1 | 0.5 |
| | 6.2 | 0.6 |

© EFFECTIVE TRAINING INC.

**FIGURE 22-8  Position Tolerance (MMC) Used to Define Hole Pattern Location and Orientation**

## Position Tolerance (MMC) Used as a Bidirectional Tolerance

In certain cases, the function of a part may require a hole to have less location tolerance in one direction and more in another direction. In these cases, a cylindrical tolerance zone would not work, and a bidirectional position tolerance is used.

A *bidirectional position tolerance* is where the location tolerance of a hole is defined with different tolerance values in two directions. This is accomplished by using two position tolerance feature control frames to indicate the direction and magnitude of each location tolerance relative to the datum specified. Bidirectional tolerances result in noncylindrical tolerance zones. The diameter symbol is not shown in the feature control frames. Figure 22-9 shows an example.

In this example, the following conditions apply:

- The axis tolerance zones are 0.2 vertical x 0.6 horizontal rectangles at MMC.
- The virtual condition is 15.4 vertical x 15 horizontal.
- The virtual conditions are perpendicular to datum plane A.
- The virtual conditions are located at true position of each hole.
- The surface of each hole must not violate its virtual condition.
- The virtual condition (acceptance boundary) limits the location, orientation, and spacing between the hole.
- Maximum bonus is 0.4 = (16.0 - 15.6).
- Datum feature shift is not allowed (datum B referenced RMB).
- Rule #1 applies to each hole.
- Each hole must be within its size limits.

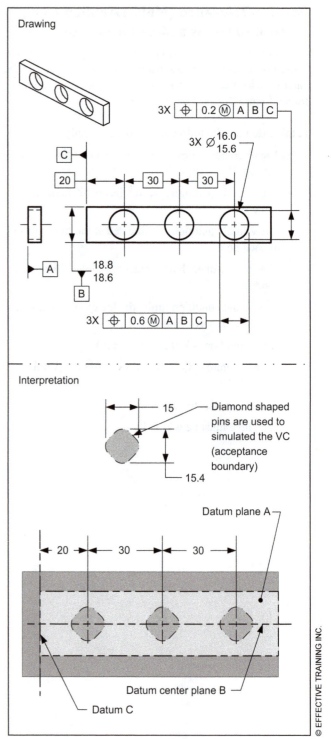

FIGURE 22-9  *Bidirectional Position Tolerances*

© EFFECTIVE TRAINING INC.

## Position Tolerance (MMC) Applied to Elongated (Slotted) Holes

A position tolerance may be applied to irregular features of size, such as elongated holes. Where a position tolerance is applied to an elongated hole, it is often applied at MMC. When tolerancing an elongated hole, two position tolerances are often used: one for the length and one for the width. The feature control frames are attached to the dimensions lines to create a bidirectional tolerance. A bidirectional tolerance results in noncylindrical tolerance zones. The diameter symbol is not shown in the feature control frame. Figure 22-10 shows an example.

In this example, the following conditions apply:

- The acceptance boundaries are 15.4 vertical x 15 horizontal virtual condition.

- The virtual conditions are perpendicular to datum plane A.

- The virtual conditions are located at the true position of each hole.

- The surface of each hole must not violate its virtual condition.

- The virtual condition limits the location, orientation, and spacing between the holes.

- Maximum location deviation for the elongated holes is 0.6 (±0.3) in the Y direction when their width is at LMC.

- Maximum location deviation for the elongated holes is 1.6 (±0.8) in the X direction when their length is at LMC.

- Datum feature shift is not allowed.

- Rule #1 applies to the length and width of each hole.

- Each hole must be within its size limits for length and width.

Where that interpretation of the position tolerance zone may not be clear for all drawing users, the keyword "BOUNDARY" may be specified along with a position tolerance.

*"BOUNDARY"* is a keyword used with a position tolerance on to denote that only a boundary interpretation exists for the position tolerance. There is no axis interpretation.

The BOUNDARY notation is a practice from prior versions of Y14.5. The 2009 edition of Y14.5 states that the use of the MMC modifier indicates that the boundary interpretation applies.

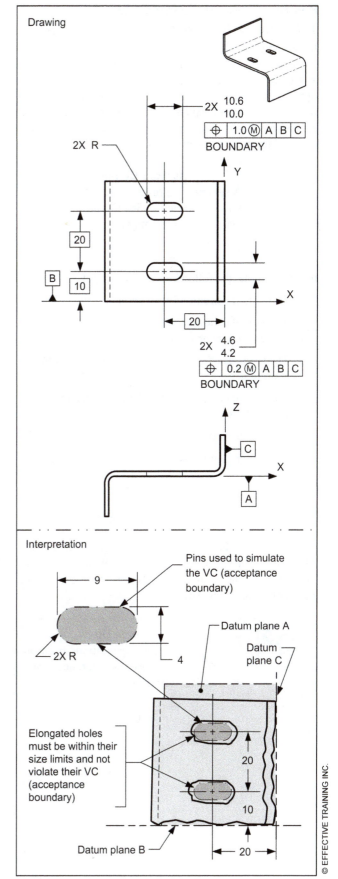

*FIGURE 22-10 Position Tolerance Applied to an Elongated Hole*

## Position Tolerance (MMC) Used to Tolerance a Symmetrical Relationship

In certain cases, the function of a part may require that the location deviation for a symmetrical relationship be defined. A symmetrical relationship is where the location deviation of the center plane of a feature of size is being defined relative to a datum center plane. In these cases, a position tolerance may be used to limit the offset between the center planes. Figure 22-10 shows an example.

**Author's Comment**
A symmetrical relationship is not the same as symmetry. See Chapter 25 for information on symmetry.

In this example, the following conditions apply:

- The virtual condition is two parallel planes 12 apart.

- The virtual condition is centered (basic zero offset) to the datum center plane.

- The surfaces of the slot must not violate its virtual condition.

- The virtual condition limits the location, and orientation of the slot.

- Max bonus is 0.4 = (12.6 - 12.2).

- Max datum feature shift is 0.4 = (18.6 - 18.2) (at MMB).

- Rule #1 applies to the width of the slot.

- The slot must be within its size limits.

The example in Figure 22-11 uses a position tolerance at MMC and MMB. The use of a position tolerance to control symmetrical relationships may also be specified at LMC, LMB, or RFS, and RMB.

**Design Tip**
In many industrial applications, position tolerances are used to tolerance symmetrical relationships.

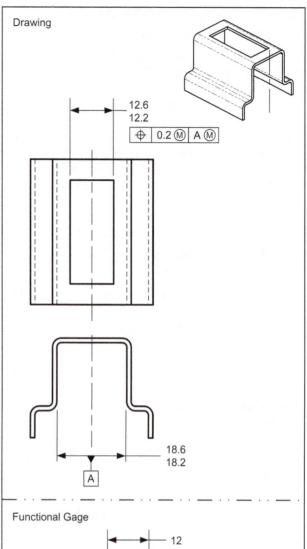

Drawing

12.6
12.2

⌖ 0.2 Ⓜ A Ⓜ

18.6
18.2

A

Functional Gage

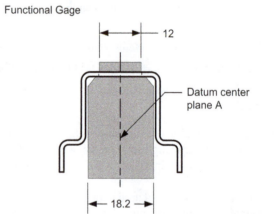

12

Datum center plane A

18.2

| Total Position Tolerance Allowed | | | |
|---|---|---|---|
| Slot Actual Mating Envelope | Datum Feature Actual Mating Envelope | | |
| | 18.2 | 18.4 | 18.6 |
| 12.2 | 0.2 | 0.4 | 0.6 |
| 12.4 | 0.4 | 0.6 | 0.8 |
| 12.6 | 0.6 | 0.8 | 1.0 |

**FIGURE 22-11  Using a Position Tolerance at MMC to Define a Symmetrical Relationship**

## Position Tolerance at LMC Applications

In certain cases, the function of a part may require a minimum distance of a part to be maintained. The LMC modifier is used in a position tolerance when the functional consideration is to tolerance a minimum distance on a part. The minimum distance can be a minimum wall thickness, a minimum part distance, or the minimum machine stock on a casting.

### Author's Comment
Caution: when specifying the LMC modifier, the effects of Rule #1 are reversed. Perfect form at LMC is required. Study the applicable sections in Y14.5 before using this concept.

The bonus tolerance for a position tolerance at LMC works the opposite of the bonus for MMC. Where the LMC modifier is used in a position tolerance, the stated tolerance applies (no bonus) when the feature of size is at LMC. A bonus tolerance is permitted when the feature of size departs from LMC towards MMC. The maximum bonus applies at MMC. Figure 22-12 shows an example of a position tolerance with the LMC modifier.

In this example, the following conditions apply:

- The tolerance zone is a virtual condition (inner) boundary cylinder of $\phi24.0 = (24.2$ LMC - 0.2).
- The VC is centered about the datum axis.
- The datum axis is the axis of the internal feature of size (RMB).
- The surface of the O.D. must not violate the LMC virtual condition (inner) boundary.
- The virtual condition limits location and orientation.
- Max bonus is 0.6 = (24.8 - 24.2) (at MMC).
- Datum feature shift is not allowed (datum A referenced RMB).
- Perfect form at LMC applies to the O.D. (invoked by position LMC).
- Perfect form at MMC applies to datum feature A.
- All features of size must be within their size limits.

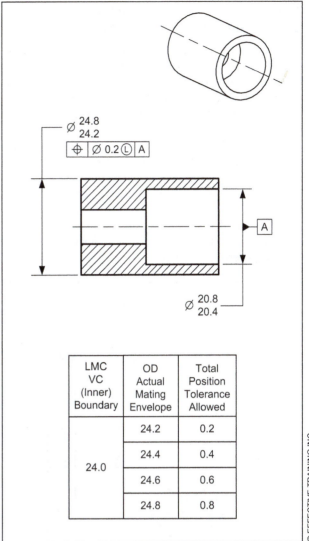

| LMC VC (Inner) Boundary | OD Actual Mating Envelope | Total Position Tolerance Allowed |
|---|---|---|
| | 24.2 | 0.2 |
| 24.0 | 24.4 | 0.4 |
| | 24.6 | 0.6 |
| | 24.8 | 0.8 |

**FIGURE 22-12** *Position Tolerance Using the LMC Modifier*

## Position Tolerance Used to Control Spacing and Orientation of a Hole Pattern

In certain cases, the function of a part may require controlling the spacing and orientation of a hole pattern, but not its location on a part. This can be accomplished by using a position tolerance with a single datum feature reference (the mounting surface), as illustrated in Figure 22-13.

In this figure, the position tolerance limits the spacing between the holes and their perpendicularity to datum plane A (the mounting surface), but it does not tolerance the location of the pattern. Often, when this type of position tolerance is used, the hole pattern is designated as the secondary datum feature that locates the part. The outside edges and the center hole of the part would be toleranced relative to the hole pattern (datum feature B).

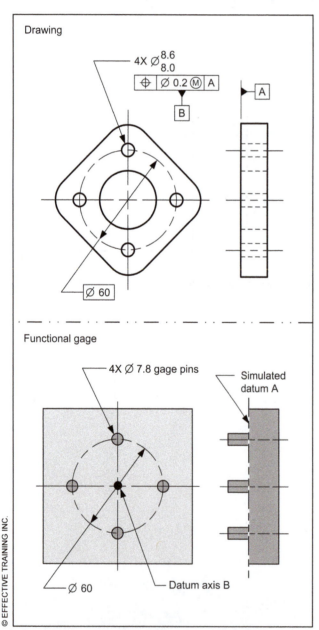

*FIGURE 22-13  Position Tolerance at MMC Used to Define Spacing and Location of a Hole Pattern*

In this example, the following conditions apply:

- The tolerance zones are four virtual condition cylinders of ⌀7.8 = (8.0 - 0.2).
- The virtual conditions are perpendicular to datum plane A.
- The virtual conditions are located at the true position of each hole.
- The surface of each hole must not violate their virtual condition.
- The virtual condition limits the orientation and spacing between the holes.
- Max bonus is 0.6 = (8.6 - 8.0).
- Datum feature shift is allowed. The datum feature shift is the same as the bonus because the position tolerance invokes an implied self-datum for the bolt holes. The part can shift on the gage to get all four holes to fit their gage pins.
- Rule #1 applies to each hole.
- Each hole must be within its size limits.

A position tolerance with a single datum feature reference is used in two applications:

1. Where a hole pattern is used as a secondary datum and the position tolerance controls spacing between the holes and their relationship to the primary datum feature (usually the mounting surface).

2. A hole pattern toleranced with a multiple single-segment or composite position tolerance where the lower segment controls the spacing between the holes and their orientation to the primary datum feature.

> ***Author's Comment***
> Using a pattern of features of size as a datum is a common practice when functionally dimensioning a part. In Figure 22-13, a profile tolerance could be used to relate the outside edges of the part to the hole pattern. However, this concept is not well understood and is often controversial.

## Zero Tolerance at MMC

How the tolerances are specified on a part affects the part function and manufacturing costs. Tight tolerances do not guarantee a quality part, only an expensive one. A dimensioning method that protects the part function and helps to reduce cost is called "zero tolerance at MMC."

Zero tolerance at MMC appears restrictive to anyone not familiar with the benefits and concepts of GD&T. However, once the method is understood, it becomes apparent that zero tolerance at MMC protects the part function and offers the maximum flexibility for manufacturing.

*Zero tolerance at MMC* is a dimensioning method that takes the tolerance stated in the feature control frame, combines it with the size tolerance, and states a zero tolerance in the feature control frame.

The zero tolerance at MMC method should be used wherever the function of the toleranced feature is assembly.

## Benefits of Zero Tolerance at MMC

There are three primary benefits of using the zero tolerance at MMC method:

1. Provides flexibility for manufacturing

2. Prevents the rejection of usable parts

3. Reduces manufacturing costs

The benefits of zero tolerance at MMC can have a significant impact in an organization. The use of zero tolerance at MMC, while still misunderstood in some companies, is considered a best practice by many other companies.

**Design Tip**
All of the benefits of zero tolerance at MMC are also available with the following geometric tolerances: straightness, flatness, parallelism, perpendicularity and angularity.

## Real-World Applications of Zero Tolerance at MMC

The zero tolerance at MMC method should be considered for use where ever the function of the toleranced feature(s) of size is assembly.

Zero tolerance at MMC is common where the function is assembly. Some examples are:

• Clearance holes for fasteners

• Mating features of size

## Zero Tolerance at MMC Application

Figure 22-14A shows a part with a conventional position tolerance of 0.3 at MMC. Figure 22-14B shows the same part dimensioned with zero tolerance at MMC. The 0.3 tolerance has been removed from the feature control frame and combined with the size tolerance. Note that the functional parameters (virtual condition 8.2) is the same for both parts. With zero tolerance at MMC, all of the hole location tolerance is derived from the size (bonus) tolerance, so manufacturing can divide the available tolerance between size and position to best suit the capability of the manufacturing process.

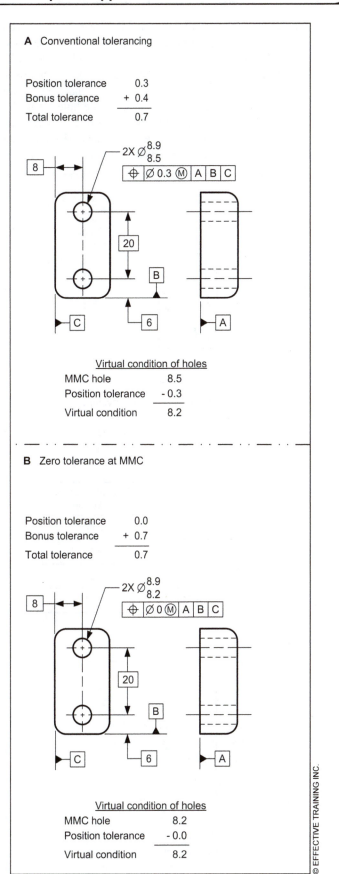

**A** Conventional tolerancing

| | |
|---|---|
| Position tolerance | 0.3 |
| Bonus tolerance | + 0.4 |
| Total tolerance | 0.7 |

Virtual condition of holes

| | |
|---|---|
| MMC hole | 8.5 |
| Position tolerance | - 0.3 |
| Virtual condition | 8.2 |

**B** Zero tolerance at MMC

| | |
|---|---|
| Position tolerance | 0.0 |
| Bonus tolerance | + 0.7 |
| Total tolerance | 0.7 |

Virtual condition of holes

| | |
|---|---|
| MMC hole | 8.2 |
| Position tolerance | - 0.0 |
| Virtual condition | 8.2 |

**FIGURE 22-14** *Comparison of Conventional Position Tolerancing at MMC and Zero Tolerance at MMC*

The effects of zero tolerance at MMC can be visualized through the use of a tolerance analysis chart. A *tolerance analysis chart* is a chart that graphically displays the available tolerances of a feature of size as defined by the drawing specifications. The tolerance analysis charts in Figure 22-15 show the available tolerance for the holes from the parts in Figure 22-14A and B. On the vertical scale, the allowable position tolerances are listed. The horizontal scale lists the hole sizes and the virtual condition.

In Figure 22-15A, the shaded area of the chart represents the acceptable parts, according to the drawing specifications. The dots labeled "1 through 4" represent the measured hole sizes and hole locations of four parts. Three of the four parts would be rejected on the basis of the hole sizes not being within the drawing specifications. Even though the parts would function, the parts are rejected because the holes do not violate their virtual conditions.

In Figure 22-15B, the shaded area of the chart represents the acceptable parts according to the drawing specifications. The dots labeled "1 through 4" represent the measured hole sizes and hole locations of the same four parts. Note that all four parts are now acceptable. The zero tolerance at MMC dimensioning method increases the zone of acceptable parts by making the MMC and virtual condition of the hole the same value. The zero tolerancing at MMC method results in all functional parts being accepted.

When dimensioning holes, if their function is assembly, the designer should consider using the zero tolerance at MMC method. With this method of dimensioning, functional parts are not rejected, and more flexibility for manufacturing results in lower costs.

**Design Tip**

When dimensioning a part, zero tolerance at MMC should be considered wherever the function of a feature of size is assembly.

The chart in Figure 22-16 shows a comparison of parts accepted and rejected using conventional tolerancing and zero tolerancing at MMC.

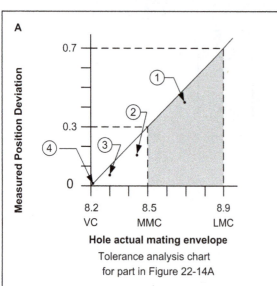

| Part | Hole Actual Mating Envelope | Measured Position Deviation |
|------|------|------|
| 1 | 8.69 | 0.42 |
| 2 | 8.45 | 0.15 |
| 3 | 8.30 | 0.05 |
| 4 | 8.21 | 0.01 |

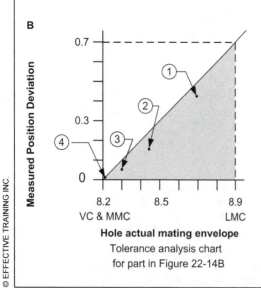

FIGURE 22-15 Tolerance Analysis Chart of Measured Position Deviations From the Parts in Figure 22-14

| Measured Hole Actual Mating Envelope | Measured Position Deviation | A ⌖ ⌀0.3Ⓜ A B C | B ⌖ ⌀0Ⓜ A B C |
|------|------|------|------|
| 8.2 | 0.0 | Rejected | Accepted |
| 8.3 | 0.1 | Rejected | Accepted |
| 8.4 | 0.2 | Rejected | Accepted |
| 8.5 | 0.3 | Accepted | Accepted |
| 8.6 | 0.4 | Accepted | Accepted |
| 8.7 | 0.5 | Accepted | Accepted |
| 8.8 | 0.6 | Accepted | Accepted |
| 8.9 | 0.7 | Accepted | Accepted |

FIGURE 22-16 Comparison of Accepted/Rejected Parts From Figure 22-14

## SUMMARY
## Key Points

- A projected tolerance zone is a tolerance zone that is projected (moved) outside the feature of size.

- One real world application of using a projected tolerance zone is in bolted joint assemblies where the thickness of the part with the clearance hole is greater than the depth of the threaded hole.

- A multiple single-segment position tolerance is two or more single-segment position tolerances applied to the same feature of size or pattern of features of size.

- A composite position tolerance is a feature control frame that contains multiple segments with a single entry of a position tolerance symbol that is applicable to all horizontal segments of the feature control frame.

- Bidirectional position tolerancing is where two different position tolerance values are applied to the same hole in two directions.

- When tolerancing an elongated hole, two position tolerances are often used: one for the length and one for the width.

- The term "BOUNDARY" indicates that only a boundary interpretation exists.

- A symmetrical relationship is where the center plane of a feature of size is located within a tolerance zone that is centered about a datum center plane.

- The bonus tolerance for a position tolerance at LMC works the opposite of a bonus tolerance at MMC.

- A position tolerance with a single datum feature reference is used in two applications:
  o Where a hole pattern is used as a secondary datum, and the position tolerance controls spacing between the holes and their orientation to the primary datum feature
  o A hole pattern toleranced with a multiple single-segment or composite position tolerance where the lower segment controls the spacing between the holes and their orientation to the primary datum feature

- Zero tolerance at MMC is a dimensioning method that takes the geometric tolerance usually stated in the feature control frame, combines it into the size tolerance, and states a zero tolerance at MMC in the feature control frame.

- There are three primary benefits of using the zero tolerance at MMC method:
  o Provides flexibility for manufacturing
  o Prevents the rejection of usable parts
  o Reduces manufacturing costs

- A tolerance analysis chart is a chart that graphically displays the available tolerances of a feature of size as defined by the drawing specifications.

## Additional Related Topics

*These topics are recommended for further study to improve your understanding of advanced position applications.*

| Topic | Source |
|---|---|
| Simultaneous requirements | *ASME Y14.5-2009, Para. 7.5.4* |
| Position tolerances at LMC | *ASME Y14.5-2009, Para. 7.3.5* |
| Composite position tolerances | *ASME Y14.5-2009, Para. 7.5.1* |
| Comparing composite and multiple single-segment position tolerancing | ETI's *Advanced Concepts of GD&T* textbook |

## QUESTIONS AND PROBLEMS

**Website Bonus Materials**
Additional questions are available at our website. To access bonus materials for this textbook, please visit:
www.etinews.com/textbookbonus

## True and False

*Indicate if each statement is true or false.*

T / F    1. A projected tolerance zone is a tolerance zone that is projected (moved) outside of the feature of size.

T / F    2. When using a projected tolerance zone, the projected tolerance zone modifier and value for the height of the projected zone may be placed inside the feature control frame.

T / F    3. Each segment of a multiple single-segment position tolerance creates a separate requirement for location, orientation, or spacing of a hole pattern.

T / F    4. When two individual position tolerance feature control frames are stacked with their left sides aligned, they are a composite position tolerance.

T / F    5. Using a zero tolerance at MMC increases manufacturing costs.

T / F    6. When interpreting a position tolerance with zero tolerance at MMC, the allowable tolerance is equal to the amount the toleranced feature of size departs from LMC.

T / F    7. A multiple single-segment position tolerance may have five segments.

T / F    8. A composite position tolerance may have three segments.

## Multiple Choice

*Circle the best answer to each statement.*

1. When applying a position tolerance to a threaded hole, the projected tolerance zone modifier should be used if:
   A. The MMC modifier is specified.
   B. The thickness of the mating part (with the clearance hole) is greater than the depth of the threaded hole.
   C. The hole is too deep to measure the full extent of the tolerance zone.
   D. The function is assembly.

2. When a position tolerance is applied with the projected tolerance zone modifier:
   A. The tolerance zone is projected into the part.
   B. The tolerance zone is projected above the part.
   C. A bonus tolerance is permissible.
   D. It increases the allowable orientation tolerance.

3. When a position tolerance contains the LMC modifier:
   A. The design intent is to protect a minimum distance.
   B. The maximum bonus occurs at LMC.
   C. A functional gage must be used to verify the position tolerance.
   D. It results in zero tolerance at MMC.

4. A zero tolerance at MMC should be used when:
   A. The function of a feature of size is a sliding fit.
   B. The function of a feature of size is assembly.
   C. The design cannot permit a tolerance.
   D. The manufacturing costs are too high.

5. A benefit of zero at MMC is:
   A. Manufacturing can select the tolerance.
   B. Zero tolerance at MMC allows flexibility for manufacturing.
   C. The tolerance is obtained from the general tolerances for the drawing.
   D. The part will have less tolerance permitted.

6. A _____ tolerance is often applied where the function of a part can permit a larger tolerance for the location of a hole pattern, but requires a smaller tolerance for the orientation and spacing of the hole pattern.
   A. Position
   B. Composite position
   C. Bidirectional
   D. Coordinate

## Application Problems

*The application problems are designed to provide practice applying the chapter concepts to situations that are similar to on-the-job conditions.*

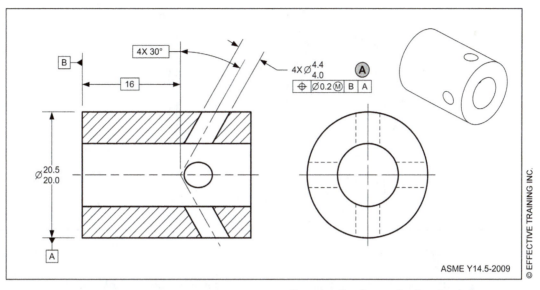

ASME Y14.5-2009

*Question 1 refers to the drawing above.*

1. Use the Significant Seven Questions to interpret the position tolerance labeled "A."

   1) Which dimensioning and tolerancing standard applies? _____

   2) What does the tolerance apply to? _____

   _____

   3) Is the specification standard-compliant? _____

   4) What are the shape and size of the tolerance zone?

   _____

   5) How much total tolerance is permitted? _____

   6) What are the shape and size of the datum simulators? _____

   7) Which geometry attributes are affected by this tolerance? _____

*Question 2 refers to the drawing below.*

2. Use the Significant Seven Questions to interpret the position tolerance labeled "A."

   1) Which dimensioning and tolerancing standard applies? _____

   2) What does the tolerance apply to? _____

   _____

   3) Is the specification standard-compliant? _____

   4) What are the shape and size of the tolerance zone?

   _____

   5) How much total tolerance is permitted? _____

   6) What are the shape and size of the datum simulators? _____

   7) Which geometry attributes are affected by this tolerance? _____

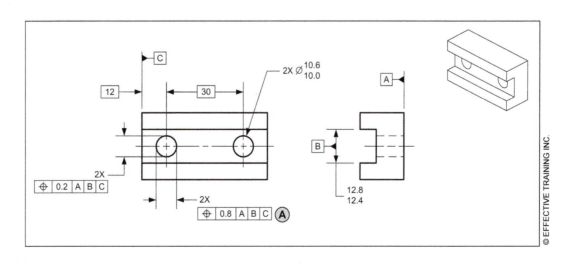

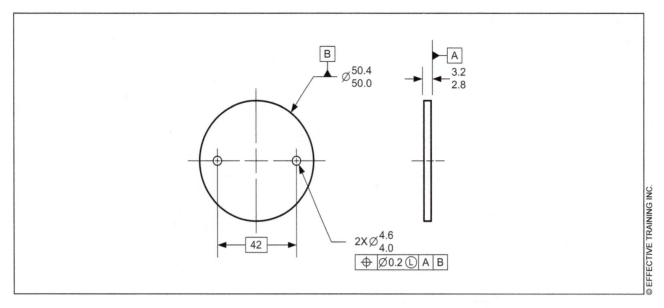

*Questions 3 & 4 refer to the drawing above.*

3.  When the holes are at MMC, how much bonus tolerance is permissible for the position tolerance? _____

   _____

4.  When the holes are at MMC, how much total tolerance is permissible for the position tolerance? _____

   _____

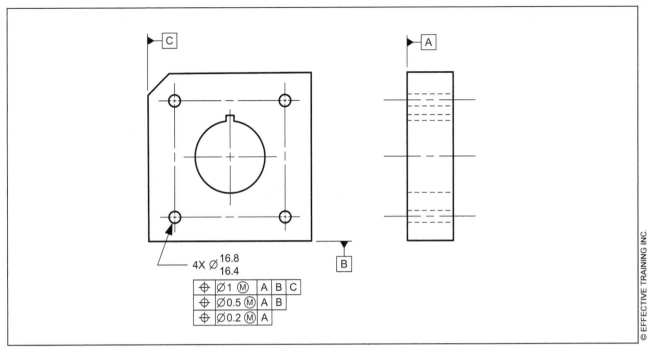

*Question 5 refers to the drawing above.*

5. Fill in the maximum allowable deviations for the conditions described. (Consider the part to be at MMC.)

   A. Location deviation relative to datum C _____

   B. Location deviation relative to datum B _____

   C. Orientation deviation relative to datum A _____

   D. Spacing deviation between the holes _____

# Position Tolerance Calculations

## Goal

Calculate position tolerance values using the fixed and floating fastener formulas

## Performance Objectives

Upon completing this chapter, you should be able to:

1. Describe a fixed fastener assembly (p.316)
2. State the general fixed fastener formula for calculating position tolerance values (with equal distribution) (p.316)
3. Calculate the position tolerance values for a fixed fastener application using the general fixed fastener formula (p. 316)
4. State the modified fixed fastener formula for calculating position tolerance values (with unequal distribution) (p.317)
5. Calculate the position tolerance values for a fixed fastener application using the modified fixed fastener formula (p.317)
6. Describe a floating fastener assembly (p.318)
7. State the formula for calculating position tolerance values for floating fasteners assemblies (p.318)
8. Calculate the position tolerance values for floating fastener applications (p.318)
9. Describe the limitations of the fixed and floating fastener formulas (p.319)

## New Terms

- Fixed fastener assembly
- Floating fastener assembly
- Floating fastener formula

- General fixed fastener formula
- Modified fixed fastener formula

## What This Chapter Is About

This is the last of four chapters about position tolerances. In this chapter, we'll look at the calculations for determining position tolerance values in bolted joint assemblies. When designing bolted joint assemblies, the fixed and floating fastener formulas are convenient design tools. They allow a designer to quickly determine the tolerance values in a position callout for mating parts.

This chapter explains how to calculate position tolerance values using the fixed and floating fastener formulas. We will cover the fastener assembly types and the formulas for each type.

## TERMS AND CONCEPTS

### Fixed Fastener Assembly

A *fixed fastener assembly* is an assembly of two or more parts where a fastener is held in place in (threaded or pressed into) one of the components of the assembly. Often, the holes in one component of the assembly are clearance holes, and the holes in the other component are threaded holes (or a press fit, like a dowel pin). This type of assembly is called a fixed fastener assembly because the fastener is "fixed" in the assembly. An example of a fixed fastener assembly is shown in Figure 23-1.

The components are assembled with four M14 screws. The cover has four clearance holes. The housing has four threaded holes. Both hole patterns are dimensioned with position tolerances.

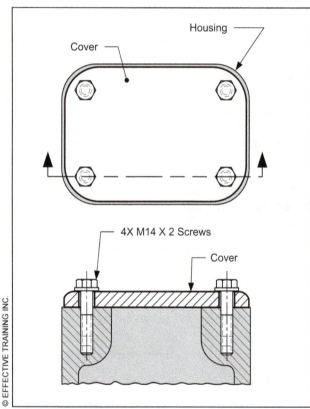

**FIGURE 23-1  Fixed Fastener Assembly**

The procedure for determining the amount of tolerance for the position tolerances in a fixed fastener application is a simple process. (The formula in this text applies when the projected tolerance zone modifier—explained in Chapter 22—is used on the position callout of the threaded holes.)

### General Fixed Fastener Assembly Formula

The *general fixed fastener formula* for equally distributed tolerances is:

$$T = \frac{H - F}{2}$$

Where:   T = total position tolerance for assembly
         H = MMC of the clearance hole
         F = MMC of the fastener

The MMC of the fastener is equal to the stated thread major diameter for the clearance holes. Since the function of the holes is assembly, the MMC modifier is used in the tolerance portion of the position feature control frame, allowing additional position tolerance as the clearance holes depart from MMC. Since a screw thread is a self-centering feature for the threaded holes, the position tolerance is specified at RFS.

## Calculating Position Tolerance Values Using the General Fixed Fastener Formula

Figure 23-2 shows an example of using the fixed fastener formula for determining the tolerance values in position callouts using the cover and housing from the assembly in Figure 23-1. Using the fixed fastener formula, the total tolerance for the clearance holes and the threaded holes is 0.4. Divided equally, the tolerance for each set of holes is 0.2.

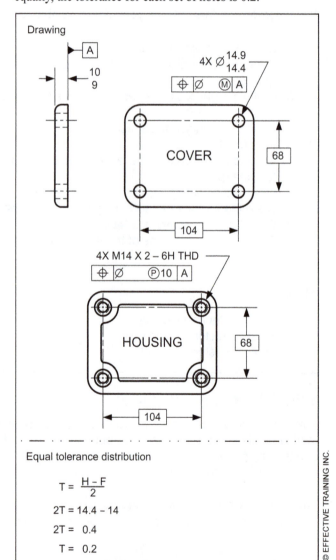

**FIGURE 23-2  Calculating Position Tolerance Values Using the General Fixed Fastener Formula**

**For More Info...**
The formulas that apply when the projected tolerance zone modifier is not specified are shown in Y14.5 Appendix B.

## Modified Fixed Fastener Formula

In some cases, it is desired to have an unequal distribution of the tolerances between the clearance holes and the threaded holes. Where the mating parts have a different tolerance distribution, the modified fixed fastener formula is used.

The *modified fixed fastener formula* for unequally distributed tolerances is:

$$H - F = T_1 + T_2$$

Where: H = MMC of the clearance hole

F = MMC of the fastener

$T_1$ = tolerance for the threaded hole

$T_2$ = tolerance for the clearance hole

## Calculating Position Tolerance Values Using the Modified Fixed Fastener Formula

Figure 23-3 shows an example of using the general fixed fastener formula for determining the tolerance values in position callouts. The cover and housing are from the assembly shown in Figure 23-1. In this case, we will split the tolerances so that the threaded holes receive 75% and the clearance holes receive 25% of the total tolerance.

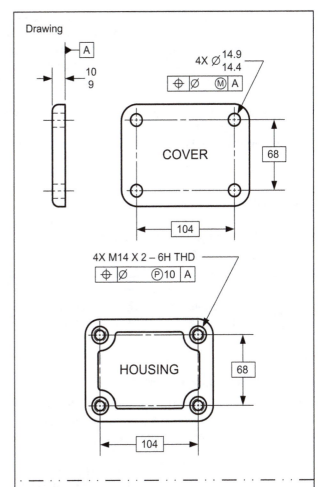

Unequal tolerance distribution
25% for clearance hole and 75% for threaded hole

Where:

H = MMC of the clerance hole

F = MMC of the fastener

$T_1$ = Positional tolerance for threaded hole

$T_2$ = Positional tolerance for clearance hole

$$H = F + T_1 + T_2$$
$$H - F = T_1 + T_2$$
$$14.4 - 14 = T_1 + T_2$$
$$0.4 = T_1 + T_2$$
$$T_1 = 0.75 \times 0.4 = 0.3$$
$$T_2 = 0.25 \times 0.4 = 0.1$$

© EFFECTIVE TRAINING INC.

*FIGURE 23-3 Calculating Position Tolerance Values Using the Modified Fixed Fastener Formula*

---

### TECHNOTE 23-1
### Fixed Fastener Assemblies

- A fixed fastener assembly is where a fastener is held in place (restrained) into one of the components of the assembly.

- The general fixed fastener formula (for equally distributed tolerance) is:

$$T = \frac{H - F}{2}$$

Where:  T = total position tolerance
H = MMC of the clearance hole
F = MMC of the fastener

- The modified fixed fastener formula (for unequally distributed tolerance) is:

$$H - F = T_1 + T_2$$

Where:  $T_1$ = tolerance for the threaded hole

$T_2$ = tolerance for the clearance hole

Where the fixed fastener formulas are used, the position tolerance values result in a zero clearance condition when the clearance holes are at MMC and located at the extremes of their VC boundaries and the threaded holes are located at their extremes within their position tolerance zones.

The examples shown in the figures specify the projected tolerance zone modifier. If the projected tolerance zone modifier is not specified, different formulas should be used.

When tolerancing a threaded hole, the projected tolerance zone modifier should be specified in the position tolerance feature control frame to indicate that the tolerance zone applies outside the hole where the fastener passes through the mating part. The projected tolerance zone modifier is explained fully in Chapter 22.

**For More Info...**
The formulas that apply when the projected tolerance zone modifier is not specified are shown in Y14.5, Appendix B.

## Floating Fastener Assembly

A *floating fastener assembly* is an assembly of two or more parts where two (or more) components are held together with fasteners (such as bolts and nuts), and both components have clearance holes for the fasteners. This type of assembly is called a floating fastener assembly because the fasteners can "float" (move) within the holes of each part of the assembly. An example of a floating fastener assembly is shown in Figure 23-4.

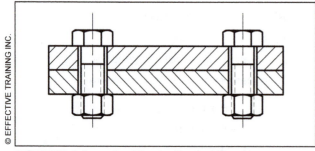

**FIGURE 23-4** *Example of Floating Fastener Assembly*

In this example, the plates are assembled with four M14 bolts and nuts. Both plates have the same size bolt clearance holes and use position to tolerance the hole locations.

The procedure for determining the tolerance values for the position tolerances in a floating fastener application is a simple process.

## Floating Fastener Formula

The *floating fastener formula* is:
$$T = H - F$$

Where: T = position tolerance zone diameter (for each part)
H = MMC of the clearance hole
F = MMC of the fastener

After the position tolerance zone value is calculated, it applies to each part in the assembly (if the clearance holes are the same hole size used in the formula). Since the function is assembly, the MMC modifier is used in the tolerance portion of the position callout. This allows additional location tolerance as a hole departs from MMC. Figure 23-5 shows an example of using the floating fastener formula for determining position tolerance values in a floating fastener application.

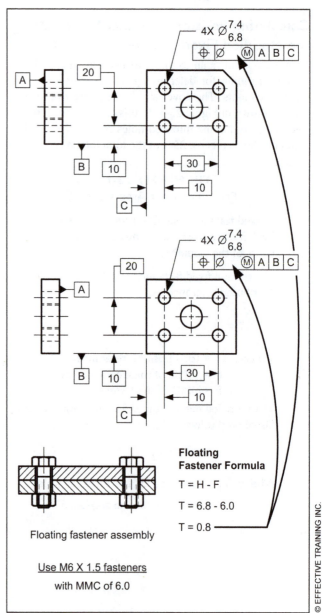

**FIGURE 23-5** *Floating Fastener Formula Example*

Where the floating fastener formula is used, the position tolerance values result in a zero clearance condition when the holes are at MMC and located at the extremes of their VC boundaries. Consideration should be given to additional geometric variations that are not accounted for in the floating fastener formula.

---

**TECHNOTE 23-2**
**Floating Fastener Assemblies**

- A floating fastener assembly is where two (or more) components are held together with fasteners (such as bolts and nuts), and both components have clearance holes for the fasteners.

- The floating fastener formula is: T = H - F

---

# FIXED AND FLOATING FASTENER APPLICATIONS

***Author's Comment***
The most cost effective method for dimensioning holes in a bolted joint application is to use a method called, "zero tolerance at MMC," as described in Chapter 22. After calculating the applicable position tolerance values, the zero tolerance at MMC method should be applied to the clearance holes to provide more flexibility and lower cost for manufacturing.

## Limitations of the Fixed and Floating Fastener Formulas

The fixed and floating fastener formulas are useful tools for a quick way to establish tolerance values. However, they do not account for the typical fastener deviations. Fasteners have two common types of variation: straightness deviation and thread to shank (body) runout (coaxiality). Often, fasteners are not perfectly straight, and threads are not perfectly coaxial to the shank diameter.

These deviations are shown in Figure 23-6. Each of these deviations can cause the fastener body to interfere with the surface of the clearance hole in a bolted assembly. Therefore, when using the fixed and floating fastener formulas, these bolt tolerances should be accounted for in the design.

***Design Tip***
When designing a bolted joint application using the fixed fastener formulas, consider the following:

- Does the position tolerance on the threaded hole contain the projected tolerance zone modifier?
- Have you considered the fastener tolerances?
- Have you considered zero tolerance at MMC?

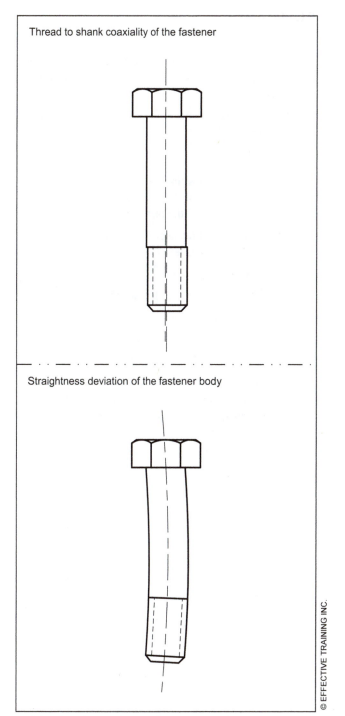

Thread to shank coaxiality of the fastener

Straightness deviation of the fastener body

© EFFECTIVE TRAINING INC.

***FIGURE 23-6 Typical Fastener Deviations***

***Author's Comment***
When using the fixed and floating fastener formulas, the results should be adjusted to account for fastener deviations.

## SUMMARY

### Key Points

- A fixed fastener assembly is where the fastener is held in place (restrained) in one of the components of the assembly.

- The general fixed fastener formula for equally distributed tolerance is:

$$T = \frac{H - F}{2}$$

Where: T = total position tolerance for assembly
H = MMC of the clearance hole
F = MMC of the fastener

- The modified fixed fastener formula for unequally distributed tolerances is:

$$H - F = T_1 + T_2$$

Where: H = MMC of the clearance hole

F = MMC of the fastener

$T_1$ = tolerance for the threaded hole

$T_2$ = tolerance for the clearance hole

- A floating fastener assembly is where two (or more) components are held together with fasteners (such as bolts and nuts), and both components have clearance holes for the fasteners.

- The floating fastener formula is:

$$T = H - F$$

Where: T = position tolerance (for each part)
H = MMC of the clearance hole
F = MMC of the fastener

- A limitation of the fixed and floating fastener formulas is that they do not account for the fastener tolerances such as:
  o Straightness deviation of the body
  o Thread to body runout

### Additional Related Topics

*These topics are recommended for further study to improve your understanding of the fixed and floating fastener formulas.*

| Topic | Source |
|---|---|
| Using the fixed fastener formula where the projected tolerance zone modifier is not specified | *ASME Y14.5-2009, Appendix B5* |
| Using position in tolerance stacks | ETI's *Tolerance Stacks Using GD&T* textbook |
| Calculating position tolerance values in complex relationships | ETI's *Tolerance Stacks Using GD&T* textbook |

## QUESTIONS AND PROBLEMS

***Website Bonus Materials***
Additional questions are available at our website. To access bonus materials for this textbook, please visit:
www.etinews.com/textbookbonus

## True and False

*Indicate if each statement is true or false*

T / F  1. In a fixed fastener assembly, the fastener is fixed in both components.

T / F  2. In a floating fastening assembly, the fastener is floating in both of the components.

T / F  3. An assembly of parts using nuts and bolts, is an example of a fixed fastener assembly.

T / F  4. The floating fastener formula requires the use of a projected tolerance zone.

T / F  5. The fixed and floating fastener formulas do consider fastener deviations in the calculation.

## Multiple Choice

*Circle the best answer to each statement.*

1. In the general fixed fastener formula, the "H" represents:
   A. The hole depth
   B. The hole location tolerance
   C. The LMC of the hole
   D. The MMC of the hole

2. In the floating fastener formula, what does "F" represent?
   A. The fastener depth
   B. The fastener location tolerance
   C. The LMC of the fastener
   D. The MMC of the fastener

3. T = H - F is the formula for _____ fastener assemblies.
   A. Fixed
   B. Floating
   C. Projected
   D. None of the above

4. The deviations of the fastener must be accounted for when using the:
   A. The general fixed fastener formula
   B. The floating fastener formula
   C. Both A and B
   D. None of the above

## Application Problems

*The application problems are designed to provide practice applying the chapter concepts to situations that are similar to on-the-job conditions.*

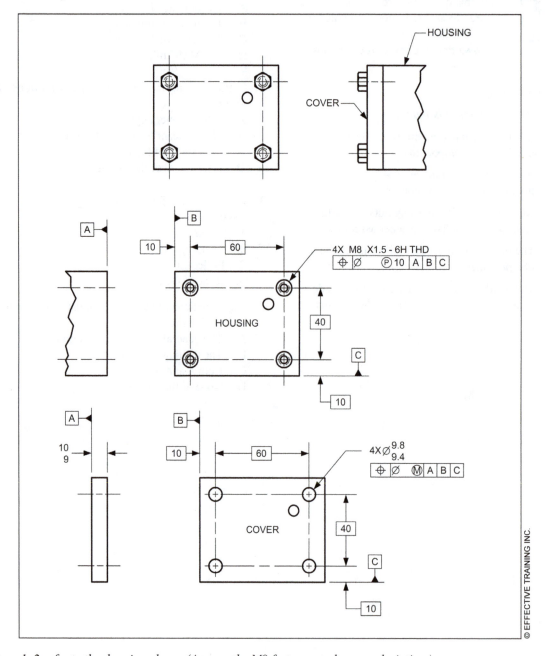

*Questions 1–2 refer to the drawing above. (Assume the M8 fasteners to have no deviation.)*

1. Using the general fixed fastener formula with equal distribution, calculate the position tolerance values for the clearance holes and the threaded holes.

    Clearance hole position tolerance value = _____

    Threaded hole position tolerance value = _____

2. Using the modified fixed fastener formula with unequal distribution or 75% for the threaded hole and 25% for the clearance hole, calculate the position tolerance values for the clearance holes and the threaded holes.

    Clearance hole position tolerance value = _____

    Threaded hole position tolerance value = _____

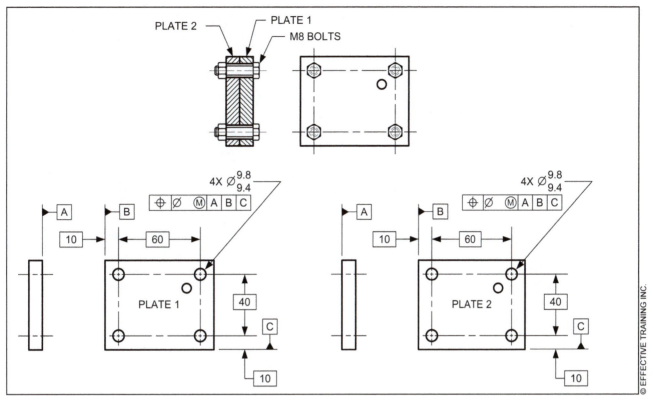

*Question 3 refers to the drawing above. (Assume the M8 fasteners to have no deviation.)*

3. Using the floating fastener formula, calculate the position tolerance values for the clearance holes and the threaded holes.

　　　Plate 1 position tolerance value = _____

　　　Plate 2 position tolerance value = _____

# Circular and Total Runout Tolerances

## Goal

Interpret the runout tolerances (circular and total)

## Performance Objectives

Upon completing this chapter, you should be able to:

1. Describe the terms "runout," "circular runout tolerance," and "total runout tolerance" (p.326)
2. Describe the geometry attributes that runout tolerances can affect (p.326)
3. Describe two tolerance zone shapes for a circular runout tolerance (p.326)
4. Describe two tolerance zone shapes for a total runout tolerance (p.326)
5. Describe three ways a datum axis can be established for a runout tolerance (p.327)
6. Explain which modifiers may be used in runout tolerances (p.328)
7. Describe the types of part deviations controlled with runout tolerances (p.328)
8. Explain the cumulative effects of runout tolerances (p.328)
9. Explain two differences between circular and total runout (p.329)
10. Evaluate if a runout tolerance specification is standard-compliant (p.330)
11. Describe three real-world applications for runout tolerances (p.331)
12. Interpret a circular runout tolerance (applied to a cylindrical surface) (p.331)
13. Interpret a total runout tolerance (applied to a cylindrical surface) (p.331)
14. Interpret a circular runout tolerance (applied to a planar surface) (p.332)
15. Interpret a total runout tolerance (applied to a planar surface) (p.333)
16. Interpret a circular runout tolerance by using the "Significant Seven Questions" (p.333)
17. Understand the verification principles for circular and total runout (p.334)
18. Explain how to verify a circular runout tolerance (p.334)
19. Explain how to verify a total runout tolerance (p.335)

## New Terms

- Circular runout tolerance
- Runout
- Total runout tolerance

## What This Chapter Is About

In this chapter, you'll learn where to use, how to interpret, and how to verify circular runout and total runout tolerances. The runout tolerances are two of the fourteen geometric tolerances. They are commonly used for defining coaxial relationships on rotating parts.

The runout tolerances are important because they define surface deviations and axis offset in coaxial relationships.

# TERMS AND CONCEPTS

## Runout

Runout tolerances are used to control the functional relationship of a surface (or surface elements) relative to a datum axis established from a datum feature at RMB. *Runout* is the high to low point deviation of the surface elements relative to a datum axis. The deviations may be on a surface of revolution coaxial to a datum axis or on a planar surface that is perpendicular to and intersects a datum axis. There are two types of runout tolerances: circular runout and total runout.

A *circular runout tolerance* is a geometric tolerance that limits the high to low point deviation (runout) of the circular elements of any surface of revolution, or the circular elements of a planar surface that is perpendicular to and intersects the datum axis. The tolerance zone (runout) applies independently at each circular element of the toleranced surface (i.e., at each cross section perpendicular to the datum axis). The tolerance zone applies normal to the surface elements.

A *total runout tolerance* is a geometric tolerance that limits the high to low point deviation (runout) of all surface elements of a cylindrical surface, or all surface elements of a planar surface that is perpendicular to and intersects the datum axis. The tolerance zone applies normal to the surface elements.

Runout tolerances control surface deviations relative to an axis of rotation. Therefore, they are always specified at RFS and always require a datum feature reference RMB. The datum feature reference(s) must use one of the three methods for establishing a repeatable datum axis, as shown on page 327.

## Geometry Attributes Affected by Runout Tolerances

A runout tolerance controls the geometry attributes of form, location and orientation. The chart in Figure 6-4 (page 66) shows the geometric tolerances and the geometry attributes that they can control.

**Author's Comment**
Keep in mind that a runout tolerance does not control size.

**Author's Comment**
Although runout tolerances apply to surfaces (circular elements, cylindrical surfaces, planar surface elements and planar surfaces) on drawings they are, on occasion, shown as being applied to a size dimension. It is preferred that runout tolerances be applied to a surface with a leader line directed to a surface.

## Runout Tolerance Zones

There are four common runout tolerance zone shapes. The shape of the tolerance zone depends upon two factors:

1. The type of runout (circular or total)

2. The type of feature to which the runout is applied (cylindrical or planar surface)

The four common tolerance zones are:

**Circular Runout Tolerance**
- The radial space between two coaxial circles
- The axial space between two coaxial circles (offset axially)

**Total Runout Tolerance**
- The radial space between two coaxial cylinders
- The axial space between two planes (offset axially)

The tolerance zones for runout tolerances are illustrated in Figures 24-1 and 24-2.

Where circular runout is applied to a cylindrical surface, the tolerance zone is two coaxial circles centered on the datum axis. The size of the outer circle is determined by the radius from the datum axis to the farthest point on the toleranced surface. The radial distance between the two circles is equal to the runout tolerance value.

We will discuss different applications of circular and total runout and how they are measured later in this chapter to gain a better understanding of the tolerance zones and the deviations that both types of runout affect.

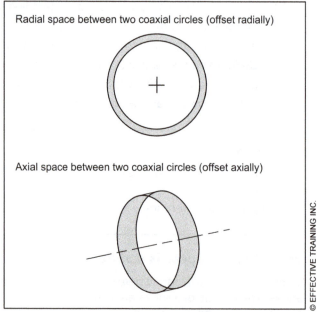

Radial space between two coaxial circles (offset radially)

Axial space between two coaxial circles (offset axially)

© EFFECTIVE TRAINING INC.

**FIGURE 24-1  Circular Runout Tolerance Zones**

Radial space between two coaxial cylinders (offset radially)

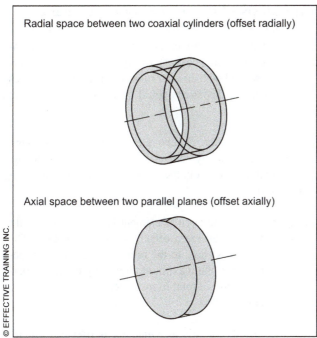

Axial space between two parallel planes (offset axially)

*FIGURE 24-2  Total Runout Tolerance Zones*

## Three Ways a Datum Axis Can Be Established for a Runout Tolerance

A runout tolerance requires a datum axis. The Y14.5 Standard specifies three methods used to establish a repeatable datum axis for a runout tolerance specification, as listed below and illustrated in Figure 24-3. The methods used to establish a repeatable datum axis are to reference one of the following datum features at RMB:

1. A single cylindrical feature of sufficient length

2. Two coaxial cylindrical features a sufficient distance apart

3. A surface (as a primary datum) and a cylindrical feature (as a secondary datum) at right angles

**Design Tip**
If a cylindrical feature is too short to orient the part relative to a datum feature simulator and establish a repeatable axis for inspection, the cylindrical feature will not be long enough to orient the part in the assembly.

Two important considerations in choosing one of the methods shown in Figure 24-3 are part shape and functional design requirements. Usually, the same part features are used to establish the datum axis and orient and locate the part in the assembly.

A single cylindrical feature is used where it orients the part in the assembly. Two coaxial cylindrical feature are used where they equally establish the orientation of the part in the assembly. A surface primary and cylindrical feature secondary are used where the surface orients the part and the cylindrical feature locates the part. Where the surface is the primary datum feature, the cylindrical feature should be short.

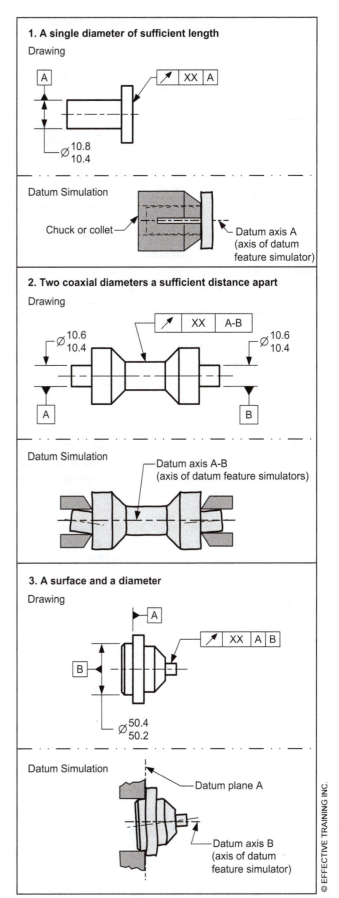

*FIGURE 24-3  Three Ways to Establish a Datum Axis for a Runout Tolerance*

## Modifiers Used With Runout Tolerances

There are two modifiers that may be used in the tolerance portion of a circular or total runout tolerance. The choice of modifier is often related to the functional requirements of the application. Figure 24-4 shows the modifiers and examples of applications where they could be used.

| Tolerance Modifier | Can Be Applied To | Effect | Functional Application |
|---|---|---|---|
| (F)* | Feature | Release the restraint requirement | Non-rigid parts with restraint notes |
| (ST)* | Feature | Requires statistical process controls | Statistically derived tolerances or tolerances used in statistical tolerance analyses |

\* These modifiers are only introduced in this text

*FIGURE 24-4 Runout Tolerance Modifiers*

## Types of Part Deviations Included in a Runout Tolerance Zone

Where circular runout is applied to circular elements constructed around a datum axis, it limits the deviations in circularity, orientation, and axis offset (location).

Where circular runout is applied to a planar surface at a right angle to a datum axis, it controls deviations of the circular elements of the plane for wobble, which includes orientation and form deviations of the circular elements.

Where total runout is applied to a cylindrical surface constructed around a datum axis, it limits the deviations in cylindricity (includes circularity, straightness, and taper), orientation, and axis offset (location).

Where total runout is applied to a surface perpendicular to a datum axis, it controls orientation (perpendicularity) and form (flatness) deviations. The types of part deviations included in a runout tolerance zone are shown in Figure 24-5.

| Runout Type | Applied to | Part Deviations Being Limited |
|---|---|---|
| Circular | Circular elements of a surface of revolution | Circularity Orientation Location |
| | Circular elements of a planar surface | Form and orientation of circular line elements (wobble) |
| Total | Cylindrical surface | Cylindricity (includes circularity, straightness, and taper) Orientation Location |
| | Planar surface | Orientation Flatness |

*FIGURE 24-5 Comparison of Circular and Total Runout Tolerances*

## Cumulative Effects of Runout Tolerances

Where a circular runout tolerance is applied to a cylindrical surface, it limits deviations in form (circularity), orientation, and location (axis offset). In Figure 24-6A, a part surface is toleranced with circular runout. The figure illustrates how the various deviations could occur within the runout tolerance zone.

In Figure 24-6B, the circular element is perfectly round (circular) and perfectly coaxial with the datum axis. As the part is rotated 360° about the datum axis, the runout deviation is zero.

In Figure 24-6C, the circular element is not round (lobed), but it is still coaxial (zero offset) to the datum axis. The runout deviation is equal to the roundness (circularity) deviation. Therefore, the maximum circularity deviation permitted by a circular or total runout tolerance is equal to the runout tolerance value.

In Figure 24-6D, the circular element is perfectly round, but its center point is offset 0.15mm above the datum axis (12 o'clock position). When the part is rotated 180°, the center point will move from the 12 o'clock position to the six o'clock position. The runout deviation is twice the axis offset (0.3mm). Therefore, the maximum axis offset permitted by a circular or total runout tolerance is one-half of the runout tolerance value.

Most parts have a combination of form and location deviations. Circular runout limits the cumulative effects of circularity (form) and axis offset (location). Total runout limits the cumulative effects of cylindricity and axis offset. (Total runout includes straightness and taper; circular runout does not.) The maximum possible axis offset in a runout application (circular or total) is equal to one-half the runout tolerance value. Circular runout limits center points. (If all the center points were offset in the same direction, it would be equivalent to an axis offset.)

The maximum possible orientation (parallelism) deviation is equal to the runout tolerance value. When the circular elements at each end of the cylindrical surface are offset in opposite directions, the orientation (parallelism) deviation is equal to the runout tolerance value.

Runout does not separate form and location deviations, so the amount each of these deviations contribute to the runout deviation will not be known. Therefore, the verification of a runout tolerance reports the cumulative effects of form and axis offset.

---

**TECHNOTE 24-1**
**Runout Tolerance Axis Offset**
Where a cylindrical surface is controlled by a runout tolerance, its maximum possible axis offset from the datum axis is equal to one-half the runout tolerance value.

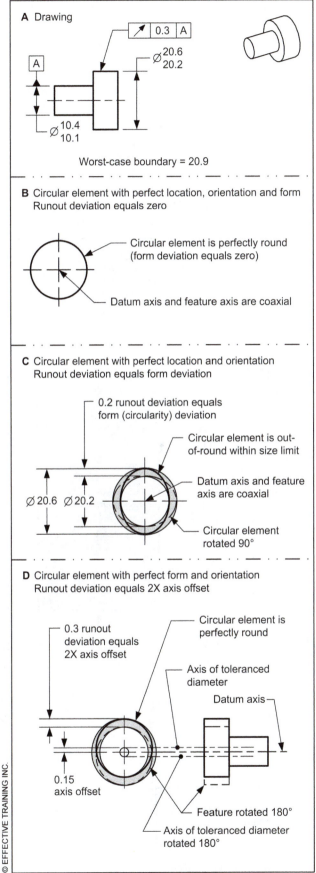

A Drawing

Worst-case boundary = 20.9

B Circular element with perfect location, orientation and form
Runout deviation equals zero

Circular element is perfectly round
(form deviation equals zero)

Datum axis and feature axis are coaxial

C Circular element with perfect location and orientation
Runout deviation equals form deviation

0.2 runout deviation equals
form (circularity) deviation

Circular element is out-
of-round within size limit

Datum axis and feature
axis are coaxial

Ø 20.6　Ø 20.2

Circular element
rotated 90°

D Circular element with perfect form and orientation
Runout deviation equals 2X axis offset

0.3 runout
deviation equals
2X axis offset

Circular element is
perfectly round

Axis of toleranced
diameter

Datum axis

0.15
axis offset

Feature rotated 180°

Axis of toleranced diameter
rotated 180°

**FIGURE 24-6** *Cumulative Effects of a Circular Runout Tolerance*

## Differences Between Circular and Total Runout Tolerances

Circular and total runout tolerances are mainly used to control deviations on coaxial cylindrical surfaces. These tolerances are similar but have a few significant differences. The chart in Figure 24-7 compares circular and total runout.

| Runout Applied to a Cylindrical Surface | | |
|---|---|---|
| Concept | Circular Runout | Total Runout |
| Tolerance zone | Two coaxial circles | Two coaxial cylinders |
| Relative cost to produce | $ | $$ |
| Relative cost to inspect | $ | $$ |
| Part characteristics being controlled | Location Orientation Circularity | Location Orientation Cylindricity |

**FIGURE 24-7** *Comparison of Circular and Total Runout Tolerances*

Since the tolerance zone shape is different between circular and total runout tolerances, a part could pass circular runout verification and fail a total runout verification. Figure 24-8 shows two parts with deviations. In each case, the flatness or straightness deviations would not be detected with a circular runout check, but would be detected with a total runout check.

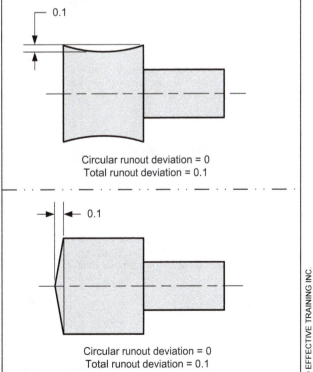

0.1

Circular runout deviation = 0
Total runout deviation = 0.1

0.1

Circular runout deviation = 0
Total runout deviation = 0.1

**FIGURE 24-8** *Comparison of Deviations Affected by Circular and Total Runout Tolerances*

## Evaluate if a Runout Control Specification is Standard-Compliant

For a runout tolerance (circular or total) to be a standard-compliant specification (legal), it must satisfy certain requirements.

In order to make it easier to remember what to look for when evaluating the correctness of a geometric control on a drawing, I created a mnemonic.

The requirements are divided into four areas represented by the initials "CARE":

> C - Check for datums
>
> A - Assess the application
>
> R - Review the modifiers
>
> E - Evaluate the tolerance

A flowchart that identifies the requirements for a standard-compliant runout tolerance is shown in Figure 24-9.

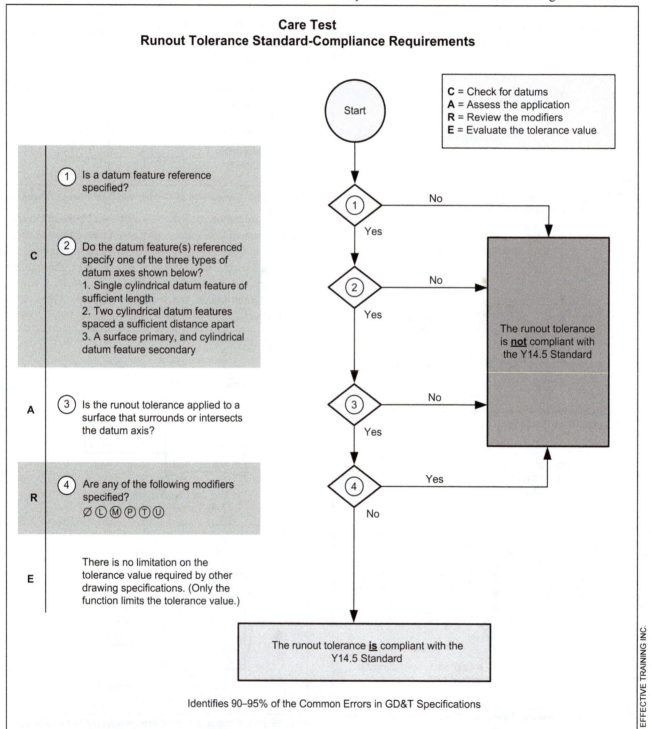

**Care Test**
**Runout Tolerance Standard-Compliance Requirements**

Start

C = Check for datums
A = Assess the application
R = Review the modifiers
E = Evaluate the tolerance value

**C**

① Is a datum feature reference specified?

② Do the datum feature(s) referenced specify one of the three types of datum axes shown below?
1. Single cylindrical datum feature of sufficient length
2. Two cylindrical datum features spaced a sufficient distance apart
3. A surface primary, and cylindrical datum feature secondary

**A**

③ Is the runout tolerance applied to a surface that surrounds or intersects the datum axis?

**R**

④ Are any of the following modifiers specified?
⌀ Ⓛ Ⓜ Ⓟ Ⓣ Ⓤ

**E**

There is no limitation on the tolerance value required by other drawing specifications. (Only the function limits the tolerance value.)

① No →
① Yes ↓

② No →
② Yes ↓

③ No →
③ Yes ↓

④ Yes →
④ No ↓

The runout tolerance is **not** compliant with the Y14.5 Standard

The runout tolerance **is** compliant with the Y14.5 Standard

Identifies 90–95% of the Common Errors in GD&T Specifications

*FIGURE 24-9 Test for a Standard-Compliant Runout Tolerance*

# RUNOUT APPLICATIONS

The following sections describe various runout applications.

## Real-World Applications of Runout Tolerances

There are many real-world applications for runout tolerances. Three common real-world applications of circular and total runout tolerance are:

1. Clearance between rotating parts or a rotating part and stationary part

2. Gear mesh

3. Rotating parts for alignment, balance, vibration, wear (wheels, axles, shafts)

## Circular Runout Tolerance Applied to a Cylindrical Surface

In certain applications, the function of a part requires limiting the circular runout deviation of a cylindrical surface relative to a datum axis.

In the example in Figure 24-10, a circular runout tolerance is applied to a cylindrical surface relative to a datum axis. This is the most common application of a circular runout tolerance. In this application, the following conditions apply:

- The cylindrical surface must meet its size requirements.

- The tolerance zone is the radial space between two coaxial circles 0.2 apart.

- The tolerance zone applies to each circular element independently.

- The runout tolerance also limits the roundness of the circular elements.

- The maximum possible axis offset is 0.1.

- The maximum parallelism deviation is 0.2.

- The worst-case boundary of the diameter is 24.8 (24.6 + 0.2).

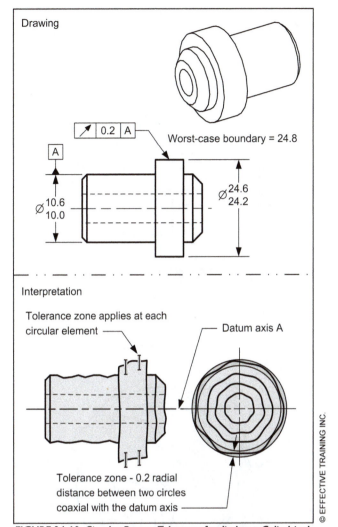

FIGURE 24-10  Circular Runout Tolerance Applied to a Cylindrical Surface

## Total Runout Tolerance Applied to a Cylindrical Surface

In certain applications, the function of a part requires limiting the total runout deviation of a cylindrical surface relative to a datum axis. In the example in Figure 24-11, a total runout tolerance is applied to a cylindrical surface relative to a datum axis. This is the most common application of a total runout tolerance. The following conditions apply:

- The cylindrical surface must meet its size requirements.

- The tolerance zone is the space between two coaxial cylinders 0.1 apart.

- The tolerance zone applies to all surface elements simultaneously.

- The runout tolerance also limits the cylindricity of the circular elements (includes straightness, taper, and circularity).

- The maximum possible axis offset is 0.05.

- The maximum parallelism deviation is 0.1.

- The worst-case boundary of the diameter is 12.9 (12.8 + 0.1).

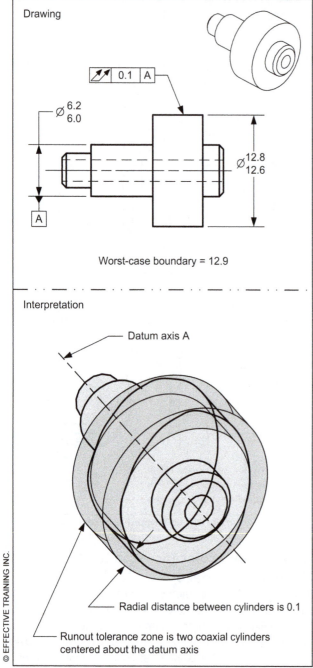

Drawing

Ø 6.2 / 6.0

Ø 12.8 / 12.6

Worst-case boundary = 12.9

Interpretation

Datum axis A

Radial distance between cylinders is 0.1

Runout tolerance zone is two coaxial cylinders centered about the datum axis

*FIGURE 24-11  Total Runout Tolerance Applied to a Cylindrical Surface*

## Circular Runout Tolerance Applied to a Planar Surface

In certain applications, the function of a part requires limiting the circular runout deviation of a planar surface perpendicular to a datum axis.

In the example in Figure 24-12, a circular runout tolerance is applied to a planar surface relative to a datum axis. Where circular runout is applied to a planar surface, the following conditions apply:

- The tolerance zone is the space between two coaxial circles offset axially 0.2 apart.
- The tolerance zone applies to all circular elements independently.
- The tolerance zone limits deviations in form and orientation of the circular elements (but not location).

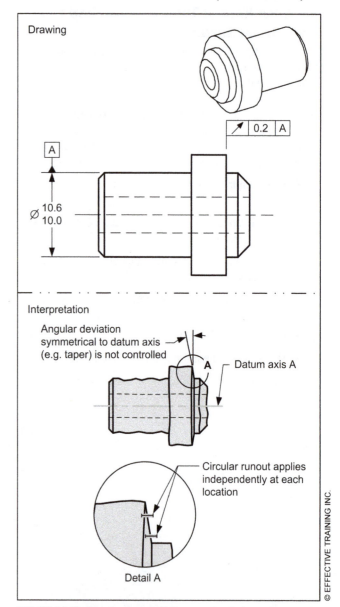

Drawing

0.2   A

A

Ø 10.6 / 10.0

Interpretation

Angular deviation symmetrical to datum axis (e.g. taper) is not controlled

A

Datum axis A

Circular runout applies independently at each location

Detail A

*FIGURE 24-12  Circular Runout Tolerances Applied to a Planar Surface*

## Total Runout Tolerance Applied to a Planar Surface

In certain applications, the function of a part requires limiting the total runout deviation of a planar surface perpendicular to a datum axis.

In the example in Figure 24-13, a total runout tolerance is applied to a planar surface relative to a datum axis. This is not a common application of a total runout tolerance. Where total runout is applied to a planar surface, the following conditions apply:

- The tolerance zone is the space between two parallel planes 0.4 apart and perpendicular to the datum axis.
- The tolerance zone applies to all surface elements simultaneously.
- The tolerance zone controls the flatness of the surface and the orientation of the surface relative to the datum axis (but not location).

*Design Tip*
If the design intent is to control the perpendicularity of a surface relative to an axis, it would be more straightforward to use a perpendicularity tolerance.

## The Significant Seven Questions

This section uses the Significant Seven Questions to interpret circular runout tolerance application. The significant seven questions are explained in Appendix B. Figure 24-14 contains a circular runout application. The Significant Seven Questions are answered in the interpretation of the figure. Notice how each answer increases your understanding of the geometric tolerance.

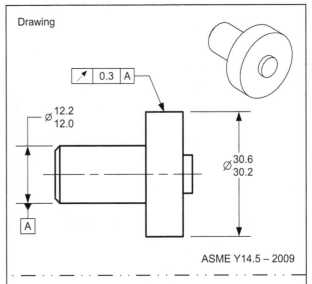

ASME Y14.5 – 2009

Interpretation

The Significant Seven Questions

1. Which dimensioning and tolerancing standard applies?
Answer – ASME Y14.5-2009

2. What does the tolerance apply to?
Answer – Circular elements along the entire cylindrical surface

3. Is the specification standard-compliant?
Answer – Yes

4. What are the shape and size of the tolerance zone?
Answer – A 0.3mm radial space between two coaxial circles

5. How much total tolerance is permitted?
Answer – 0.3

6. What are the shape and size of the datum simulators?
Answer – A contracting device like a chuck or collet

7. Which geometry attributes are affected by this tolerance?
Answer – Form, orientation, and location

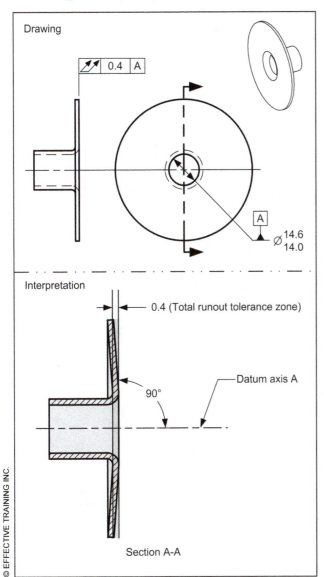

*FIGURE 24-13 Total Runout Tolerance Applied to a Planar Surface*

*FIGURE 24-14 Using the Significant Seven Questions to Interpret a Circular Runout Tolerance*

# VERIFICATION PRINCIPLES AND METHODS

## Verification Principles for Runout Tolerances

This section contains a simplified explanation of the verification principles and methods that can be used to inspect a circular or total runout tolerance applied to a cylindrical surface. Verifying a circular runout tolerance applied to a cylindrical surface has two parts:

1. Establish the datum axis.

2. Verify the high to low point deviation (runout) of each individual circular element (at each cross section independently) is less than the specified circular runout tolerance.

Verifying a total runout tolerance applied to a cylindrical surface has two parts:

1. Establish the datum axis.

2. Verify the high to low point deviation (runout) of all surface elements (for full length and circumference).

The datum axis RMB may be established by mounting the datum feature in a datum feature simulator (such as a chuck or collet) or by using a CMM to collect a cloud of surface points to establish the axis from the unrelated actual mating envelope.

Verifying the radial deviation of the surface points can be done with a CMM or a dial indicator. If a CMM is used, the difference between the closest and furthest point from the datum axis is calculated and compared to the runout tolerance zone. If a chuck or collet is used to establish the datum axis, a dial indicator is often used to measure the radial deviation of the surface elements. The dial indicator reading (full indicator movement) is compared to the runout tolerance value.

Since a circular runout tolerance applies to each circular element of the part surface individually, several separate runout measurements should be taken at different locations (cross sections) along the length of the toleranced surface. Each runout measurement is independent. Since a total runout tolerance applies to the entire surface simultaneously, a single runout measurement is taken for the full length and circumference of the toleranced surface. The examples in this section use a dial indicator to verify the runout tolerances.

## Verifying Circular Runout Applied to a Cylindrical Surface

There are many ways to verify a circular runout tolerance. In Figure 24-15, a circular runout tolerance is applied to a cylindrical surface. When verifying this cylindrical surface, three separate checks should be made: the size, the Rule #1 boundary, and the circular runout deviations of the surface. Chapter 7 explained how to verify the size and Rule #1 boundary. Now we'll look at how to verify the circular runout requirement.

One way a circular runout tolerance could be verified is shown in the lower half of Figure 24-15. The part is mounted in a device similar to a chuck or collet to establish datum axis A. Then a dial indicator is placed perpendicular to the surface being inspected. The part is rotated 360°, and the full indicator movement (FIM) is the runout deviation for that circular element. The dial indicator is moved to another location on the cylindrical surface, reset to zero, and another indicator reading is obtained. The number of circular elements checked is determined during the inspection planning process.

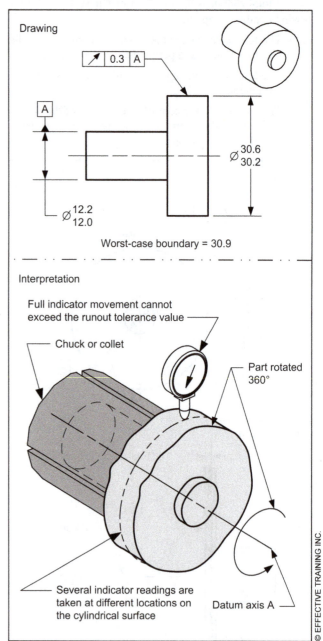

**FIGURE 24-15 Verifying a Circular Runout Tolerance**

## Verifying Total Runout Applied to a Cylindrical Surface

There are many ways to verify a total runout tolerance. In Figure 24-16, a total runout tolerance is applied to a diameter. When verifying this cylindrical surface, three separate checks should be made: the size, the Rule #1 boundary, and the total runout deviations of the surface. Chapter 7 explained how to verify the size and Rule #1 boundary. Now we'll look at how to verify the total runout requirement.

One way a total runout tolerance could be verified is shown in the lower half of Figure 24-16. The part is mounted in a chuck or collet to establish the datum axis. Then a dial indicator is placed perpendicular to the surface being inspected. The part is rotated around the datum axis while the dial indicator is moved along the cylindrical surface. The full indicator movement (FIM) is the total runout deviation. The number of rotations and rate of movement of the dial indicator is determined during the inspection planning process.

---

**TECHNOTE 24-2**
**Verifying a Total Runout Tolerance**

When verifying a total runout tolerance, the dial indicator is moved axially along the surface while the part is rotated about the datum axis. The dial indicator reading is the runout deviation of the surface.

---

**Author's Comment**
If the dial indicator is zeroed at a point that is not the lowest or highest point, the full indicator movement will be the total difference between the greatest positive value reading and the greatest negative value indicator reading.

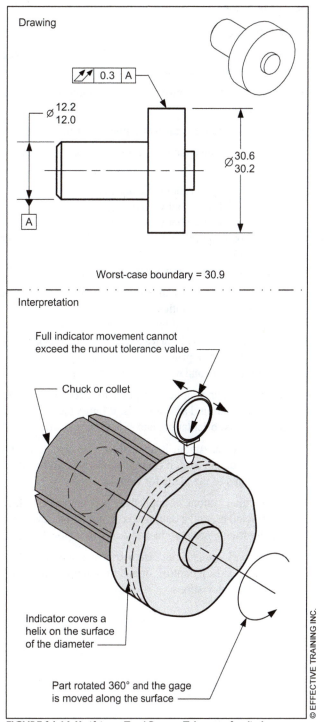

Worst-case boundary = 30.9

**FIGURE 24-16** *Verifying a Total Runout Tolerance Applied to a Cylindrical Surface*

## SUMMARY

### Key Points

- Runout is the high to low point deviation of the surface elements relative to a datum axis.

- Runout tolerances control surface deviations relative to a datum axis.

- A runout tolerance is always applies at RFS.

- A runout tolerance datum feature reference is always at RMB.

- The four common runout tolerance zones are:
  - o The radial space between two coaxial circles
  - o The radial space between two coaxial cylinders
  - o The axial space between two coaxial circles (offset axially)
  - o The axial space between two planes (offset axially)

- There are three ways to establish a datum axis for a runout tolerance specification:
  - o A single cylindrical feature of sufficient length
  - o Two coaxial cylindrical features a sufficient distance apart
  - o A surface (as a primary datum) and a cylindrical feature (as a secondary datum) at right angles

- Where circular runout is applied to circular elements constructed around a datum axis, it limits the deviations in circularity, orientation, and axis offset (location).

- Where circular runout is applied to a surface at a right angle to a datum axis, it controls deviations of the circular elements of the plane for wobble.

- Where total runout is applied to a cylindrical surface constructed around a datum axis, it limits the deviations in cylindricity, orientation, and axis offset (location).

- Where total runout is applied to a surface at a right angle to a datum axis, it controls orientation (perpendicularity) and flatness deviations.

- Where a cylindrical feature is controlled by a circular or total runout tolerance, its maximum possible axis offset from the datum axis is equal to one-half the runout tolerance value.

- Three common real-world applications of circular and total runout tolerance are:
  - o Clearance between rotating parts or a rotating part and stationary part
  - o Gear mesh
  - o Rotating parts for alignment, balance, vibration, wear (wheels, axles, shafts)

- When verifying a circular runout tolerance, several measurements, at different locations along the surface, are made. Each measurement is independent.

- When verifying total runout, the dial indicator is moved axially along the surface while the part is rotated about the datum axis. The dial indicator reading (full indicator movement) is the runout deviation of the surface.

### Additional Related Topics

*These topics are recommended for further study to improve your understanding of circular and total runout.*

| Topic | Source |
|---|---|
| Using runout to control deviations on conical surfaces | *ASME Y14.5-2009, Para. 6.4.3* |
| Contoured surfaces of revolution | ETI's *Advanced Concepts of GD&T* textbook |

# QUESTIONS AND PROBLEMS

***Website Bonus Materials***
Additional questions are available at our website. To access bonus materials for this textbook, please visit:
www.etinews.com/textbookbonus

## True and False

*Indicate if each statement is true or false.*

T / F   1.   A circular runout tolerance can be applied to circular elements and planar surface elements that are perpendicular to the datum axis.

T / F   2.   A total runout tolerance must reference a datum axis.

T / F   3.   Runout tolerances must be specified at RFS.

T / F   4.   Runout tolerances can only have a primary datum feature reference.

T / F   5.   A total runout tolerance may have a bonus tolerance.

T / F   6.   A datum axis may be established from two cylindrical features a sufficient distance apart.

T / F   7.   Two tolerance zone shapes for a total runout tolerance are the space between two coaxial cylinders and the space between two parallel planes.

T / F   8.   A total runout tolerance can control the geometry attribute of size.

T / F   9.   Runout runout tolerance can never use the MMC modifier.

T / F   10.   A circular runout tolerance affects the cylindricity of a cylindrical surface.

## Multiple Choice

*Circle the best answer to each statement.*

1. When a circular runout tolerance is applied to a surface of revolution:
   A. The tolerance zone applies to all circular elements of the surface simultaneously.
   B. The tolerance zone applies to one circular element of the surface.
   C. The tolerance zone applies only once anywhere on the indicated surface.
   D. The tolerance zone applies to each circular element of the surface independently.

2. Which of the following is a Y14.5 compliant method of establishing a datum axis?
   A. A surface primary and a cylindrical feature, 90° to the surface, secondary
   B. A diameter primary and a surface , 90° to the cylindrical feature, secondary
   C. A pattern of holes
   D. Any single cylindrical feature

3. Which surface deviations may be limited by total runout applied to a surface perpendicular to the datum axis?
   A. Circularity and straightness
   B. Circularity and flatness
   C. Perpendicularity and flatness
   D. Location, perpendicularity, and straightness

4. When a total runout tolerance is applied to a cylindrical surface:
   A. Rule #1 is overridden.
   B. The tolerance zone applies to the entire surface.
   C. The diameter modifier should be applied to the tolerance value.
   D. The tolerance zone applies to each circular element of the surface of revolution independently.

5. What type of deviations are limited by total runout but not circular runout?
   A. Circularity
   B. Cylindricity
   C. Parallelism
   D. None of the above

6. The "cumulative effect" of runout tolerances applied to a cylindrical surface refers to the fact that a runout tolerance affects both form and_____ .
   A. Location deviations
   B. Bonus tolerances
   C. Size deviations
   D. None of the above

7. A real-world application of a runout tolerance is:
   A. Assembly
   B. Rotating parts or alignment, balance, vibration, etc.
   C. Bolted joint applications
   D. None of the above

8. A total runout tolerance may be verified with a:
   A. Functional gage
   B. Dial indicator
   C. Set of calipers
   D. All of the above

## Application Problems

*The application problems are designed to provide practice applying the chapter concepts to situations that are similar to on-the-job conditions.*

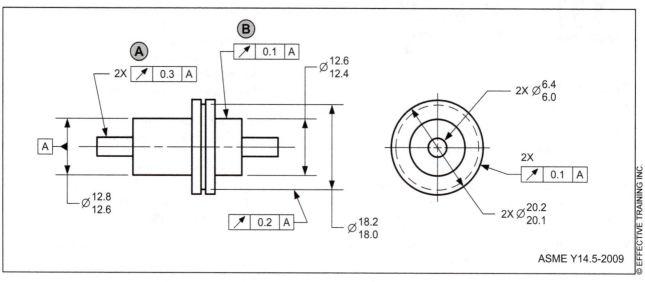

ASME Y14.5-2009

*Questions 1 and 2 refer to the figure above.*

1. Use the Significant Seven Questions to interpret the runout tolerance labeled "A."

   1) Which dimensioning and tolerancing standard applies?

   _____

   2) What does the tolerance apply to?_____

   _____

   3) Is the specification standard-compliant?_____

   4) What are the shape and size of the tolerance zone?

   _____

   5) How much total tolerance is permitted?_____

   6) What are the shape and size of the datum simulators?

   _____

   7) Which geometry attributes are affected by this tolerance?

   _____

2. Use the Significant Seven Questions to interpret the runout tolerance labeled "B."

   1) Which dimensioning and tolerancing standard applies?

   _____

   2) What does the tolerance apply to?_____

   _____

   3) Is the specification standard-compliant?_____

   4) What are the shape and size of the tolerance zone?

   _____

   5) How much total tolerance is permitted?_____

   6) What are the shape and size of the datum simulators?

   _____

   7) Which geometry attributes are affected by this tolerance?

   _____

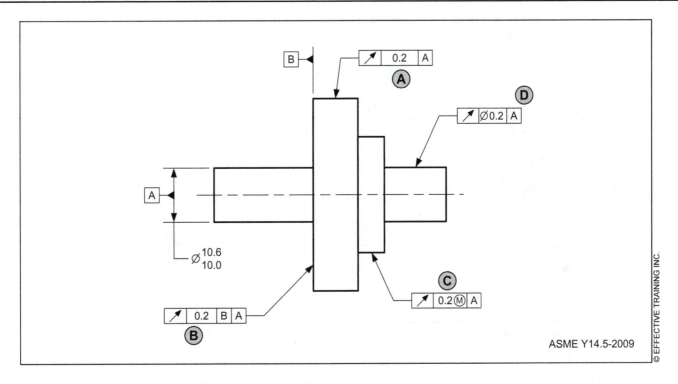

ASME Y14.5-2009

© EFFECTIVE TRAINING INC.

3.  Indicate if each runout tolerance below is standard-compliant. If it is not, describe why. (Hint: use the CARE test.)

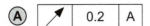  _____

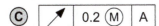  _____

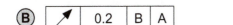  _____

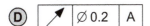  _____

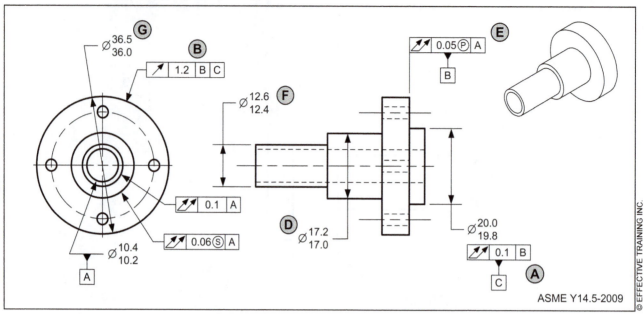

ASME Y14.5-2009

*Use the drawing above to answer questions 4 and 5.*

4.  Fill in the chart.

| QUESTION | APPLIES TO | |
| --- | --- | --- |
| | DIA **F** | DIA **G** |
| The size of the diameter is limited to. . . | | |
| The roundness of the diameter is limited to. . . | | |
| The maximum offset between the diameter axis and datum axis A is. . . | | |
| Describe the tolerance zone for the runout tolerance applied to the surface. | | |
| What is the outer boundary of this diameter? | | |

5.  Indicate if each runout tolerance below is standard-compliant. If it is not, describe why. (Hint: use the CARE test.)

**B**  | ↗ | 1.2 | B | C |  _____

**C**  | ↗↗ | 0.4 | A |  _____

**D**  | ↗↗ | 0.06 ⓈS | A |  _____

**E**  | ↗↗ | 0.05 Ⓟ | A |  _____

# Concentricity and Symmetry Tolerances

## Goal

Interpret the concentricity and symmetry tolerances

## Performance Objectives

Upon completing this chapter, you should be able to:

1. Describe the terms "concentricity," "symmetry," "median point," "concentricity tolerance," and "symmetry tolerance" (p.342)
2. Describe the tolerance zone for a concentricity tolerance (p.342)
3. Describe the tolerance zone for a symmetry tolerance (p.342)
4. Describe the geometry attributes that concentricity and symmetry tolerances can affect (p.343)
5. List the modifiers that can be used with a concentricity and symmetry tolerance (p.343)
6. Evaluate if a concentricity tolerance specification is standard-compliant (p.344)
7. Evaluate if a symmetry tolerance specification is standard-compliant (p.345)
8. Describe one real-world application for a concentricity tolerance (p.346)
9. Describe one real-world application for a symmetry tolerance (p.346)
10. Interpret a concentricity tolerance applied to a cylindrical feature of size (p.346)
11. Explain two differences between concentricity and total runout tolerances (p.347)
12. Explain two differences between concentricity and position (RFS) tolerances (p.347)
13. Interpret a symmetry tolerance applied to a width (p.347)
14. Explain two differences between symmetry and position (RFS) tolerances (p.349)
15. Interpret a concentricity tolerance by using the "Significant Seven Questions" (p.349)
16. Understand the verification principles for concentricity and symmetry tolerances (p.350)
17. Explain how to verify a concentricity tolerance (p.350)
18. Explain how to verify a symmetry tolerance (p.351)

## New Terms

- Concentricity
- Concentricity tolerance
- Median point
- Symmetry
- Symmetry tolerance

## What This Chapter Is About

In this chapter, you'll learn where to use, how to interpret, and how to verify concentricity and symmetry tolerances. Concentricity and symmetry tolerances are two of the fourteen geometric tolerances. Concentricity tolerances are used to control coaxial relationships, and symmetry tolerances are used to control a symmetrical relationship of a planar feature of size relative to a datum center plane or axis.

Although concentricity and symmetry tolerances are not common, it is important to learn about them because you need to be able to interpret these tolerances when they appear on drawings.

# TERMS AND CONCEPTS

## Concentricity, Symmetry, and Median Point

*Concentricity* is the condition where the median points of all diametrically opposed elements of a surface of revolution (or the median points of correspondingly located elements of two or more radially disposed features) are congruent with a datum axis (or center point).

A *median point* is the mid-point of a two-point measurement. Each two-point measurement (actual local size) divided by two yields a median point. In Figure 25-1 shows median points for a circular element and for a width. Median points can be derived from a surface of revolution, radially disposed features (e.g., a hex, or square, etc.), or a width.

## Concentricity and Symmetry Tolerances

A *concentricity tolerance* is a geometric tolerance that defines the permissible deviation that all median points of a surface of revolution (or correspondingly located elements of two or more radially disposed features) are permitted to vary from a datum axis or center point.

A concentricity tolerance is considered a location tolerance. A concentricity tolerance:

- Is one of the three direct location controls

- Must be applied to a surface of revolution or radially disposed surfaces (e.g., hex)

- Must always use a datum reference

- Always applies at RFS, and the datum feature at RMB

- The tolerance zone is always cylindrical

An example of a concentricity tolerance is shown in Figure 25-2.

*Symmetry* is the condition where all the median points of all opposed correspondingly located elements of two or more feature surfaces are congruent with a datum axis or center plane.

A *symmetry tolerance* is a geometric tolerance that defines the permissible deviation that median points of all opposed correspondingly located elements of two or more feature surfaces are permitted to vary from a datum axis or center plane.

A symmetry tolerance is considered a location tolerance. A symmetry tolerance:

- Is one of the three direct location controls

- Must be applied to a width

- Must always use a datum reference

- Always applies at RFS, and the datum feature at RMB

- The tolerance zone is always two parallel planes

An example of a symmetry tolerance is shown in Figure 25-2.

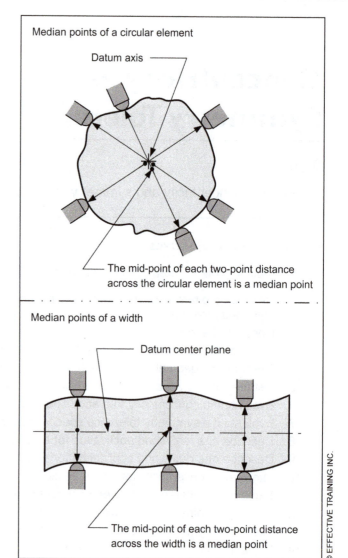

Median points of a circular element

Datum axis

The mid-point of each two-point distance across the circular element is a median point

Median points of a width

Datum center plane

The mid-point of each two-point distance across the width is a median point

© EFFECTIVE TRAINING INC.

**FIGURE 25-1  Median Points**

## Concentricity and Symmetry Tolerance Zones

The tolerance zone shape for a concentricity tolerance is the space within a cylinder. The tolerance zone is centered on the datum axis. Since the tolerance zone is always cylindrical, the diameter symbol modifier is always specified in a concentricity tolerance feature control frame.

The tolerance zone shape for a symmetry tolerance is the space between two parallel planes. The tolerance zone is centered on a datum axis or datum center plane. Since the tolerance zone is always two parallel planes, the diameter symbol is never specified in a symmetry tolerance.

The tolerance zones for concentricity and symmetry tolerances are illustrated in Figure 25-2.

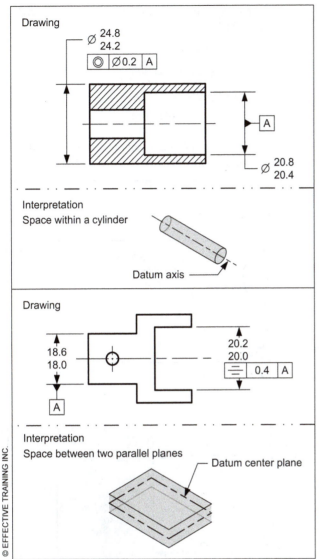

FIGURE 25-2 Concentricity and Symmetry Tolerance Zones

The types of geometry attributes affected by a concentricity or symmetry tolerance are shown in Figure 25-3.

| Tolerance | Applied to | Geometry Attributes Controlled |
|---|---|---|
| Concentricity | Surface of revolution or radially disposed features | Location of median points<br>Orientation of derived median line<br>Form (straightness) of derived median line |
| Symmetry | Planar feature of size (e.g., slot & tab) | Location of median points<br>Orientation of derived median plane<br>Form (flatness) of derived median plane |

FIGURE 25-3 Geometry Attributes Controlled With Concentricity and Symmetry Tolerances

## Modifiers Used With Concentricity and Symmetry Tolerances

Concentricity and symmetry tolerances are always specified at RFS; however, there are a few modifiers that may be used in the feature control frame. The choice of modifier is often related to the functional requirements of the application. The chart in Figure 25-4 shows the common modifiers and examples of applications where they could be used.

| Tolerance Modifier | Can Be Applied to | Effect | Application |
|---|---|---|---|
| Ⓕ * | Feature or feature of size | Releases the restraint requirement | Non-rigid parts with restraint notes |
| ⟨ST⟩ * | Feature or feature of size | Requires statistical process control | Statistically derived tolerances or tolerances used in statistical tolerance analysis |
| ⌀ (Concentricity only) | Feature or feature of size | Invokes a cylindrical tolerance zone | Required in concentricity tolerances** |

\* These modifiers are only introduced in this text

\*\* Except where the spherical diameter is specified

FIGURE 25-4 Modifiers Used with Concentricity and Symmetry Tolerances

## Geometry Attributes Affected by a Concentricity or Symmetry Tolerance

Where concentricity or symmetry are specified, the deviations of the location of the median points of the toleranced feature are limited. Concentricity and symmetry tolerances affect orientation and form of the axis or center plane of the toleranced feature of size (e.g., parallelism, perpendicularity, straightness or flatness).

A concentricity or symmetry tolerance can control the geometry attributes of form, location and orientation. The chart in Figure 6-4 (page 68) shows the geometric tolerances and the geometry attributes that they can control.

---

**TECHNOTE 25-1**
**Concentricity**

The tolerance zone for a concentricity tolerance is the space within a cylinder centered on the datum axis. The median points of the toleranced surface must be located within the tolerance zone.

---

**Author's Comment**

Although rarely used, the tolerance zone for a concentricity tolerance may also be a sphere centered about a datum center point.

## Evaluate if a Concentricity Tolerance Specification is Standard-Compliant

For a concentricity tolerance to be a standard-compliant specification (legal), it must satisfy certain requirements. In order to make it easier to remember what to look for when evaluating the correctness of a geometric control on a drawing, I created a mnemonic.

The requirements are divided into four areas represented by the initials "CARE":

C - Check for datums

A - Assess the application

R - Review the modifiers

E - Evaluate the tolerance

A flowchart that identifies the requirements for a standard-compliant concentricity tolerance is shown in Figure 25-5.

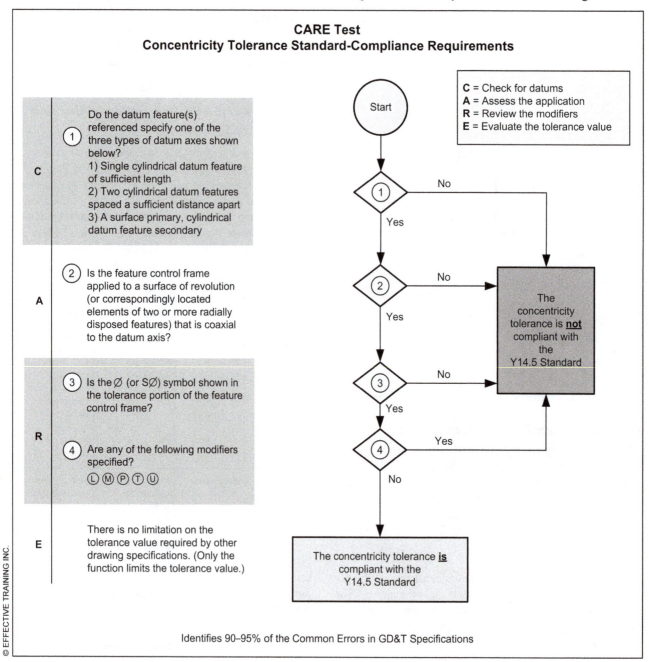

**CARE Test**
**Concentricity Tolerance Standard-Compliance Requirements**

C = Check for datums
A = Assess the application
R = Review the modifiers
E = Evaluate the tolerance value

**C**

(1) Do the datum feature(s) referenced specify one of the three types of datum axes shown below?
1) Single cylindrical datum feature of sufficient length
2) Two cylindrical datum features spaced a sufficient distance apart
3) A surface primary, cylindrical datum feature secondary

**A**

(2) Is the feature control frame applied to a surface of revolution (or correspondingly located elements of two or more radially disposed features) that is coaxial to the datum axis?

**R**

(3) Is the Ø (or SØ) symbol shown in the tolerance portion of the feature control frame?

(4) Are any of the following modifiers specified?
(L) (M) (P) (T) (U)

**E**

There is no limitation on the tolerance value required by other drawing specifications. (Only the function limits the tolerance value.)

Start

(1) — No / Yes
(2) — No / Yes
(3) — No / Yes
(4) — Yes / No

The concentricity tolerance is **not** compliant with the Y14.5 Standard

The concentricity tolerance **is** compliant with the Y14.5 Standard

Identifies 90–95% of the Common Errors in GD&T Specifications

© EFFECTIVE TRAINING INC.

*FIGURE 25-5 Test for a Standard-Compliant Concentricity Tolerance*

## Evaluate if a Symmetry Tolerance Specification is Standard-Compliant

For a symmetry tolerance to be a standard-compliant specification (legal), it must satisfy certain requirements. In order to make it easier to remember what to look for when evaluating the correctness of a geometric control on a drawing, I created a mnemonic.

The requirements are divided into four areas represented by the initials "CARE":

   C - Check for datums

   A - Assess the application

   R - Review the modifiers

   E - Evaluate the tolerance

A flowchart that identifies the requirements for a standard-compliant symmetry tolerance is shown in Figure 25-6.

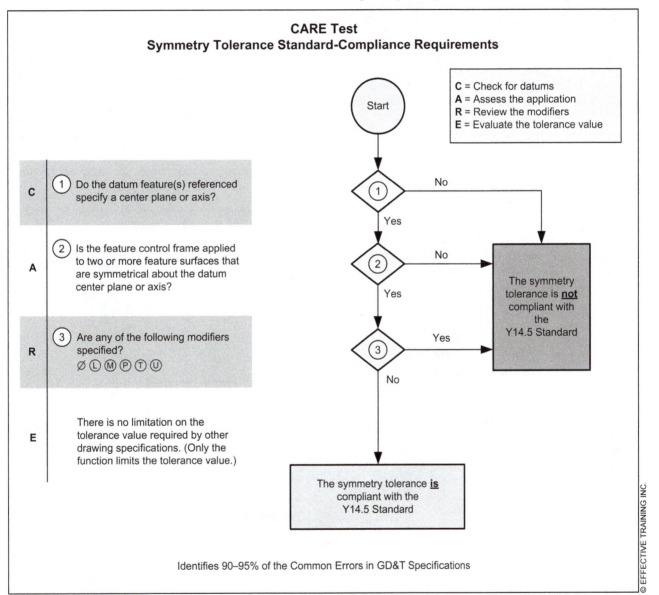

**CARE Test**
**Symmetry Tolerance Standard-Compliance Requirements**

C = Check for datums
A = Assess the application
R = Review the modifiers
E = Evaluate the tolerance value

**C**   (1)  Do the datum feature(s) referenced specify a center plane or axis?

**A**   (2)  Is the feature control frame applied to two or more feature surfaces that are symmetrical about the datum center plane or axis?

**R**   (3)  Are any of the following modifiers specified? ∅ Ⓛ Ⓜ Ⓟ Ⓣ Ⓤ

**E**   There is no limitation on the tolerance value required by other drawing specifications. (Only the function limits the tolerance value.)

Start

The symmetry tolerance is **not** compliant with the Y14.5 Standard

The symmetry tolerance **is** compliant with the Y14.5 Standard

Identifies 90–95% of the Common Errors in GD&T Specifications

© EFFECTIVE TRAINING INC.

*FIGURE 25-6 Test for a Standard-Compliant Symmetry Tolerance*

# CONCENTRICITY AND SYMMETRY TOLERANCE APPLICATIONS

The following sections describe concentricity and symmetry tolerance applications.

## Real-World Applications of Concentricity and Symmetry Tolerances

Real-world applications of concentricity and symmetry tolerance are rare. This section describes two applications of where concentricity or symmetry tolerances could be used on a drawing. The first application is using concentricity tolerances to limit the deviations of median points of radially disposed non-cylindrical surfaces relative to a datum axis for balance of a high-speed rotating part (equal distribution of mass of a high-speed rotating part). The second application is using symmetry tolerances for precise balance to control oscillating motions (e.g., a tuning fork or pendulum).

### Author's Comment

Concentricity and symmetry tolerances are rarely used on drawings. These tolerances are expensive to verify. Runout or position tolerances, along with balance notes, are often used instead.

### Design Tip

Where an assembly requires precision balance (such as an airplane propeller), the use of concentricity or symmetry tolerances at the component level is expensive and impractical. A more practical approach is to use runout or position tolerances with notes that require the parts or assembly to be balanced.

## Concentricity Applied to a Cylindrical Feature of Size

In certain applications, the function of a part requires defining the concentricity deviation of a cylindrical feature of size relative to a datum axis.

In the example in Figure 25-7, the concentricity deviation of a cylindrical feature of size is being limited relative to a datum axis. In this application, the following conditions apply:

- The feature of size must meet its size requirements.

- The tolerance zone is the space within a 0.2mm diameter cylinder.

- All median points of the diametrically opposed elements of the surface must be located within the tolerance zone.

- The maximum distance between median points of the toleranced feature of size and the datum axis is half the concentricity tolerance value.

- The worst case-boundary of the toleranced feature of size is a 24.8mm feature of size.

- The concentricity tolerance also limits the straightness of the derived median line.

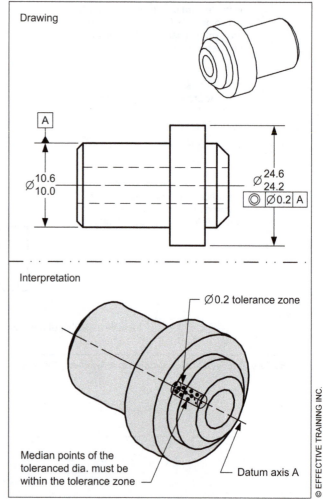

**FIGURE 25-7** *Concentricity Tolerance Application*

© EFFECTIVE TRAINING INC.

## Comparison of Coaxial Diameter Tolerances

When dimensioning coaxial features of size, several geometric tolerances may be used. The choices are position, runout, profile, and concentricity tolerances. In this section, I will compare concentricity with two common alternatives.

Concentricity, total runout, and position tolerances are mainly used to control deviations on coaxial features of size. These controls are similar, but have a few significant differences. The chart in Figure 25-8 compares concentricity, total runout, and position (RFS).

| Concept | Geometric Tolerance | | |
|---|---|---|---|
| | Concentricity | Total Runout | Position (RFS) |
| Tolerance zone shape | Cylinder | Two coaxial cylinders | Cylinder |
| Tolerance zone applies to... | Median points | Surface | Axis of unrelated actual mating envelope |
| Relative cost to produce | $$ | $$$ | $ |
| Relative cost to verify | $$$ | $$ | $ |
| Deviations being controlled | Location and orientation | Location, orientation, and form | Location and orientation |

**FIGURE 25-8** *Comparison of Concentricity, Total Runout, and Position Tolerances Applied to Coaxial Features of Size*

© EFFECTIVE TRAINING INC.

---

**TECHNOTE 25-2**
**Comparison of Concentricity,
Total Runout, and Position Tolerances**

Two differences between runout and concentricity tolerances are:

1. The shape of the tolerance zone

2. A runout tolerance affects circularity

One difference between position (RFS) and concentricity tolerances is that a position tolerance controls the location of the axis of the unrelated actual mating envelope, and a concentricity tolerance controls the location of a cloud of median points.

---

## Symmetry Tolerance Applied to a Width

In certain applications, the function of a part requires defining the symmetry deviation of a width feature of size relative to a datum center plane or datum axis.

In the example in Figure 25-9, the symmetry deviation of a width feature of size is being limited relative to a datum center plane. In this application, the following conditions apply:

- The width must meet its size requirements.

- The tolerance zone is the space within two parallel planes centered on the datum plane.

- All median points of the opposed elements of the feature of size must be located within the tolerance zone.

- The maximum distance between median points of the toleranced width and the datum center plane is half the symmetry tolerance value.

- The worst-case boundary of the toleranced width is 21.6mm.

- The symmetry tolerance also limits the flatness of the derived median plane.

***Design Tip***
Consider using a position tolerance (RFS) when defining a symmetrical relationship of a width. A position tolerance is less expensive to produce and inspect.

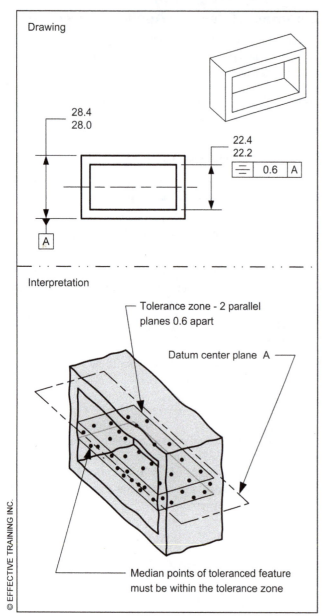

**FIGURE 25-9  Symmetry Tolerance Application**

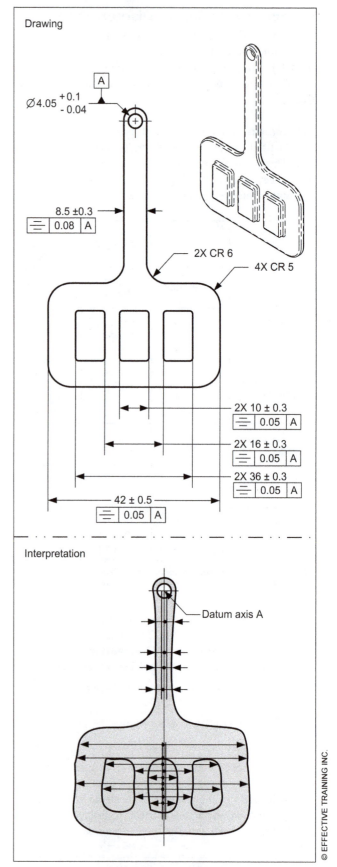

**FIGURE 25-10  Symmetry Applied to a Pendulum**

Precision balance is a functional requirement of a pendulum. A planar width symmetrical to a datum axis (or center plane) may vary greatly in size as long as its median points remain close to the datum axis (or center plane). As the location of the median points vary from the datum, the balance is affected.

In Figure 25-10, several symmetry tolerances are applied to keep the median points of various pendulum elements symmetrical about the datum axis (or center plane).

The variations in size do not affect the balance providing the median points remain near the datum axis. Depending upon the level of precision required, hundreds of median points may be required to ensure the balance requirement.

## Comparison Between Symmetry and Position (RFS) Tolerances

When dimensioning symmetrical relationships, several geometric tolerances may be used. The choices are position, profile, and symmetry. In this section, I will compare symmetry with position at RFS.

Where used to control symmetrical relationships, symmetry and position tolerances are similar but have a few significant differences. The chart in Figure 25-11 compares symmetry and position (RFS) tolerances.

| Concept | Geometric Tolerance | |
|---|---|---|
| | **Symmetry** | **Position (RFS)** |
| Tolerance zone shape | Two parallel planes | Two parallel planes |
| Tolerance zone applies to... | Median points | Center plane of the unrelated actual mating envelope |
| Part deviations being controlled | Location and orientation | Location and orientation |
| Relative cost to produce | $$ | $ |
| Relative cost to verify | $$$ | $$ |

*FIGURE 25-11  Comparison Between Symmetry and Position Tolerances*

---

**TECHNOTE 25-3**
**Comparison of Symmetry and Position Tolerances at RFS**

Two similarities between position and symmetry are:

1. The shape of the tolerance zone
2. Both tolerances limit deviations in location and orientation

One difference between position (RFS) and symmetry tolerances is that a position tolerance controls the location of the center plane of the unrelated actual mating envelope, and a symmetry tolerance controls the location of a cloud of median points.

---

## The Significant Seven Questions

This section uses the Significant Seven Questions to interpret concentricity tolerance application. The significant seven questions are explained in Appendix B.

Figure 25-12 contains a concentricity tolerance application. The Significant Seven Questions are answered in the interpretation of the figure. Notice how each answer increases your understanding of the geometric tolerance.

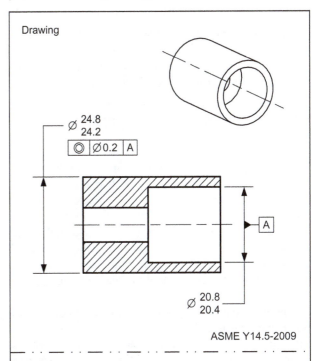

Drawing

⌀ 24.8 / 24.2

◎ ⌀0.2 A

A

⌀ 20.8 / 20.4

ASME Y14.5-2009

Interpretation

The Significant Seven Questions

1. Which dimensioning and tolerancing standard applies?
   Answer – ASME Y14.5-2009

2. What does the tolerance apply to?
   Answer – The median points of the toleranced feature

3. Is the specification standard-compliant?
   Answer – Yes

4. What are the shape and size of the tolerance zone?
   Answer – The space within a 0.2 diameter cylinder

5. How much total tolerance is permitted?
   Answer – 0.2

6. What are the shape and size of the datum simulators?
   Answer – A expanding cylinder such as a mandrel

7. Which geometry attributes are affected by this tolerance?
   Answer – Location and orientation

*FIGURE 25-12 The Significant Seven Questions Applied to a Concentricity Tolerance*

# VERIFICATION PRINCIPLES AND METHODS

## Verification Principles for Concentricity and Symmetry Tolerances

This section contains a simplified explanation of the verification principles and methods that can be used to inspect a concentricity or symmetry tolerance. Verifying a concentricity or symmetry tolerance has two parts:

1. Establishing the datum axis (or center plane) at RMB

2. Verifying that the location of the median points are within the tolerance zone; depending upon the level of precision desired, hundreds of median points may be required to ensure the balance requirement

The datum axis RMB may be established by mounting the datum feature in a datum feature simulator, such as a chuck or collet.

A concentricity tolerance can be verified with dial indicators, a CMM, or by other methods. If dial indicators are used, two diametrically opposed dial indicators are placed on either side of the feature and positioned equidistant from the datum axis. The dial indicators are rotated, and several readings are obtained at each section along the entire length of the feature of size. The mid-point of each measurement is calculated and compared to the tolerance zone.

A symmetry tolerance can also be verified with dial indicators using a method similar to concentricity. However, the dial indicators are not rotated; they are moved to various points on the surfaces. The size of the feature of size and the Rule #1 requirement also have to be verified. The examples in this section use dial indicators to verify concentricity and symmetry tolerances.

## Verifying a Concentricity Tolerance Applied to a Cylindrical Feature of Size

There are many ways to verify a concentricity tolerance. In Figure 25-13, a concentricity tolerance is applied to a cylindrical feature of size. One way the concentricity tolerance could be verified is shown in the lower half of the figure. The part is mounted in a chuck or collet to establish datum axis A. Then a height gage is used to measure the distance from the datum axis to two diametrically opposed points on the cylindrical surface. The median point is calculated, and its distance from the datum axis is determined.

The part is rotated, another median point is determined, and its location is checked. The dial indicator is moved axially to another location on the cylindrical feature of size, and several sets of median points are obtained. Their location relative to the datum axis is calculated. The number of median points checked should be determined during inspection planning. Depending upon the level of precision desired, hundreds of median points may be required to ensure the balance requirement.

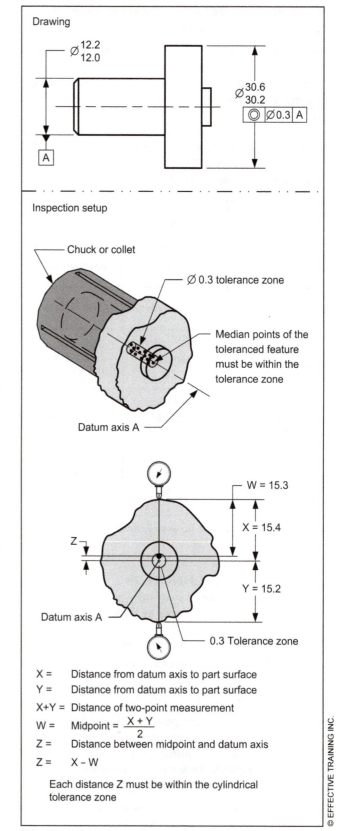

Drawing

Inspection setup

Chuck or collet

Ø 0.3 tolerance zone

Median points of the toleranced feature must be within the tolerance zone

Datum axis A

W = 15.3

X = 15.4

Z

Y = 15.2

Datum axis A

0.3 Tolerance zone

X = Distance from datum axis to part surface
Y = Distance from datum axis to part surface
X+Y = Distance of two-point measurement
W = Midpoint = $\dfrac{X+Y}{2}$
Z = Distance between midpoint and datum axis
Z = X – W

Each distance Z must be within the cylindrical tolerance zone

**FIGURE 25-13** *Verifying a Concentricity Tolerance*

Inspecting a concentricity tolerance is very difficult and costly. Concentricity can be inspected, for acceptance only, by using a runout verification. A runout tolerance controls all points on the surface, while a concentricity tolerance only controls opposing points. Therefore, a cylindrical feature may pass a concentricity check, but fail a runout check because it is not round.

Although the dial indicator method shown in Figure 25-9 is similar to methods used to check runout, it is more difficult to perform. A concentricity tolerance is actually a "looser" tolerance than a runout tolerance. If a concentricity tolerance is inspected using a single dial indicator, as a runout check, and it passes the runout check, then the median points will be within the concentricity tolerance zone.

If a part passes ⌿ 0.3 A , then it also passes ◎ Ø0.3 A . Therefore, the part would only need to be inspected to the "looser" concentricity tolerance if it failed the more restrictive runout tolerance check.

## Verifying a Symmetry Tolerance Applied to a Width

There are many ways to verify a symmetry tolerance. In Figure 25-14, a symmetry tolerance is applied to a width. One way the symmetry tolerance could be verified is shown in the lower half of the figure. The part is mounted in a set of parallel jaws to establish datum center plane A. A height gage is used to measure the distance from the datum center plane to two opposed points on each surface. The median point is calculated and its distance from the datum axis is determined. Additional median points are found. Their location relative to the datum center plane is calculated. The number of median points checked should be determined during inspection planning.

Depending upon the level of precision desired, hundreds of median points may be required to ensure the balance requirement.

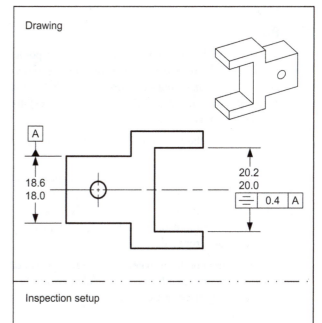

Drawing

18.6 / 18.0

20.2 / 20.0 ⬌ 0.4 A

Inspection setup

Height gage with opposed indicators
Variable jaws
Part
Surface plate

1. Center the indicators to the datum center plane
2. Zero the indicators to the nominal distance from the datum center plane
3. Measure the deviation of the opposed points from the datum center plane
4. Calculate the distance of the median point from the datum center plane (see Figure 25-12)
5. Repeat steps 1 thru 4 if necessary

*FIGURE 25-14 Verifying a Symmetry Tolerance*

## SUMMARY
### Key Points

- Concentricity is the condition where the median points of all diametrically opposed elements of a surface of revolution (or the median points of correspondingly located elements of two or more radially disposed features) are congruent with a datum axis.

- A median point is the mid-point of a two-point measurement.

- A concentricity tolerance is a geometric tolerance that defines the permissible deviation that median points of a surface of revolution (or correspondingly located elements of two or more radially disposed features) are permitted to vary from a datum axis.

- Symmetry is the condition where the median points of all opposed or correspondingly located elements of two or more feature surfaces are congruent with a datum axis or center plane.

- A symmetry tolerance is a geometric tolerance that defines the permissible deviation that median points of all opposed or correspondingly located elements of two or more feature surfaces are permitted to vary from a datum axis or center plane.

- A concentricity tolerance zone is a cylinder centered on the datum axis.

- A symmetry tolerance zone is two parallel planes centered on a datum axis or datum center plane.

- Concentricity and symmetry tolerances limit the location of median points relative to a datum.

- The only modifiers permitted in a symmetry tolerance are Ⓕ ⟨ST⟩.

- The only modifiers permitted in a concentricity tolerance are Ⓕ ⟨ST⟩ ∅ S∅.

- Due to the type of application, the concentricity tolerance is rarely used on drawings. A concentricity tolerance is expensive to verify. Runout or position tolerances are often used to replace concentricity tolerances.

- Consider using a position tolerance when tolerancing the symmetrical relationship of a width. A position tolerance is less expensive to produce and inspect.

- A real-world application of a concentricity or symmetry tolerance balance of a high speed rotating part.

### Additional Related Topics

*These topics are recommended for further study to improve your understanding of concentricity and symmetry.*

| Topic | Source |
|---|---|
| Using concentricity tolerances to control deviations on non-cylindrical surfaces | ETI's *Advanced Concepts of GD&T* textbook |

# QUESTIONS AND PROBLEMS

**Website Bonus Materials**
More questions are available at our website. To access bonus materials for this textbook, please visit:
www.etinews.com/textbookbonus

## True and False

*Indicate if each statement is true or false.*

T / F    1. A concentricity tolerance must be applied with datum feature references.

T / F    2. The tolerance zones for a concentricity tolerance and a total runout tolerance are the same.

T / F    3. Inspection of concentricity tolerance is more expensive than inspection of total runout tolerance.

T / F    4. Concentricity tolerances require datum feature references at RMB.

T / F    5. A symmetry tolerance is used where precision balance is important.

T / F    6. A symmetry tolerance may use the MMC modifier.

T / F    7. The tolerance zone for a  concentricity tolerance is the space between two concentric cylinders.

T / F    8. The tolerance zone for a symmetry tolerance is the space between two parallel planes

## Multiple Choice

*Circle the best answer to each statement.*

1. The midpoint of a two-point measurement is known as:
   A. A center point
   B. A midpoint
   C. A median point
   D. A datum point

2. Concentricity is the condition where:
   A. Two or more cylindrical features are shown on the same center line
   B. The axis of the related actual mating envelope is related to the datum axis
   C. Median points of diametrically opposed elements are congruent with a datum axis
   D. The axis of an unrelated actual mating envelope is aligned to a datum axis

3. Two types of geometry attributes affected by a concentricity or symmetry tolerance are:
   A. Location and orientation
   B. Size and form
   C. Concentricity and symmetry
   D. None of the above

4. The modifiers that can be specified in a concentricity tolerance are:
   A. Ⓕ Ⓜ and ⟨ST⟩
   B. ∅ Ⓟ and ⟨ST⟩
   C. Ⓕ ∅ and ⟨ST⟩
   D. All of the above

5. A real-world application for a concentricity tolerance is:
   A. Assembly
   B. Support
   C. Balance of a high-speed rotating part
   D. Seal on a rotating part

6. A difference between a concentricity tolerance and a position tolerance at RFS is:
   A. The shape of the tolerance zones
   B. What the tolerance zone applies to
   C. The part deviations being controlled
   D. None of the above

7. Verifying a concentricity tolerance has two parts: establishing the datum axis and verifying that the _____ is within the tolerance zone.
   A. Surface
   B. Axis
   C. Median points
   D. None of the above

8. The tolerance zone for a symmetry tolerance is:
   A. The space between two parallel planes
   B. The space within a cylinder
   C. A worst-case boundary
   D. None of the above

## Application Problems

*The application problems are designed to provide practice applying the chapter concepts to situations that are similar to on-the-job conditions.*

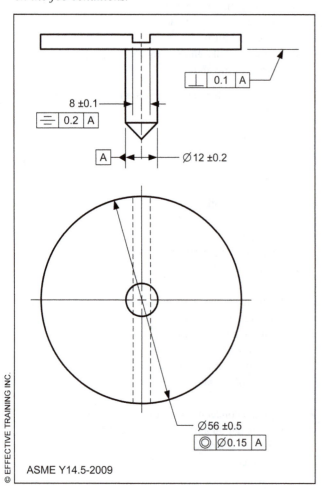

8 ±0.1

Ø 12 ±0.2

0.1 | A

0.2 | A

A

Ø 56 ±0.5

◎ | Ø0.15 | A

ASME Y14.5-2009

© EFFECTIVE TRAINING INC.

*Use the figure above to answer questions 1-4.*

2. The geometry attributes controlled by concentricity tolerance are location and:
   A. Size
   B. Form
   C. Orientation
   D. All of the above

3. The maximum possible distance between a median point of the slot and the datum plane passing through datum axis A is:
   A. 0.05
   B. 0.2
   C. 0.4
   D. None of the above

4. The tolerance zone for the symmetry tolerance is:
   A. The space within a 0.2 diameter cylinder.
   B. The space between two parallel planes 0.2 apart
   C. The space between two parallel lines 0.2 apart
   D. None of the above

1. Use the Significant Seven Questions to interpret the concentricity tolerance.

   1) Which dimensioning and tolerancing standard applies?
   _____

   2) What does the tolerance apply to? _____
   _____

   3) Is the specification standard-compliant? _____

   4) What are the shape and size of the tolerance zone?
   _____

   5) How much total tolerance is permitted? _____

   6) What are the shape and size of the datum simulators?
   _____

   7) Which geometry attributes are affected by this tolerance?
   _____

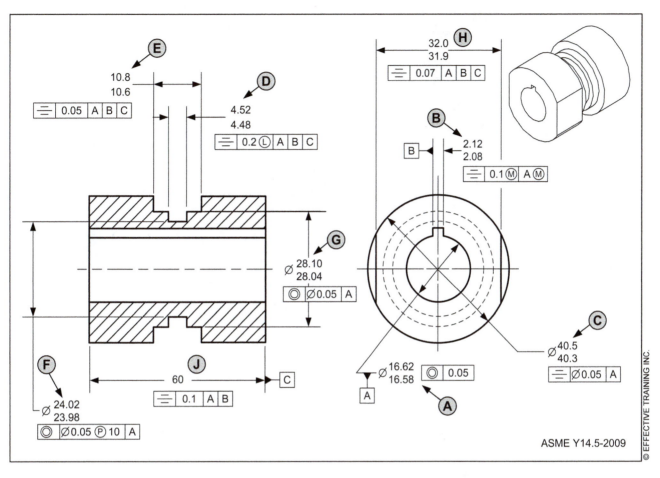

ASME Y14.5-2009

© EFFECTIVE TRAINING INC.

5. Indicate if each symmetry or concentricity tolerance below is standard-compliant. If it is not, describe why. (Hint: use the CARE test.)

A. ⌖ 0.05 _____

B. ⌰ 0.1 Ⓜ A Ⓜ _____

C. ⌰ ⌀0.05 A _____

D. ◎ 0.2 Ⓛ A B C _____

E. ⌰ 0.05 A B C _____

F. ◎ ⌀0.05 Ⓟ 10 A _____

G. ◎ ⌀ 0.05 A _____

H. ⌰ 0.07 A B C _____

J. ⌰ 0.1 A B _____

# Profile Tolerance - Basic Concepts

## Goal

Interpret the profile tolerance

## Performance Objectives

Upon completing this chapter, you should be able to:

1. Describe the terms "profile," "true profile," "profile of a surface tolerance," and "profile of a line tolerance" (p.358)
2. Describe the tolerance zone shape for profile of a surface and profile of a line tolerances (p.359)
3. Describe the geometry attributes that a profile tolerance can affect (p.359)
4. Explain the effect of using profile with or without datum references (p.359)
5. List the modifiers used with profile tolerances (p.360)
6. Describe bilateral, unilateral, and non-uniform profile tolerance zones (p.360)
7. Describe the effect of the unequally disposed profile symbol (p.361)
8. Describe the default condition for the extent a profile control tolerance zone applies to a part (p.362)
9. Describe the effect of the "between" symbol (p.362)
10. Describe the effect of the "all around" symbol (p.362)
11. Describe the effects of the "all over" symbol (p.362)
12. List three advantages of using profile tolerances (p.363)
13. Evaluate if a profile tolerance specification is standard-compliant (p.364)

## New Terms

- All around symbol
- All over symbol
- Between symbol
- Bilateral equally disposed tolerance zone
- Bilateral unequally disposed tolerance zone
- Non-uniform tolerance zone

- Profile
- Profile of a line tolerance
- Profile of a surface tolerance
- True profile
- Unequally disposed profile symbol
- Unilateral tolerance zone

## What This Chapter Is About

There are two types of profile tolerances: profile of a surface and profile of a line. In this chapter, you'll learn the basic concepts of the profile tolerances, and in the next chapter, you'll learn about profile tolerance applications. The basic concepts include default conditions for a profile tolerance zone, profile tolerance zone options, modifiers used with profile tolerances, and the advantages of using profile tolerances.

Profile tolerances are important because they are considered the most versatile and powerful tolerances in the geometric tolerancing system.

# TERMS AND CONCEPTS

## Profile and True Profile

A *profile* is an outline of a surface, a shape made up of one or more features, or a two-dimensional element of one or more features. A *true profile* is a profile located and defined by basic radii, basic angular dimensions, basic coordinate dimensions, basic size dimensions, profile to drawings, formulas, or mathematical data, including design models.

Figure 26-1 shows examples of profile and true profile.

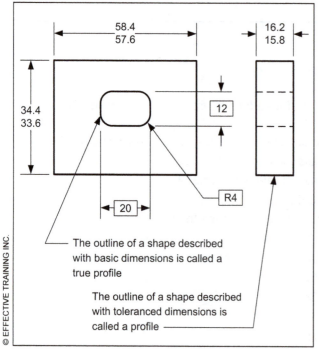

FIGURE 26-1  *Profile and True Profile*

## Profile of a Surface Tolerance

A *profile of a surface tolerance* is a geometric tolerance that establishes a three-dimensional tolerance zone that extends along the length and width (or circumference) of the considered feature or features. The tolerance zone is centered along the true profile. The shape of the tolerance zone is the same as the true profile of the toleranced surface.

A profile of a surface tolerance applies to the entire surface simultaneously. Profile tolerances may be used with or without a datum feature reference. The actual part surface is allowed to have any type of deviation within the tolerance zone. The tolerance zone for a profile of a surface tolerance is illustrated in Figure 26-2.

## Profile of a Surface Tolerance Zone

When a profile of a surface tolerance is specified, the tolerance zone is the space within a uniform 3D boundary centered about the true profile. The tolerance zone applies to the full length, width, and depth of the surface. The part surface is allowed to have any type of deviation within the tolerance zone. An example is shown in Figure 26-2.

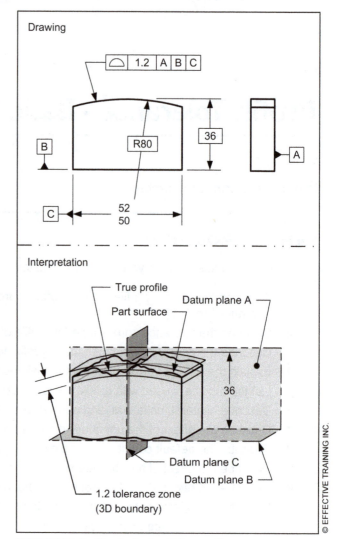

FIGURE 26-2  *Profile of a Surface Tolerance Zone*

## Profile of a Line Tolerance

A *profile of a line tolerance* is a geometric tolerance that establishes a two-dimensional tolerance zone that is normal to the true profile at each line element. The tolerance zone is centered along the true profile. The shape of the tolerance zone is the same as the true profile of the toleranced line element.

A profile of a line tolerance applies to each line element of a surface independently. It may be used with or without a datum feature reference.

Since profile of a line tolerance zone is two-dimensional, it only applies in the view in which it is shown. The actual line elements of the part surface are allowed to have any type of deviation within the tolerance zone. The tolerance zone for a profile of a line tolerance is illustrated in Figure 26-3.

## Profile of a Line Tolerance Zone

When a profile of a line tolerance is specified, the tolerance zone is the space within a uniform 2D boundary centered about the true profile. The tolerance zone applies independently at each line element of the surface. An example is shown in Figure 26-3

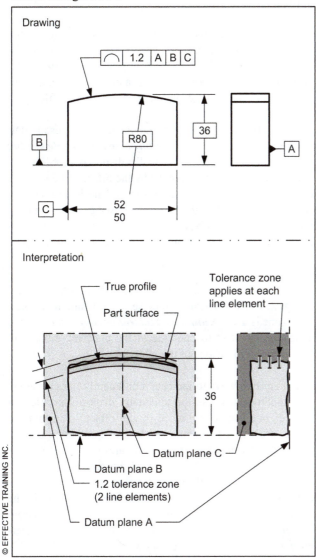

FIGURE 26-3  *Profile of a Line Tolerance Zone*

## Geometry Attributes Affected by Profile Tolerances

Profile tolerances are versatile controls. A profile tolerance can be used to limit deviations of line elements or a whole surface. A profile tolerance, depending upon how it is applied, can affect four types of geometry attributes:

1. Size
2. Location
3. Orientation
4. Form

The chart in Figure 6-4 (page 66) shows the geometric tolerances and the geometry attributes that they can control.

## Profile Tolerances With and Without Datum References

A unique aspect of profile tolerances is that they can be specified with or without datum references. All applications of profile tolerances control form, and (depending upon the application) a profile tolerance may control coplanarity (or alignment) or size. In addition, depending upon the datums referenced, profile tolerances can control orientation and location.

Where datums are referenced in a profile tolerance, they orient and locate the tolerance zone. The result is that profile tolerances (profile of a surface, in particular) often control orientation and location of the toleranced surface(s).

A profile tolerance with no datum references, based on the application, may be a size and/or form control. Where a profile tolerance does not contain any datum references, the tolerance zone is oriented and located by the toleranced surface.

***Author's Comment***
A profile tolerance always applies to a surface or line element. Therefore, a profile tolerance should never be attached to a basic dimension.

### TECHNOTE 26-1
**Profile Tolerances**

A profile tolerance:

• Must be applied to a true profile

• Without datum references, controls form

• Depending upon application, may control coplanarity or alignment

• Depending upon application, may control size

• With datum references, also controls location and/ or orientation

• Depending upon application, it may affect all four geometry attributes of size, location, orientation, and form

## Modifiers Used With Profile Tolerances

There are several modifiers that are used in the tolerance portion of a profile feature control frame. The choice of modifier is often related to the functional requirements of the application. The chart in Figure 26-4 shows the common modifiers and examples of applications where they could be used.

| Tolerance Modifier | Effect | Functional Application |
|---|---|---|
| Ⓕ * | Release the restraint requirement | Non-rigid parts with restraint notes |
| ⟨ST⟩ * | Requires statistical process controls | When statistical tolerance stacks have been used to determine the tolerance value |
| NON-* UNIFORM | Invokes a non-uniform tolerance zone | Where a surface requires a tighter tolerance in some areas but not for an entire surface |
| Ⓤ | Invokes an offset tolerance zone | Where tolerances cannot be applied as an equal ± distribution |
| Ⓣ | Releases the indirect flatness requirement on planar surfaces | Where the flatness deviation may be larger than the location and orientation requirements of a profile tolerance of a surface |
| * These modifiers are only introduced in this text | | |

© EFFECTIVE TRAINING INC.

**FIGURE 26-4** *Modifiers Used With Profile Tolerances*

## Bilateral, Unilateral, and Non-Uniform Profile Tolerance Zones

There are five types of tolerance zones that may be used with a profile tolerance. They are listed below and explained in the following paragraphs:

- Bilateral - equally disposed (default)
- Bilateral - unequally disposed (option)
- Unilateral - outside of material (option)
- Unilateral - inside of material (option)
- Non-uniform (option)

Unless otherwise indicated, where the leader line of a profile tolerance (surface or line) is directed to a surface or its extension line, the tolerance zone is bilateral (equally disposed about the true profile).

A *bilateral equally disposed tolerance zone* is a tolerance zone that is equally divided bilaterally to both sides of the true profile. Where an equally disposed (centered) bilateral tolerance zone is intended, the profile tolerance has a leader line directed to the surface or an extension line of the surface. An equally disposed bilateral tolerance zone is shown in Figure 26-5.

In certain applications, the function of a part may require a profile tolerance zone that is bilateral but unequally disposed (not centered) about the true profile. A *bilateral unequally disposed tolerance zone* is a uniform boundary tolerance zone that is not centered about the true profile.

There are two ways to indicate an unequally disposed bilateral tolerance zone. The *unequally disposed profile symbol* indicates a unilateral or unequally disposed profile tolerance. Use this symbol followed by the offset value (outside of material) in the tolerance portion of the feature control frame. The second method (for 2D drawings only) is to use phantom lines and a basic dimension to indicate the offset of the tolerance zone. Both methods are shown in Figure 26-5.

In certain cases, the function of a part may require a unilateral profile tolerance zone. In these cases, a unilateral tolerance zone is used. A *unilateral tolerance zone* is a tolerance zone that is offset to one side of the material, either "all outside" or "all inside" of the material.

There are two ways to indicate a unilateral tolerance zone. One method is to use the unequally disposed profile tolerance zone symbol followed by the offset value (outside of material) in the tolerance portion of the feature control frame. An offset value of zero indicates that all of the tolerance zone exists to the inside of the material. An offset value equal to the full tolerance value indicates that all of the tolerance zone exists to the outside of the material. The second method (for 2D drawings only) is to use a phantom line and a basic dimension to indicate the offset of the tolerance zone. Both methods are shown in Figure 26-5.

In certain cases, the function of a part may require a profile tolerance zone that is not a uniform distance from the true profile. In these cases, a non-uniform tolerance zone is used. A *non-uniform tolerance zone* is a maximum material boundary and least material boundary of unique shape that encompasses the true profile. The method to indicate a non-uniform tolerance zone is use the term "NON-UNIFORM" to replace the tolerance value in the feature control frame. The feature control frame is directed to the surface. The tolerance zone boundary may be defined with phantom lines and basic dimensions on the 2D drawing or with supplemental geometry in the 3D CAD model. This method is shown in Figure 26-6.

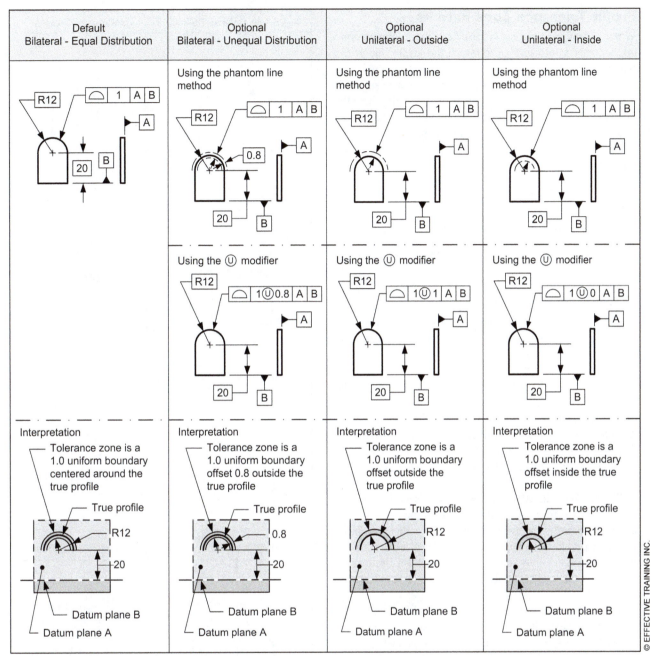

**FIGURE 26-5  Bilateral and Unilateral Profile Tolerance Zones**

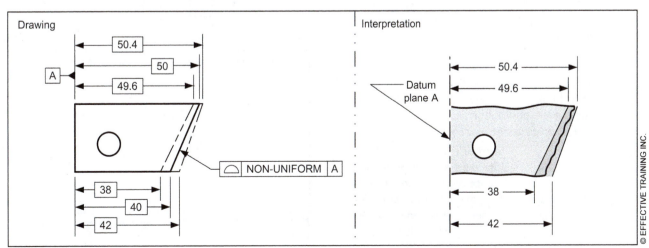

**FIGURE 26-6  Non-Uniform Profile Tolerance Zone**

© EFFECTIVE TRAINING INC.

## Profile Tolerance Zone Extents

When a profile tolerance is specified, the tolerance zone applies only to the surface to which the leader line of the feature control frame is directed. A surface is considered to end where the geometry changes shape. For example, if the toleranced surface is a planar surface, it ends where the planar surface comes to a corner or to a tangent point with another surface. If desired, a profile tolerance zone can be extended to include additional surfaces by using one of four methods:

1. The between symbol
2. The all around symbol
3. The all over symbol
4. A note associated with the profile tolerance feature control frame

In Figure 26-7A, the profile tolerance zone only applies to the radius surface to which it is directed. The tolerance zone ends where the radius become tangent to the planar surfaces. This is the default condition for the tolerance zone for a profile tolerance.

In Figure 26-7B, the between symbol is used to extend the tolerance zone to include all surfaces on the part. The *between symbol* is the symbolic means of indicating that a profile tolerance zone applies to all surfaces between two points. The symbol for "between" is a double-ended arrow with letters at each end to identify the extents of the tolerance zone.

In Figure 26-7C, the all around symbol is used to extend the tolerance zone to include all surfaces on the part. The *all around symbol* is the symbolic means of indicating that a profile tolerance zone applies to all surfaces, all around the true profile in the view shown. The symbol is a circle at the junction of the leader line from the feature control frame.

In Figure 26-7D, the all over symbol is used to extend the tolerance zone to include several surfaces. The *all over symbol* is the symbolic means of indicating that a profile tolerance zone applies to all surfaces, all over the three-dimensional profile of a part. The "all over" symbol is a double circle at the junction of the leader line from the feature control frame.

In Figure 26-7E, a notation is used to extend the tolerance zone to include several surfaces. The notation can be in several different forms. In this case, the notation "2X" is used to indicate that the profile tolerance zone applies to both radius surfaces on each side of the gap.

### Author's Comment
The Y14.5 Standard is not explicit about profile tolerance zone extents. The information in this text is one interpretation that many experts agree upon. However, the words in the standard also leave room for other interpretations.

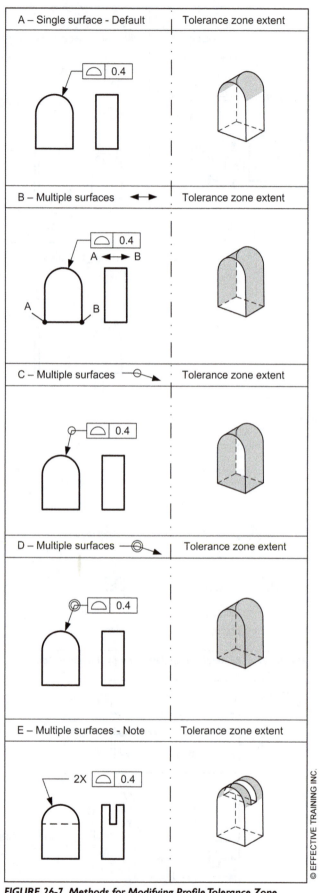

*FIGURE 26-7 Methods for Modifying Profile Tolerance Zone Extents*

© EFFECTIVE TRAINING INC.

## TECHNOTE 26-2
### Profile Tolerance Zone Coverage

A profile tolerance zone (surface or line) applies only to the surface to which the feature control frame is directed, unless it is extended with the between symbol, the all around symbol, the all over symbol, or a note.

## Advantages of Using Profile Tolerances

Profile tolerancing is very flexible and can be used in many places on an engineering drawing. It also offers many advantages over coordinate tolerancing.

One way to recognize the advantages of profile tolerancing is to compare profile tolerancing with coordinate tolerancing. Three important advantages of profile tolerances are:

- Profile tolerances provide a clear definition of the tolerance zone.
- Profile tolerancing communicates how to hold the part for repeatable measurements.
- Profile tolerancing eliminates tolerance accumulation.

In Figure 26-8, a part is dimensioned with profile tolerancing in panel A and with coordinate tolerancing in panel B. Let's look at the profile tolerancing first.

With the profile tolerance, the tolerance zone for the top radius surface of the part is clearly defined. It is a 0.4-wide boundary centered on the true profile. The datum references communicate part functional requirements (location and orientation) and the inspection setup. The top surface of the part is affected by only one tolerance: the profile tolerance.

In Figure 26-8B, the part is dimensioned with coordinate tolerances. With coordinate tolerances, the tolerance zone for the top surface of the part is not clear. Since the radius comes from the center of the hole, both hole location tolerances affect the location of the radius. Also, the inspection setup, (which surfaces to use, and in which order) is not defined.

The accumulation of tolerances for the hole location and the assumption required to hold the part for measurement result in a complex tolerance zone shape for the top surface of the part.

### Author's Comment

Profile tolerancing is often a good tool to replace coordinate tolerancing. Profile tolerances can also be used as general tolerances on a drawing.

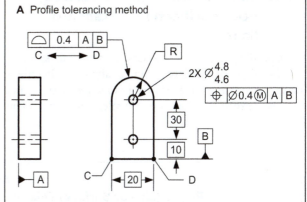

**A** Profile tolerancing method

- Radius tolerance zone 0.4 wide uniform boundary
- Datum sequence specified
- No tolerance accumulation

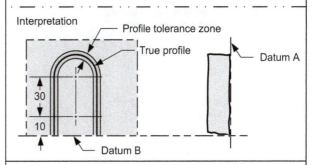

Interpretation

**B** Coordinate tolerancing method

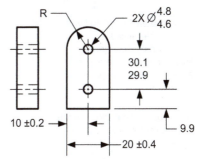

- Radius tolerance zone shape unclear
- Must assume a datum sequence
- Hole tolerances accumulate and may affect radius location

Interpretation

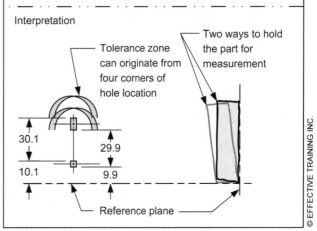

*FIGURE 26-8 Advantages of Profile Tolerancing*

## Evaluate if a Profile (Line or Surface) Tolerance Specification is Standard-Compliant

For a profile tolerance to be a standard-compliant specification (legal), it must satisfy certain requirements. In order to make it easier to remember what to look for when evaluating the correctness of a geometric tolerance on a drawing, I created a mnemonic.

The requirements are divided into four areas represented by the initials "CARE":

    C - Check for datums

    A - Assess the application

    R - Review the modifiers

    E - Evaluate the tolerance

A flowchart that identifies the requirements for a standard-compliant profile tolerance is shown in Figure 26-9.

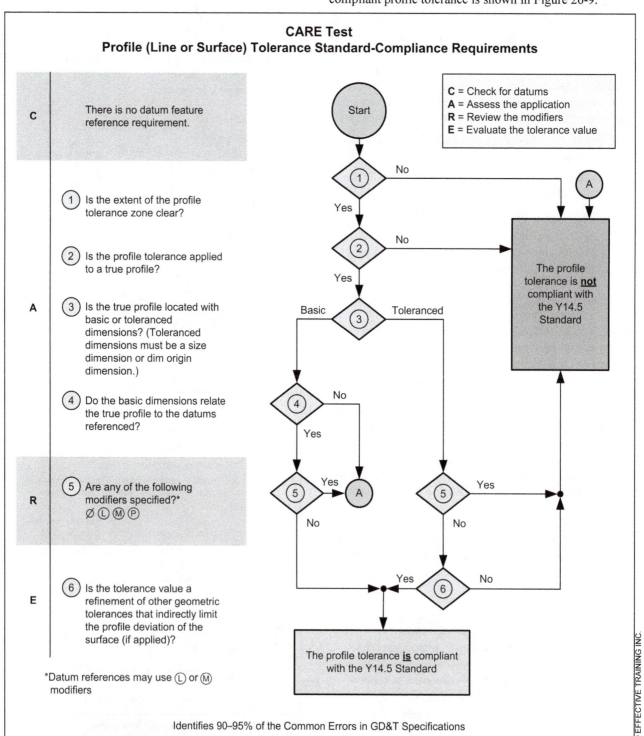

**CARE Test**
**Profile (Line or Surface) Tolerance Standard-Compliance Requirements**

C    There is no datum feature reference requirement.

C = Check for datums
A = Assess the application
R = Review the modifiers
E = Evaluate the tolerance value

(1) Is the extent of the profile tolerance zone clear?

(2) Is the profile tolerance applied to a true profile?

A    (3) Is the true profile located with basic or toleranced dimensions? (Toleranced dimensions must be a size dimension or dim origin dimension.)

(4) Do the basic dimensions relate the true profile to the datums referenced?

R    (5) Are any of the following modifiers specified?*
Ø Ⓛ Ⓜ Ⓟ

E    (6) Is the tolerance value a refinement of other geometric tolerances that indirectly limit the profile deviation of the surface (if applied)?

*Datum references may use Ⓛ or Ⓜ modifiers

The profile tolerance is **not** compliant with the Y14.5 Standard

The profile tolerance **is** compliant with the Y14.5 Standard

Identifies 90–95% of the Common Errors in GD&T Specifications

*FIGURE 26-9 Test For a Standard-Compliant Profile Tolerance*

## SUMMARY

### Key Points

- A profile is an outline of a surface, a shape made up of one or more features, or a two-dimensional element of one or more features.

- A true profile is a profile defined by basic radii, basic angular dimensions, basic coordinate dimensions, basic size dimensions, undimensioned drawings, formulas, or mathematical data, including design models.

- A profile of a surface tolerance is a geometric tolerance that establishes a three-dimensional tolerance zone that extends along the length and width (or circumference) of the considered feature or features.

- A profile of a line tolerance is a geometric tolerance that establishes a line element two-dimensional tolerance zone that is normal to the true profile at each line element.

- A profile tolerance can affect location, orientation, form (or coplanarity), and size.

- All profile tolerances control form deviations of a surface or line elements.

- Depending upon the application, profile tolerances may control form.

- Where a profile tolerance contains datum references, it is an orientation and/or location tolerance.

- A profile tolerance with no datum references is a form and/or coplanarity control.

- There are five types of tolerance zones that may be used with a profile tolerance:
  - Bilateral - equally disposed (default)
  - Bilateral - unequally disposed (option)
  - Unilateral - outside of material (option)
  - Unilateral - inside of material (option)
  - Non-uniform (option)

- When a profile tolerance is specified, the tolerance zone applies only to the surface to which the feature control frame is directed.

- A non-uniform tolerance zone is a maximum material boundary and least material boundary of unique shape that encompasses the true profile.

- The only modifiers permitted in a profile tolerance are Ⓕ ⓈⓉ Ⓤ Ⓣ.

- Four methods to extend the coverage of a profile tolerance zone:
  - The between symbol
  - The all around symbol
  - The all over symbol
  - A note associated with the profile tolerance

- The between symbol is the symbolic means of indicating that a profile tolerance zone applies to all surfaces between two points. It is a double-ended arrow with letters at each end to identify the extents of the tolerance zone.

- The all around symbol is the symbolic means of indicating that a profile tolerance zone applies to all surfaces all around the true profile in the view shown. It is a circle at the junction of the leader line from the feature control frame.

- The all over symbol is the symbolic means of indicating that a profile tolerance zone applies to all surfaces all over the three-dimensional profile of a part. It is a double circle at the junction of the leader line from the feature control frame.

- Three important advantages of profile tolerances are:
  - Profile tolerances provide a clear definition of the tolerance zone.
  - Profile tolerancing communicates how to hold the part for repeatable measurements.
  - Profile tolerancing eliminates tolerance accumulation.

### Additional Related Topics

*These topics are recommended for further study to improve your understanding of profile.*

| Topic | Source |
|---|---|
| Limited segments of a profile | *ASME Y14.5-2009, Para. 8.3.1.5* |
| All over | *ASME Y14.5-2009, Para. 8.2.1.1* |
| Non-uniform tolerance zones | *ASME Y14.5-2009, Para. 8.3.2* |
| Composite profile tolerances | *ASME Y14.5-2009, Para. 8.6* |
| Profile with datum feature shift | ETI's *Advanced Concepts of GD&T* textbook |

## QUESTIONS AND PROBLEMS

**Website Bonus Materials**
More questions are available at our website. To access bonus materials for this textbook, please visit:
www.etinews.com/textbookbonus

### True and False

*Indicate if each statement is true or false.*

T / F    1. A true profile is the outline of one or more features.

T / F    2. A profile of a surface tolerance establishes a three-dimensional tolerance zone.

T / F    3. All applications of profile tolerances require datum references.

T / F    4. A profile of a line tolerance establishes a three-dimensional tolerance zone.

T / F    5. A profile tolerance can limit deviations in the geometry attribute of size.

T / F    6. A unequal bilateral profile tolerance can be centered about the true profile.

T / F    7. When the "all over" symbol is used, the profile tolerance applies all over the entire part.

T / F    8. When the "all around" symbol is used, the profile tolerance applies all around the entire part.

### Multiple Choice

*Circle the best answer to each statement.*

1. Which geometry attribute deviation(s) can be limited with a profile of a surface tolerance?
   A. Location        C. Form
   B. Orientation     D. All of the above

2. A true profile may be defined with:
   A. Basic radii
   B. Basic angles
   C. Formulas
   D. All of the above

3. Which modifier may be applied to the profile tolerance value?
   A. Ⓤ                      C. Ⓢ
   B. Ⓜ                      D. All of the above

4. The default tolerance zone for a profile tolerance is:
   A. Non-uniform
   B. Unilateral
   C. Bilateral - equal distribution
   D. Bilateral - unequal distribution

5. An advantage of using a profile tolerance in place of a coordinate tolerance is:
   A. A bonus tolerance is permitted
   B. A datum feature sequence may be specified
   C. A profile tolerance always controls size
   D. All of the above

6. The shape of the tolerance zone for a profile tolerance is:
   A. Two parallel planes
   B. The same as the true profile of the toleranced surface
   C. Equal bilateral
   D. Cylindrical when the diameter symbol is specified in the tolerance portion of the feature control frame

7. When a profile tolerance does not contain any datum feature references, it controls:
   A. Form
   B. Location
   C. Orientation
   D. All of the above

8. When the Ⓤ modifier is used in a profile tolerance it indicates that the tolerance zone is _____ disposed.
   A. Unilaterally
   B. Universally
   C. Unequally
   D. Unambiguously

## Application Problems

*The application problems are designed to provide practice applying the chapter concepts to situations that are similar to on-the-job conditions.*

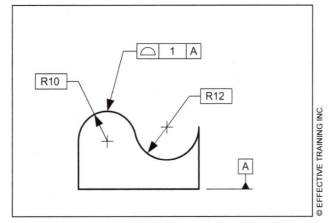

1. Draw the profile tolerance zone to indicate the extent of the part that the profile tolerance applies to on the drawing above.

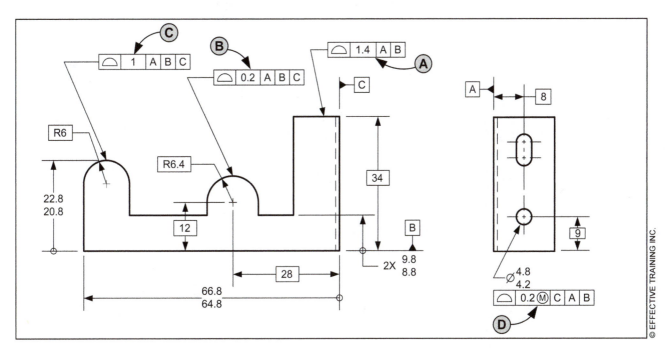

*Question 2 refers to the drawing above.*

2. Indicate if each profile tolerance below is standard-compliant. If it is not, describe why. (Hint: use the CARE test.)

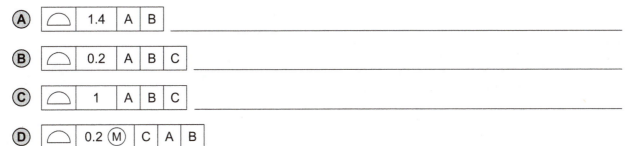

# Profile Tolerance Applications

## Goal

Interpret profile tolerance applications

## Performance Objectives

Upon completing this chapter, you should be able to:

1. Describe real-world applications for a profile of a surface tolerance (p.370)
2. Understand the verification principles for profile tolerances (p.370)
3. Interpret a profile of a surface tolerance applied to a planar surface (p.370)
4. Describe how to inspect a profile of a surface tolerance that is used to locate a planar surface (p.371)
5. Interpret profile of a surface applied to a closed polygon (p.372)
6. Interpret profile of a surface applied to a conical surface (p.373)
7. Interpret a profile tolerance applied to coplanar surfaces (p.374)
8. Describe what a multiple single-segment profile tolerance is (p.376)
9. Interpret a multiple single-segment profile control (p.376)
10. Describe what a composite profile tolerance is (p.378)
11. Interpret a composite profile tolerance (p.378)
12. Interpret a profile of a surface tolerance using the "Significant Seven Questions" (p.380)
13. Describe real-world applications for a profile of a line tolerance (p.380)
14. Interpret a profile of a line tolerance used with a dimension origin dimension to locate the surface (p.381)
15. Interpret a profile of a line tolerance application used to control orientation and form (p.382)

## New Terms

- Composite profile tolerance
- Indicator zero block
- Multiple single-segment profile tolerance

## What This Chapter Is About

In this chapter, we will look at profile tolerances in selected applications. The applications will provide you with a good understanding of the flexibility of profile tolerances.

Understanding profile tolerance applications is important because profile tolerances are used to define many different types of surfaces. Profile tolerances can be used to define the location, orientation, form, and size of any surface, from a simple plane to complex geometry. Therefore, profile tolerances are common on engineering drawings.

# TERMS, CONCEPTS, APPLICATIONS, AND VERIFICATION PRINCIPLES

This chapter covers a selection of position tolerance concepts and applications. To make the applications easier to learn, the application and verification information is combined with the terms and concepts.

## Real-World Applications of Profile of a Surface

There are many common real-world applications for profile tolerances. These applications limit the location, size, orientation, and form deviations of surfaces on parts. Examples of common profile of a surface applications are:

- Sealing between contoured surfaces

- Controlling relationship between mounting surfaces

- Relating dimensions to datum reference frames for measurement repeatability

- Controlling form of complex shapes such as cam profiles, turbine blades, hip and knee joint replacements, aircraft wings, etc.

## Verification Principles for Profile of a Surface Tolerances

This section provides a simplified description of the verification principles and methods that can be used to inspect a profile of a surface tolerance.

Verifying a profile tolerance that controls size and form:

- Collect a set of surface points.

- Use a method (fitting routine) to determine if the set of surface points fit within the tolerance zone.

Verifying a profile tolerance where it controls coplanarity (alignment):

- Establish a reference plane that is tangent to the high points of the toleranced surfaces.

- Measure the distance the surface points are from the reference plane.

- Compare the value of the measured points to the profile tolerance zone.

Verifying a profile tolerance where it controls location or orientation:

- Establish the relationship between the part and the datum reference frame.

- Establish the profile of a surface tolerance zone.

- Measure the surface points (collect the location of a set of surface points).

- Determine that the surface points are within the profile tolerance zone.

Verifying a profile tolerance where it controls orientation (and form):

- Establish the relationship between the part and the datum reference frame.

- Establish the profile of a surface tolerance zone.

- Measure the surface points (collect the location of a set of surface points).

- Determine that the surface points are within the profile tolerance zone. (The set of surface points may translate along X, Y, and Z to optimize the points within the tolerance zone).

A profile of a surface tolerance can be verified with a CMM, dedicated variable gage, or other methods as long as the measurement method complies with the verification principles.

---

### Author's Comment

A profile tolerance can control the geometry attribute of size. However, unlike a size measurement, a profile tolerance is not a two point measurement across opposed points. This is because a profile tolerance zone imposes perfect form at MMC and LMC.

Using a two-point size measurement to verify a profile tolerance is a compromise based on cost and expedience. However, the inspector should be aware that there is a risk of accepting a bad part. A part may pass a two-point measurement method but fail a profile measurement.

---

## Surface Location Application

In certain cases, the function of a part may require that a surface be located relative to a datum reference frame. This may be accomplished with a profile of a surface tolerance. In Figure 27-1, a profile of a surface tolerance is used to control the location, orientation, and form of a surface. In this application, profile is applied to a planar surface, and the following conditions apply:

- The profile tolerance is applied to a true profile.

- The true profile is related to the datum reference frame with basic dimensions.

- The datum reference frame is three orthogonal planes.

- The tolerance zone is a boundary 1.0mm wide centered about the true profile.

- There are no additional tolerances available.

- All surface elements must be located within the tolerance zone.

- The tolerance zone limits deviations in the geometry attributes of location, orientation, and form.

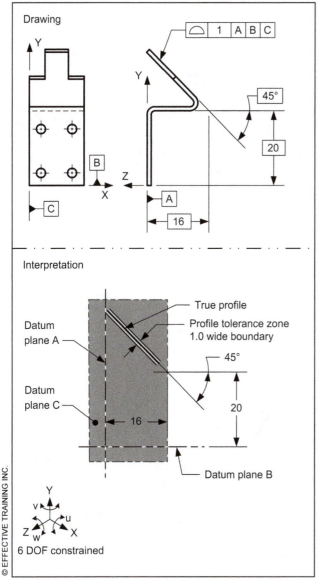

FIGURE 27-1 *Profile of a Surface Tolerance Used to Locate a Surface*

## Verifying a Profile Tolerance Application

There are many ways to verify a profile tolerance. Using a CMM is a common method, but not the only way. To inspect the profile tolerance for the part in Figure 27-1, we decided to use the dedicated gage shown in Figure 27-2.

The gage consists of datum feature simulators for datum features A, B, and C. The gage also has a gage detail with a surface parallel to the true profile of the toleranced surface and located at a basic distance from the true profile. An indicator is mounted against the gage surface so it is perpendicular to the toleranced surface. The gage detail is mounted on a precision slide or dovetail guides so it can move in the X axis. The indicator is located in a slot so it can move in the Y-Z plane parallel to the true profile. This allows the indicator to cover the whole surface.

An indicator zero block is used to set (zero) the indicator to the basic distance that the gage surface is located from the true profile of the toleranced surface. An *indicator zero block* is a gage detail that has two parallel surfaces set at a set distance apart. In this case, the basic distance is equal to the gage surface is from the true profile.

The part is brought into contact with the datum feature simulators, being careful to follow the datum sequence of the profile tolerance. The dial indicator is mounted against the gage surface, then moved along the surface. Since the indicator is zeroed to the true profile, it is measuring a plus/minus bilateral tolerance that is half of the profile tolerance (±0.5). If the indicator reading exceeds this ±0.5 tolerance as it moves along the surface, the part is rejected.

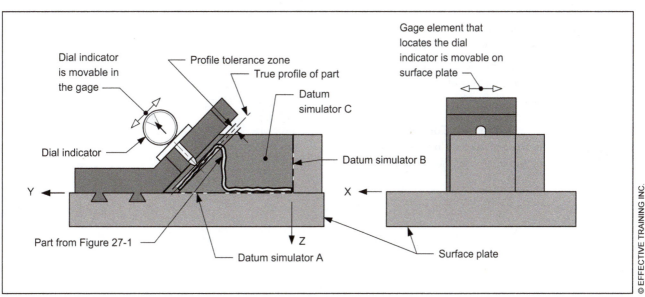

FIGURE 27-2 *Verifying Profile of a Surface Tolerance With a Dedicated Gage*

## Profile Tolerance Applied to a Closed Polygon

In certain cases, the function of a part may require controlling size, location, orientation and form deviations of a closed polygon. This may be accomplished with a profile of a surface tolerance with datum feature references. Where a profile of surface tolerance with the all around symbol is applied to a closed polygon, it limits deviations in size and form. Where datums are referenced, it also controls the location and orientation deviations of the polygon.

In Figure 27-3, a profile of a surface tolerance is used to control the size, location, orientation, and form of a closed polygon. In this application, the following conditions apply:

- A true profile is established by the basic dimensions.

- The true profile is located to datum axis A by an implied basic zero.

- The datum reference frame is an axis and two orthogonal planes (rotation unconstrained).

- The all around symbol extends the tolerance zone to all sides of the polygon.

- The tolerance zone is 3-dimensional.

- The tolerance zone is 0.4mm wide boundary centered about the true profile (two extruded coaxial hexes centered about the datum axis).

- All surface elements of the flats must lie within the profile tolerance zone.

- The profile tolerance limits deviations in the geometry attributes of size, location, orientation, and form.

- The max radius specification limits the allowable radius on the on the corners of the polygon, not the profile tolerance.

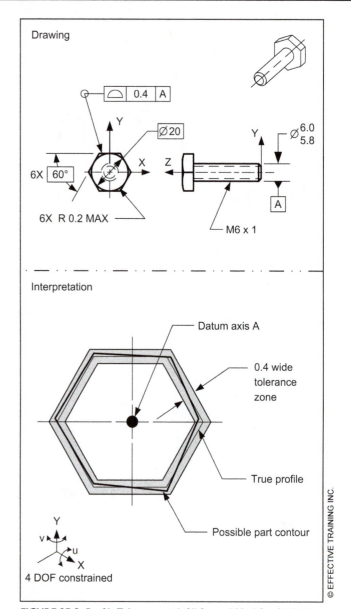

FIGURE 27-3 *Profile Tolerance with All Around Modifier Applied to a Polygon*

© EFFECTIVE TRAINING INC.

## Profile Tolerance Applied to a Conical Surface

In certain applications, the form and taper angle of a cone may need to be controlled. This may be accomplished with a profile of surface tolerance. The profile tolerance may be applied without a datum feature reference to limit deviations in form and taper angle only. It may also be applied with datum feature references to limit deviations in orientation, and location.

In Figure 27-4, a profile of surface tolerance is used to limit deviations in form, taper angle, orientation, and location of a cone. In this application, the following conditions apply:

- A true profile is established by the basic taper angle.

- The datum reference frame contains three orthogonal planes, two intersecting through the datum axis and one at the end of the part (rotation unconstrained).

- The tolerance zone applies to the full circumference of the surface.

- The tolerance zone is 3-dimensional.

- The tolerance zone is a 0.2mm wide boundary that applies to the full circumference of the surface (i.e., two coaxial cones 0.2mm radial distance between them).

- All surface elements must lie within the profile tolerance zone.

- Since a toleranced size dimension applies to the cone, the profile tolerance does not limit the size of the cone.

- The tolerance zone (two coaxial cones) may change in size within the size limits specified while maintaining the 0.2mm gap between the cones.

- The profile tolerance limits deviations in the geometry attributes of form (circularity and straightness), orientation, and location.

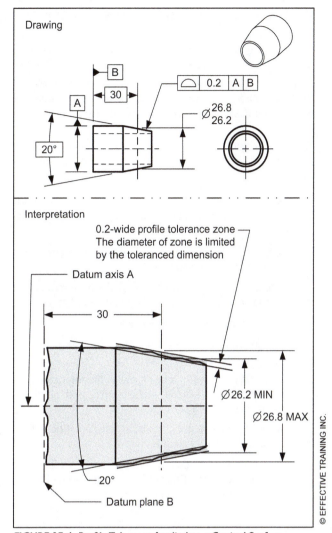

**FIGURE 27-4  Profile Tolerance Applied to a Conical Surface**

## Profile Tolerance Applied to Coplanar Surfaces

In certain cases, the function of a part may require controlling coplanar surfaces relative to each other (see Chapter 13). This may be accomplished with a profile of a surface tolerance with no datum references. The profile tolerance would limit deviations in coplanarity and form.

Where a profile tolerance is specified without any datum feature references to control coplanarity (alignment), the tolerance zone is unilateral.

An example is shown in Figure 27-5. In this application, profile is applied to coplanar surfaces, and the following conditions apply:

- A true profile is established by the implied basic zero offset between the two surfaces.

- The number of surfaces to which the profile tolerance applies is indicated above the feature control frame.

- The tolerance zone is 3-dimensional.

- The tolerance zone is a unilateral (or a total) zone of two parallel planes 0.4 apart.

- The first plane (reference plane) of the tolerance zone is an implied self-datum plane tangent to the high points across both surfaces.

- All surface elements of both surfaces must lie within the profile tolerance zone.

- The profile tolerance limits deviations in the geometry attributes of form (flatness) and coplanarity (alignment to each other).

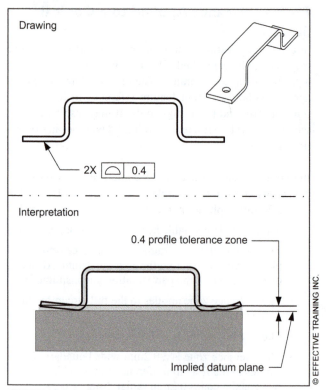

**FIGURE 27-5** *Profile Tolerance Applied to Coplanar Surfaces*

In applications where there are several coplanar surfaces, it may be desirable to control the surfaces relative to each other or relative to a datum reference frame. This may be accomplished with two profile tolerances: one with a datum feature reference and one with no datum references. Depending upon how they are specified, the profile tolerances would limit deviations in location, orientation, coplanarity and form.

An example is shown in Figure 27-6. In this application, there are two profile tolerances applied to the coplanar surfaces.

For the 0.3 profile tolerance applied to datum features B and C, the following conditions apply:

- A true profile is established by an implied basic zero between the two surfaces and an implied basic 90° to datum plane A.

- The number of surfaces to which the profile tolerance applies is indicated below the feature control frame.

- There is an implied self-secondary datum plane perpendicular to the primary datum plane and tangent to the highest point on each surface.

- The tolerance zone is a unilateral two parallel planes 0.3 apart perpendicular to the primary datum plane starting at the high points across both surfaces projecting into the surface.

- All surfaces elements of both surfaces must lie within the tolerance zone.

- The profile tolerance limits deviations in form, coplanarity, and orientation of both surfaces.

For the 0.8 profile tolerance applied to the middle two surfaces, the following conditions apply:

- A true profile is established by an implied basic zero between the two surfaces and an implied basic 90° to datum plane A.

- The true profile is oriented to datum plane A by an implied basic 90° angle.

- The true profile is located to datum plane B-C by an implied basic zero.

- The number of surfaces to which the profile tolerance applies is indicated below the feature control frame.

- The tolerance zone is two parallel planes 0.8 apart perpendicular to datum plane A and centered about the B-C plane.

- All surface elements of both surfaces must lie within the profile tolerance zone.

- The profile tolerance limits deviations of the geometry attributes of form, orientation, and location.

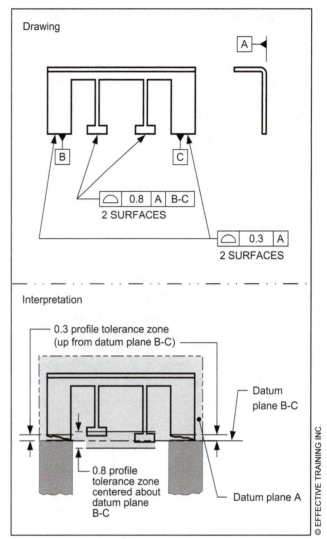

**FIGURE 27-6  Profile Tolerance Applied to Multiple Sets of Coplanar Surfaces**

© EFFECTIVE TRAINING INC.

## Multiple Single-Segment Profile Tolerance

In certain applications, the function of a feature may require different amounts of deviation for location, orientation, form, and size. These deviations are normally combined when using a profile tolerance. A multiple single-segment profile tolerance allows different tolerance values to be assigned to the various geometry attributes.

A *multiple single-segment profile tolerance* is where two or more profile tolerances are applied to a surface (or collection of surfaces) relative to different datums. When using multiple single-segment profile, different levels of control of part geometry attributes can be achieved by using different tolerance values and adding, removing, or changing datum references. The top segment of a multiple single-segment profile tolerance is typically used for location, and the lower segments refine the orientation, form, and size (if applicable). An example of a multiple single-segment profile tolerance is shown in Figure 27-7.

For example, if a profile tolerance with no datum references is applied to a surface, it will control deviations in form only. If three datum references are added to the profile tolerance, it will control deviations in location, orientation, size (if applicable) and form.

## Multiple Single-Segment Profile Tolerance Interpretation

A multiple single-segment feature control frame typically contains two or three segments, but it may contain as many segments as needed to convey the design requirements. In Figure 27-8, a multiple single-segment profile of a surface tolerance is used to control the size, location, orientation, and form of a closed polygon. In this application, the following conditions apply:

- A true profile is established by the basic dimensions that define the size and shape of the polygon.

- The all around symbol extends the profile tolerance zone to apply to all surfaces of the polygon.

- The top segment of the profile tolerance, with three datum references, limits deviations in location. The tolerance zone is a 1mm wide zone centered about the true profile.

- The 1mm wide zone is fixed in location and orientation relative to the datum reference frame (all six degrees of freedom are constrained).

- The middle segment of the profile tolerance, with one datum reference, limits deviations in orientation relative to datum plane A, and location and orientation to datum plane B.

- The 0.4mm wide zone is centered about the true profile and floats within the 1mm zone (three degrees of freedom are constrained).

- The bottom segment of the profile tolerance, with no datum references, limits deviations in size and form.

- The 0.2mm wide zone is centered about the true profile and floats within the 0.4 zone.

- The 0.2 zone is not constrained to any datum.

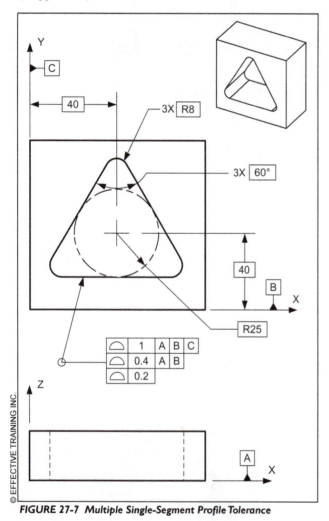

*FIGURE 27-7 Multiple Single-Segment Profile Tolerance*

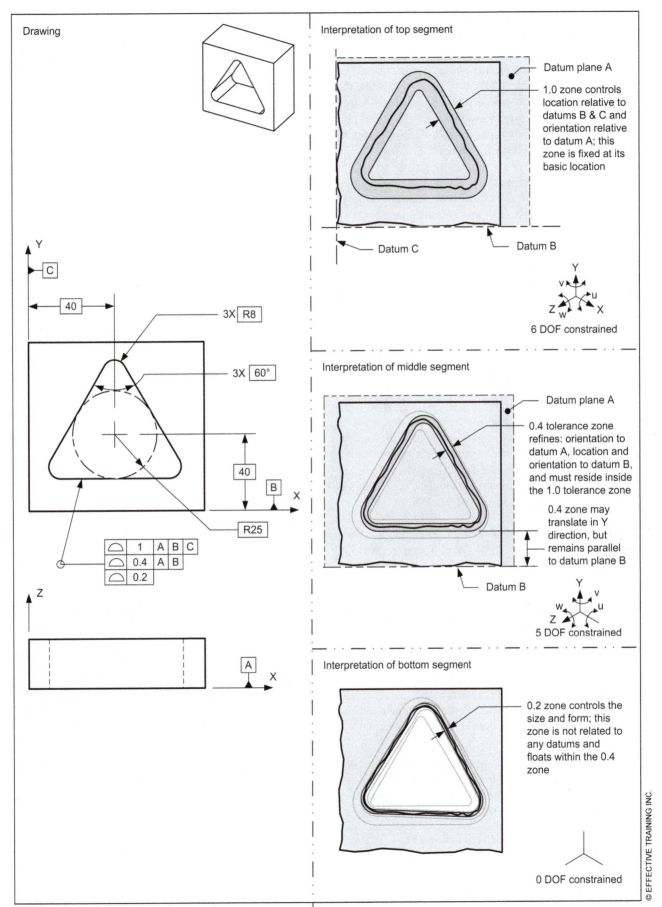

**Drawing**

3X R8

3X 60°

R25

| ⌓ | 1 | A | B | C |
| ⌓ | 0.4 | A | B | |
| ⌓ | 0.2 | | | |

**Interpretation of top segment**

Datum plane A

1.0 zone controls location relative to datums B & C and orientation relative to datum A; this zone is fixed at its basic location

Datum C

Datum B

6 DOF constrained

**Interpretation of middle segment**

Datum plane A

0.4 tolerance zone refines: orientation to datum A, location and orientation to datum B, and must reside inside the 1.0 tolerance zone

0.4 zone may translate in Y direction, but remains parallel to datum plane B

Datum B

5 DOF constrained

**Interpretation of bottom segment**

0.2 zone controls the size and form; this zone is not related to any datums and floats within the 0.4 zone

0 DOF constrained

© EFFECTIVE TRAINING INC.

*FIGURE 27-8  Interpretation of a Multiple Single-Segment Profile Tolerance*

## Composite Profile Tolerance

Where a pattern of surfaces requires their location tolerance to be larger than their orientation tolerance to a datum reference frame, a composite profile tolerance is specified. The composite profile tolerance may also be used to refine the spacing or alignment tolerance between surfaces within a pattern, as well as their form and size (if applicable). This section introduces how to specify multiple levels of controls using composite profile tolerances.

Composite profile tolerances look similar to multiple single-segment profile tolerances, but they have different interpretations and requirements. Composite profile tolerances are a complex topic and are only introduced in this text.

A *composite profile tolerance* is a feature control frame that contains a single entry of a profile symbol that is applicable to all horizontal segments of the feature control frame. A composite profile tolerance may have more than two horizontal segments, each with rules for how to specify the tolerance value, modifiers, and datums. See Figure 27-9.

A composite profile tolerance can only be applied to patterns of surfaces. Each segment creates different requirements for allowable deviations in location, orientation, and/or spacing or alignment, size, and form of the pattern. The uppermost segment is the pattern-locating control. It specifies a location tolerance for the pattern of surfaces relative to the datums references specified. Depending upon the datums referenced, the uppermost segment can constrain translation and rotational degrees of freedom.

The lower segments cannot control location, they can only constrain rotational degrees of freedom. Therefore, the lower segments can only control orientation, spacing or alignment, size and form of the surfaces in a pattern.

If datum references are not specified in a lower segment, the segment can only control size and form, in addition to spacing or alignment between surfaces in the pattern.

A major difference between multiple single-segment and composite profile tolerances is shown below:
- In a multiple single-segment profile tolerance, each segment can constrain translational and rotational degrees of freedom.
- In a composite profile tolerance, only the uppermost segment can constrain translational and rotational degrees of freedom. The lower segments can only control rotational degrees of freedom.

**Author's Comment**
This text only introduces the topic of composite profile tolerances. To learn about the requirements and limitations of composite profile tolerances see ETI's *Advanced Concepts of GD&T.*

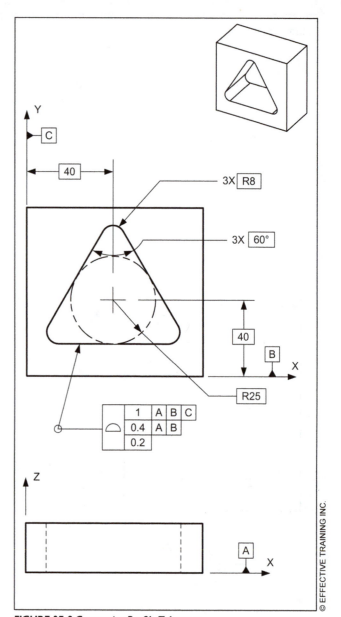

FIGURE 27-9 Composite Profile Tolerance

## Composite Profile Tolerance Interpretation

In Figure 27-10, the uppermost segment of the composite profile tolerance controls the location of the hole (the pattern of surfaces). The middle segment only controls the orientation of the hole to datums A and B (not location). The bottom segment controls the size and form of the hole.

Notice that when verifying the middle segment, only the rotational degrees of freedom are constrained relative datums A and B. Simply put, the basic dimensions relative to the datums do not apply to the middle segment.

Each segment of the composite feature control frame represents a separate requirement of a multiple interrelated requirement. The actual surface of the part must be within each of the profile tolerance zones.

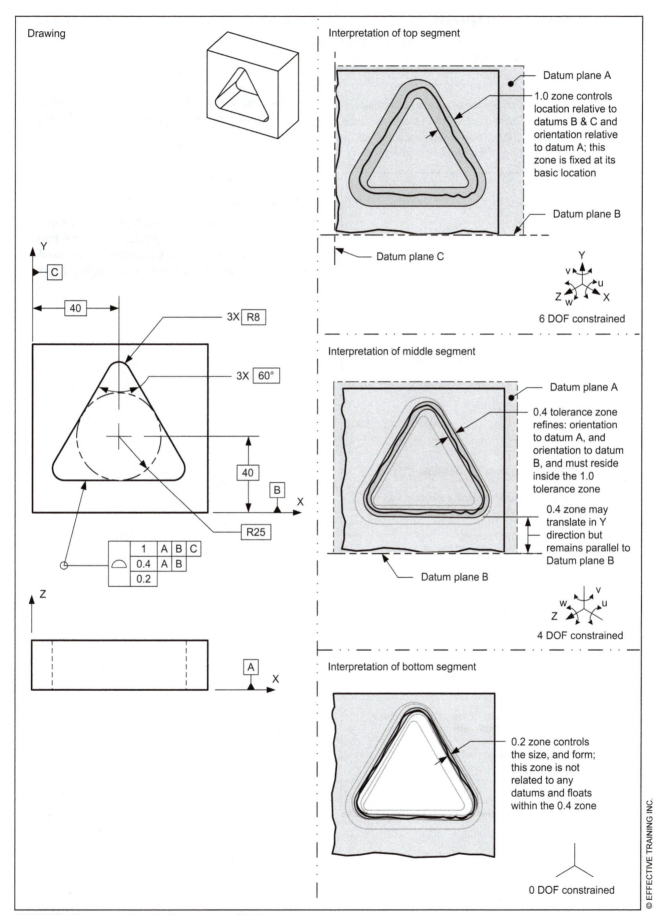

Drawing

Interpretation of top segment

— Datum plane A

1.0 zone controls location relative to datums B & C and orientation relative to datum A; this zone is fixed at its basic location

— Datum plane B

— Datum plane C

6 DOF constrained

Y

C

40

3X R8

3X 60°

40

B

X

R25

| 1 | A | B | C |
|---|---|---|---|
| 0.4 | A | B | |
| 0.2 | | | |

Z

A    X

Interpretation of middle segment

— Datum plane A

0.4 tolerance zone refines: orientation to datum A, and orientation to datum B, and must reside inside the 1.0 tolerance zone

0.4 zone may translate in Y direction but remains parallel to Datum plane B

— Datum plane B

4 DOF constrained

Interpretation of bottom segment

0.2 zone controls the size, and form; this zone is not related to any datums and floats within the 0.4 zone

0 DOF constrained

© EFFECTIVE TRAINING INC.

*FIGURE 27-10 Interpretation of a Composite Profile Tolerance*

## The Significant Seven Questions

This section uses the Significant Seven Questions to interpret a profile of a surface tolerance. The questions are explained in Appendix B.

Figure 27-11 contains a straightness tolerance application. The Significant Seven Questions are answered in the interpretation section of the figure. Notice how each answer increases your understanding of the geometric tolerance.

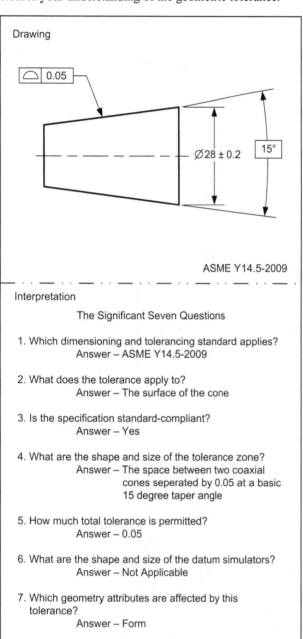

Drawing

⌒ 0.05

Ø28 ± 0.2    15°

ASME Y14.5-2009

Interpretation

The Significant Seven Questions

1. Which dimensioning and tolerancing standard applies?
   Answer – ASME Y14.5-2009

2. What does the tolerance apply to?
   Answer – The surface of the cone

3. Is the specification standard-compliant?
   Answer – Yes

4. What are the shape and size of the tolerance zone?
   Answer – The space between two coaxial cones seperated by 0.05 at a basic 15 degree taper angle

5. How much total tolerance is permitted?
   Answer – 0.05

6. What are the shape and size of the datum simulators?
   Answer – Not Applicable

7. Which geometry attributes are affected by this tolerance?
   Answer – Form

FIGURE 27-11 *The Significant Seven Questions Applied to a Profile Tolerance*

## Real-World Applications of Profile of a Line

There are several real-world applications for a profile of a line tolerance on a drawing. Traditionally, profile of a line has been applied to aerodynamic, hydrodynamic, and ergonomic surfaces to facilitate inspection of the surface at specified cross sections. However, profile of a surface may also be applied without compromising the design requirements. Applications may include contours of:

- Automobile seats
- Propeller blades
- Electric razor body

Profile of a line may be used with profile of a surface as one segment of a multiple single feature control frame or as a refinement to a surface that is located with a coordinate toleranced dimension with a dimension origin symbol.

## Interpretation of Profile of a Line Tolerance Used With the Dimension Origin Symbol

In certain cases, a surface may be located using a coordinate toleranced dimension with a dimension origin symbol, and the orientation and form of the line elements controlled using a profile of a line tolerance.

In Figure 27-12, a coordinate toleranced dimension with a dimension origin symbol is applied to the height of the part, and a profile of line tolerance, with two datum references, is applied to the top surface. In this application the following conditions apply:

- The toleranced dimension locates the true profile of the top surface, (i.e., controls the height of the part).

- The datum reference frame is two orthogonal planes.

- The profile of line tolerance applies to each individual line element independently.

- The tolerance zone is a 2D zone that applies in the view shown only.

- The tolerance zone is the space between two line elements 0.2 apart.

- The tolerance zone is allowed to move within the tolerance (15.8-16.6) height limits.

- The profile of line tolerance controls deviations in the geometry attributes of orientation (parallelism), and form (straightness).

**Author's Comment**
The use of profile of a line is rare on engineering drawings. There are very few functional applications that require profile of a line.

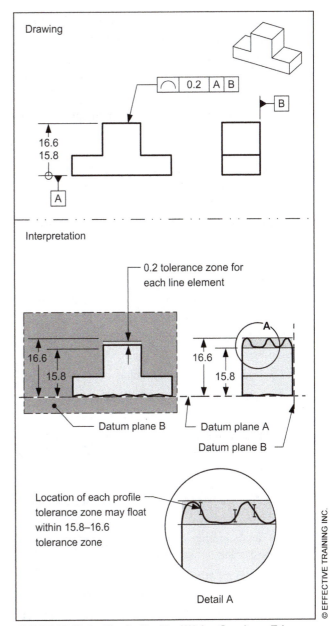

**FIGURE 27-12** *Profile of a Line Used With a Coordinate Tolerance*

© EFFECTIVE TRAINING INC.

**Author's Comment**
A common misconception about profile of a line is that it should be applied to thin parts (e.g., a gasket).

## Interpretation of Profile of a Line Tolerance Used to Control Orientation and Form

Figure 27-13 shows an example of a profile of line tolerance with one datum reference and a profile of surface tolerance with two datum references applied to a conical surface.

For the profile of a surface tolerance, the following conditions apply:

- The datum reference frame is two orthogonal planes passing through datum axis A and a third plane perpendicular to the first two, tangent to the highest point on datum feature B.

- A true profile is established by the basic diameter and basic angle.

- The true profile is located to datum axis A by an implied basic zero.

- The true profile is located from datum plane B by the basic 30 dimension.

- The tolerance zone applies to the full circumference of the cone.

- The tolerance zone is a 3D zone.

- The tolerance zone is a 0.8 wide space between two coaxial cones.

- The profile of surface tolerance controls deviations in the geometry attributes of size, location, and circularity.

For the profile of a line tolerance, the following conditions apply:

- The datum reference frame is a single orthogonal plane passing through datum axis A.

- The tolerance zone is a 2D zone that applies to each line element individually in the view shown.

- The tolerance zone is a 0.2 space between two parallel lines at a basic 10° angle from datum axis A.

- The tolerance zone may move within the 0.8 zone.

- The profile of line tolerance controls deviations in the geometry attributes of orientation to datum axis A and straightness of the line elements.

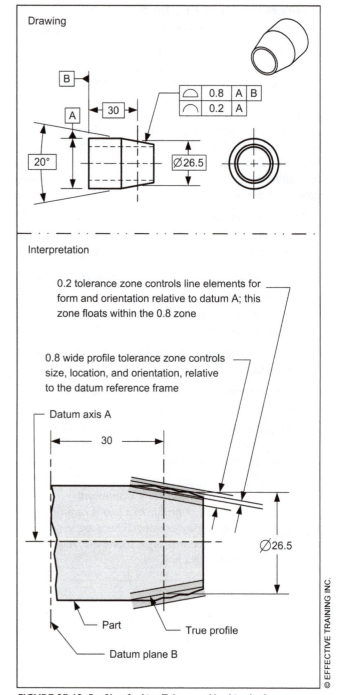

© EFFECTIVE TRAINING INC.

**FIGURE 27-13** *Profile of a Line Tolerance Used in the Lower Segment of a Multiple Single-Segment Profile Tolerance*

## SUMMARY

### Key Points

- Examples of profile of a surface applications are:
  - o Sealing between contoured surfaces
  - o Controlling relationship between mounting surfaces
  - o Relating dimensions to datum reference frames for measurement repeatability
  - o Controlling form of complex shapes such as cam profiles, turbine blades, hip and knee joint replacements, aircraft wings, etc.

- A multiple single-segment profile tolerance is where two or more profiles tolerances are applied to a surface (or collection of surfaces) relative to different datums.

- A composite profile tolerance is a feature control frame that contains a single entry of a profile symbol that is applicable to all horizontal segments of the feature control frame.

- When a profile tolerance is applied to coplanar surfaces with no datum references:
  - o A true profile is established by the implied basic zero offset between the two surfaces.
  - o The tolerance zone is unilateral (or a total) zone.

- An indicator zero block is a gage detail that has two parallel surfaces at a set distance apart. There are only a few real-world applications for profile of a line tolerance on drawings.

- An example of a profile of a line application is to control form deviations of surface line elements.

### Additional Related Topics

*These topics are recommended for further study to improve your understanding of profile tolerance applications.*

| Topic | Source |
|---|---|
| Profile tolerances with datum feature shift | ETI's *Advanced Concepts of GD&T* textbook |
| Composite profile tolerances | *ASME Y14.5-2009, Para. 8.6* |

## QUESTIONS AND PROBLEMS

***Website Bonus Materials***
More questions are available at our website. To access bonus materials for this textbook, please visit:
www.etinews.com/textbookbonus

## True and False

*Indicate if each statement is true or false.*

T / F    1. One real-world application for a profile of a surface tolerance is to relate dimensions to a datum reference frame for measurement repeatability.

T / F    2. When a profile tolerance is applied to a planar surface without a datum reference, it can simultaneously control the location, orientation, and form of the surface.

T / F    3. When verifying a profile tolerance that controls deviations in size and form, the values of the measured distances must be compared to the basic dimensions locating the true profile.

T / F    4. A multiple single-segment profile of a surface tolerance cannot have more than two segments.

T / F    5. When a profile tolerance is specified without any datum feature references to control coplanarity, the tolerance zone is automatically unilateral.

T / F    6. A composite profile tolerance is a feature control frame that contains a single entry of a profile symbol that is applicable to all horizontal segments of the feature control frame

T / F    7. The top segment of multiple single segment profile tolerance is typically used to control the form of the toleranced surface.

## Multiple Choice

*Circle the best answer to each statement.*

1. When a profile of surface tolerance with three datum references and the all around symbol is applied to a closed polygon shape, it may control the deviations of _____ for the shape.
   A. Size
   B. Location
   C. Orientation
   D. All of the above

2. When inspecting a profile of a surface tolerance that is used to locate a planar surface, the inspector should establish the relationship between _____ and the datum reference frame.
   A. The part
   B. The high points of the tolerance surface
   C. The datum planes
   D. All of the above

3. Where a profile of a surface tolerance is used to control location of a planar surface, it controls which other geometric deviations?
   A. Size
   B. Form
   C. Coplanarity
   D. All of the above

4. Where a profile of a surface tolerance (with the all-around symbol) is applied to a closed polygon without datum references, it controls deviations in the geometry attribute(s) of:
   A. Size
   B. Coplanarity
   C. Orientation
   D. All of the above

5. A real-world application of a profile of a line tolerance is _____.
   A. Complex geometry of aerodynamic or ergonomic surfaces
   B. Coplanar surfaces
   C. Size dimensions
   D. Thin surfaces (e.g., a gasket)

6. A profile of a surface tolerance without datum references applied to two coplanar surfaces controls deviations in the geometry attributes of:
   A. Flatness of each surface
   B. Alignment between surfaces
   C. Straightness of each line element on each surface
   D. All of the above

7. A profile of a line tolerance applied to a surface may be used to control deviations of:
   A. Coplanarity
   B. Cylindricity
   C. Straightness
   D. All of the above

8. A profile of a surface tolerance may be verified using a:
   A. CMM
   B. Surface plate and height gage
   C. A dedicated gage
   D. All of the above

## Application Problems

*The application problems are designed to provide practice on applying the chapter concepts to situations that are similar to on-the-job conditions.*

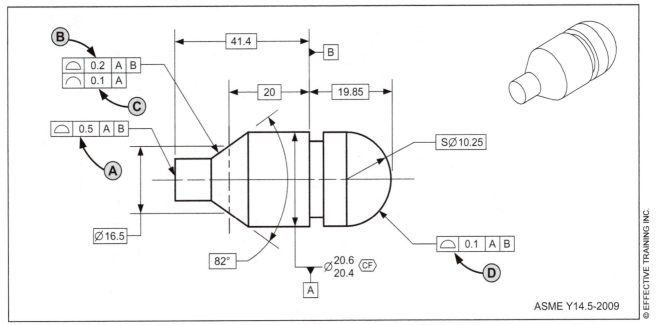

ASME Y14.5-2009

© EFFECTIVE TRAINING INC.

*Use the drawing above to answer questions 1 and 2.*

1. List the geometry attributes controlled by each profile tolerance.

    (A) _____

    _____

    (B) _____

    _____

    (C) _____

    _____

2. Use the Significant Seven Questions to interpret the profile of a surface tolerance labeled "D."

    1) Which dimensioning and tolerancing standard applies?

    _____

    2) What does the tolerance apply to?_____

    _____

    3) Is the specification standard-compliant?_____

    4) What are the shape and size of the tolerance zone?

    _____

    5) How much total tolerance is permitted?_____

    6) What are the shape and size of the datum simulators?

    _____

    7) Which geometry attributes are affected by this tolerance?

    _____

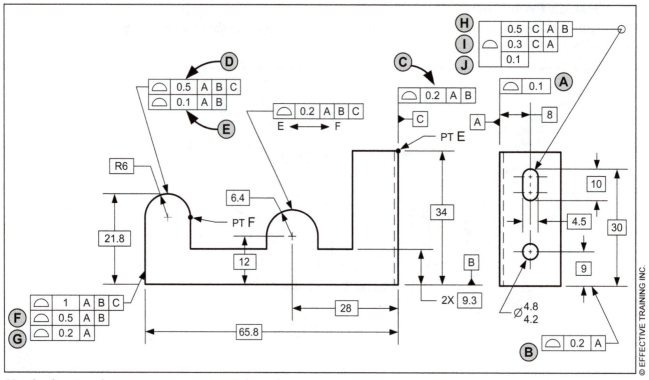

*Use the drawing above to answer questions 3 through 5.*

3. List the geometry attributes controlled by each profile tolerance.

(A) _____

(B) _____

(C) _____

(D) _____

(E) _____

(F) _____

(G) _____

(H) _____

(I) _____

(J) _____

4. Describe which feature control frames are a composite profile tolerance.

_____

_____

_____

_____

5. Describe which feature control frames are a multiple single-segment tolerance.

_____

_____

_____

_____

# Appendices

# APPENDIX A - THE HISTORY OF GD&T

As long as people have made things, they have used measurements, drawing methods, and drawings. Drawings existed as far back as six thousand B.C., when a unit of measure in the Nile and Chaldean civilizations was a "royal cubit." For thousands of years it fluctuated anywhere from 18 to 19 inches in length. Then, around four thousand B.C., the royal cubit was standardized at 18.24 inches. This set a pattern that has held true for nearly six thousand years. As long as there are measurements, drawing methods and drawings, there will be controversies, committees, and standards.

Manufacturing as we know it began with the Industrial Revolution in the 1800s. There were, of course, drawings, but these drawings were very different from the ones we use today. A typical drawing from the 1800s was a neatly inked, multi-viewed artistic masterpiece that portrayed the part with almost pictorial precision. Occasionally, the designer would write in a dimension, but generally such things were considered unnecessary.

Why? They were unnecessary because the manufacturing process was different then. There were no assembly lines, no widely dispersed departments or corporate units scattered across the nation or even worldwide as there are today. In those days, manufacturing was a cottage industry employing artisans who did it all, from parts fabrication to final assembly. These craftsmen passed their hard-won skills down from generation to generation. To them, there was no such thing as variation. Nothing less than perfection was good enough.

Of course there was variation, but back then the measuring instruments were not precise enough to identify it. When misfits and assembly problems occurred (which they routinely did), the craftsmen would simply cut-and-try, file-and-fit until the assembly worked perfectly. The total process was conducted under one roof, and communication among craftsmen was immediate and constant: "Keep that on the high side." "That edge has plenty of clearance." "That fit is OK now."

You can see that manufacturing back then was a quality process, but also slow, laborious and consequently quite an expensive one. The advent of the assembly line and other improved technologies revolutionized manufacturing. The assembly line created specialists to take the place of artisans, and these people did not have the time or skills for "file-and-fit."

Improved methods of measurement also helped to do away with the myth of "perfection." Now, engineers understand that variation is unavoidable. Moreover, in every dimension of every part in every assembly, some variation is acceptable without impairing the function of the assembly, as long as the limit of that variation—the "tolerance"—is identified, understood and controlled. This variation led to the development of the plus-minus (or coordinate) system of tolerancing, and to the determination that the logical place to record these tolerances and other information was on the engineering or design drawing.

With this development, drawings became more than just pretty pictures of parts; they became the main means of communication among manufacturing departments that were increasingly less centralized, more specialized, and subject to stricter demands.

## Engineering Drawing Standards

To improve the quality of drawings, an effort was made to standardize them. In 1935, after years of discussion, the American Standards Association (ASA) published the first recognized standard for drawings, "American Drawing and Drafting Room Practices." Of its eighteen short pages, just five discussed dimensioning; tolerancing was covered in just two paragraphs.

It was a beginning, but its deficiencies became obvious with the start of World War II. In Britain, wartime production was seriously hampered by high scrap rates due to parts that would not assemble properly. The British determined that this was caused by weakness in the plus-minus system of coordinate tolerancing, and more critically, by the absence of full and complete information on engineering drawings.

Driven by the demands of war, the British innovated and standardized. Stanley Parker of the Royal Torpedo Factory in Alexandria, Scotland, created a positional tolerancing system that called for cylindrical (rather than square) tolerance zones. The British went on to publish a set of pioneering drawings' standards in 1944, and in 1948 they published "Dimensional Analysis of Engineering Design." This was the first comprehensive standard that used fundamental concepts of true position tolerancing.

## GD&T in the United States

In the United States, Chevrolet published the Draftsman's Handbook in 1940, the first publication with any significant discussion of position tolerancing. In 1945, the U.S. Army published an ordinance manual on dimensioning and tolerancing that introduced the use of symbols (rather than notes) for specifying form and positioning tolerances.

Even so, the second edition of the American Standard Association's "American Standard Drawing and Drafting Room Practice," published in 1946, made minimal mention of tolerancing. That same year; however, the Society of Automotive Engineers (SAE) expanded coverage of dimensioning practices as applied in the aircraft industry in its "SAE Aeronautical Drafting Manual." An automotive version of this standard was published in 1952.

In 1949, the U.S. military followed the lead of the British by publishing the first standard for dimensioning and tolerancing, known as MIL-STD-8. Its successor, MIL-STD-8A, published in 1953, authorized seven basic drawing symbols and introduced a methodology of functional dimensioning.

As a result, there were three different groups in the United States publishing standards for drawings: the ASA, the SAE, and the military. This led to years of turmoil about the inconsistencies among the standards and resulted in slow, but measured progress in uniting those standards.

In 1957, the ASA (in coordination with the British and Canadians) approved the first American standard devoted to dimensioning and tolerancing. The 1959 MIL-STD-8B brought the military standards closer to ASA and SAE standards, and in 1966—after years of debate—the first united standard was published by the American National Standards Institute (ANSI), successor to the ASA. It was known as ANSI Y14.5. This first standard was updated in 1973 to replace notes with symbols in all tolerancing, and an updated standard was also published in 1982. The Y14.5 standard was published again in 1994. The current standard was published in 2009.

# APPENDIX B - THE SIGNIFICANT SEVEN QUESTIONS

Throughout this text you will be interpreting many geometric tolerances. In order to say you can interpret a geometric tolerance, you need to be able to answer the significant seven questions. The significant seven questions are explained here and are used throughout the textbook.

The seven significant questions are:

1. Which dimensioning and tolerancing standard applies?

2. What does the tolerance apply to?

3. Is the specification standard-compliant?

4. What are the shape and size of the tolerance zone?

5. How much total tolerance is permitted?

6. What are the shape and size of the datum simulators?

7. Which geometry attributes are affected by this tolerance?

When you see a geometric tolerance on a drawing, you should be able to answer these seven questions.

| Question | Description |
|---|---|
| 1. Which dimensioning and tolerance standard applies? | In order to interpret a geometric tolerance, you need to understand which standards apply. Different standards or versions of a standard often result in different interpretations of a geometric tolerance. |
| 2. What does the tolerance apply to? | A drawing is a legal document, If the specification is not legal, it should be corrected before it is interpreted. Don't assume that you know what was intended. |
| 3. Is the specification standard-compliant? | Does the geometric tolerance apply to:<br>• An element?<br>• A surface?<br>• A feature of size?<br>• A pattern? |
| 4. What are the shape and size of the tolerance zone? | What are the shape, size, orientation and location of the tolerance zone? |
| 5. How much total tolerance is permitted? | What is the stated tolerance? Is there a bonus tolerance permitted? Is there a datum shift tolerance permitted?<br>Check for the effects simultaneous requirements |
| 6. What are the shape and size of the datum simulators? | Sketch the datum simulators.<br>Determine if they are fixed or movable.<br>Where are the datums?<br>How is the part held relative to the datums? |
| 7. Which geometry attributes are affected by this tolerance? | Which geometry attributes are controlled by this geometric tolerance? Size, location, orientation, form, or surface texture? |

# APPENDIX C - STANDARD TERMS AND PART GEOMETRY

The figure below contains a visual representation of the part geometry that terms in this text refer to.

      Yellow boxes  =  terms from this text
      Green boxes   =  part geometry types
      Pink boxes    =  part geometry and dimensions used in combination

Each term applies to all of the part geometry types of the lower level branches of the tree that connect with the term.

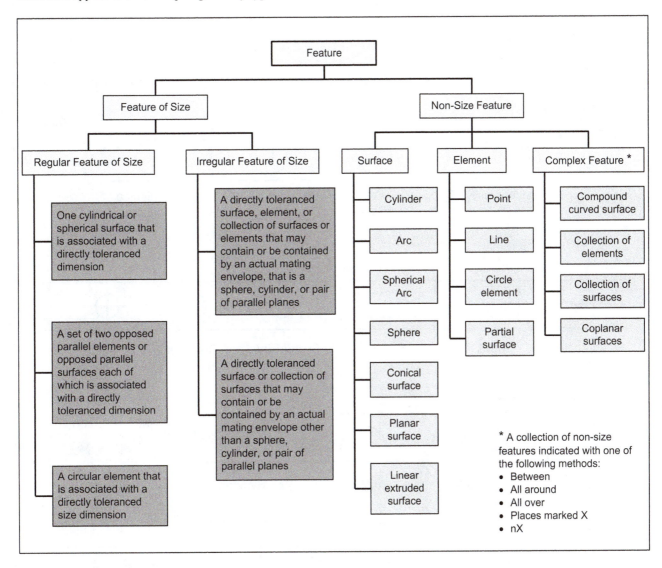

# APPENDIX D - FEATURE CONTROL FRAME PROPORTION CHART

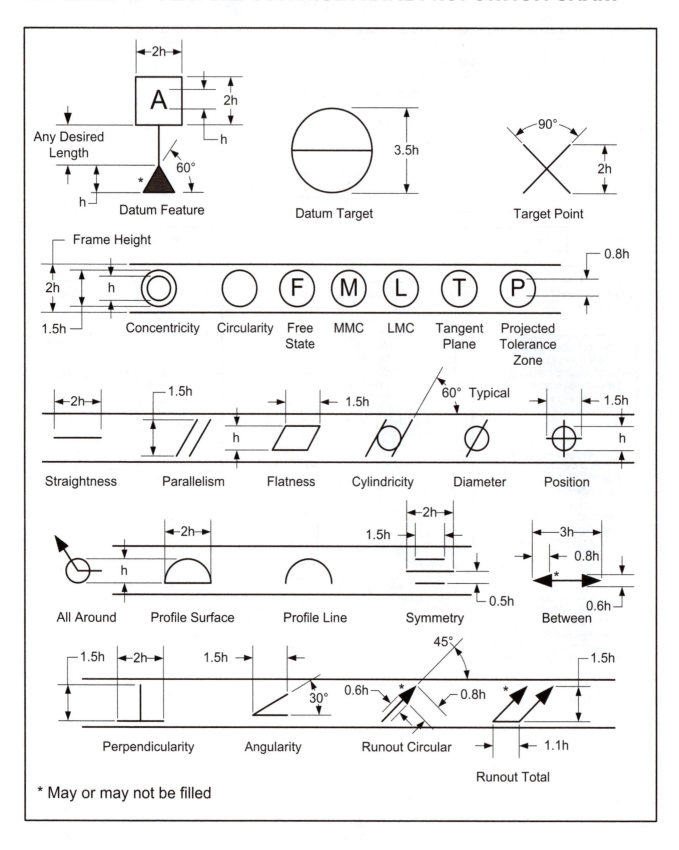

* May or may not be filled

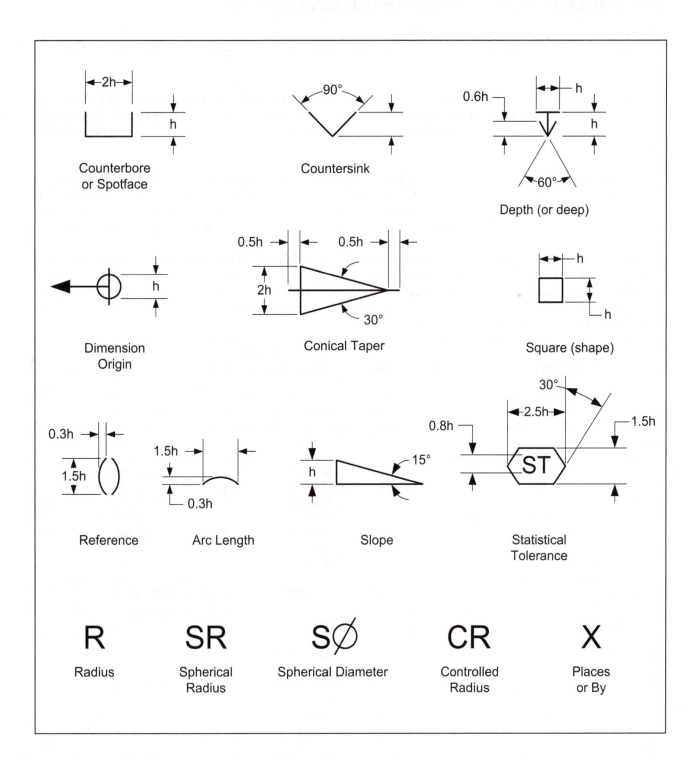

Counterbore or Spotface

Countersink

Depth (or deep)

Dimension Origin

Conical Taper

Square (shape)

Reference

Arc Length

Slope

Statistical Tolerance

R
Radius

SR
Spherical Radius

SØ
Spherical Diameter

CR
Controlled Radius

X
Places or By

# APPENDIX E - ASME/ISO COMPARISON CHARTS

Many people believe that ISO and ASME GD&T standards are nearly identical, with a few symbols and concepts that differ. Comparison charts in most reference books give this impression, including the charts in the ASME Standard on pages 197 and 198.

However, when you look beyond the shape of the symbols and consider the whole tolerance frame, the ISO/ASME standards contain numerous differences in the use and interpretation of the symbols. Some of the differences are significant and affect part acceptance criteria. Even where two symbols are the same in ISO and ASME, there are often differences in their use and interpretation.

Charts 1-4 compare the symbols from ISO 1101 and companion standards with those from the ASME Y14.5-2009 Standard. As the charts illustrate, some of the symbols are the same and are always used and interpreted in the same manner ("yes" in the chart). More often, however, the same symbol may only sometimes share the same specification/use and interpretation ("sometimes" in the chart). In a few cases, the same symbol is never interpreted in the same manner ("no" in the chart).

The comparisons in this section highlight the problems encountered when someone knowledgeable in the ISO standards attempts to interpret a technical drawing prepared to the ASME standards. Even though the symbol is the same, the tolerance specification often doesn't make sense, or its interpretation is different from that in the ISO standards. The tables listed in the comparison charts are too extensive to reproduce in this text. For details on the tables see *Alex Krulikowski's ISO Geometrical Tolerancing Reference Guide*, pages 311—330.

| Description | ISO | ASME | Same Specification / Use | Table | Same Interpretation | Table |
|---|---|---|---|---|---|---|
| Straightness | — | — | Sometimes | 1 | Sometimes | 1 |
| Flatness | ▱ | ▱ | Sometimes | 2 | Sometimes | 2 |
| Roundness[1] | ○ | ○ | Sometimes | 3 | Sometimes | 3 |
| Cylindricity | ⌭ | ⌭ | Sometimes | 4 | Sometimes | 4 |
| Profile of any line[2] | ⌒ | ⌒ | Sometimes | 5 | Sometimes | 5 |
| Profile of any surface[3] | ⌓ | ⌓ | Sometimes | 5 | Sometimes | 5 |
| Parallelism | // | // | Sometimes | 6 | Sometimes | 6 |
| Perpendicularity | ⊥ | ⊥ | Sometimes | 6 | Sometimes | 6 |
| Angularity | ∠ | ∠ | Sometimes | 6 | Sometimes | 6 |
| Position | ⊕ | ⊕ | Sometimes | 7 | Sometimes | 7 |
| Concentricity/Coaxiality[4] | ◎ | ◎ | Sometimes | 8 | No | 8 |
| Symmetry | ⹀ | ⹀ | Sometimes | 8 | No | 8 |
| Circular run-out | ↗ | ↗ | Sometimes | 9 | Sometimes | 9 |
| Total run-out | ↗↗ | ↗↗ | Sometimes | 10 | Sometimes | 10 |

1 ASME: Circularity

2 ASME: Profile of a Line

3 ASME: Profile of a Surface

4 ISO: Concentricity applies to center point of cross-section; coaxiality applies to axis.

   ASME: Concentricity applies to median points.

| Description | ISO | ASME | Same Specification/ Use | Table | Same Interpretation | Table |
|---|---|---|---|---|---|---|
| Maximum material requirement[1] | Ⓜ | Ⓜ | Sometimes | 11 | Sometimes | 11 |
| Least material requirement[2] | Ⓛ | Ⓛ | Sometimes | 11 | Sometimes | 11 |
| Reciprocity requirement | Ⓡ | -- | No | 13 | No | 13 |
| Regardless of feature size | Default | Default | Sometimes | 12 | Sometimes | 12 |
| Free state condition | Ⓕ | Ⓕ | Yes | - | Yes | - |
| Projected tolerance zone | Ⓟ | Ⓟ | No | 16 | Yes | 16 |
| Envelope requirement[3] | Ⓔ | Default | No | 13 | Sometimes | 13 |
| Independency | Default | Ⓘ | No | 14 | Sometimes | 14 |
| Tangent plane | -- | Ⓣ | No | 14 | No | 14 |
| Theoretically exact dimension[4] | 50 | 50 | Sometimes | 17 | Yes | 17 |
| Diameter | ∅ | ∅ | Yes | - | Yes | - |
| Radius | R | R | Yes | - | Sometimes | 18 |
| Controlled radius | -- | CR | No | 14 | No | 14 |
| Spherical radius | SR | SR | Yes | - | Yes | - |
| Spherical diameter | S∅ | S∅ | Yes | - | Yes | - |
| All around |  (all around symbol) |  (all around symbol) | Yes | - | Yes | - |
| All over | -- |  (all over symbol) | No | 14 | No | 14 |
| Non-uniform | -- | NON-UNIFORM | No | 14 | No | 14 |
| Profile unequally disposed | -- | Ⓤ | No | 14 | No | 14 |

1 ASME: "Maximum material condition" after tolerance; "maximum material boundary" after datum

2 ASME: "Least material condition" after tolerance; "least material boundary" after datum

3 ASME: Rule #1

4 ASME: Basic dimension

| Description | ISO | ASME | Same Specification/ Use | Table | Same Interpretation | Table |
|---|---|---|---|---|---|---|
| Counter bore | -- | ⊔ | No | 14 | No | 14 |
| Spot face | -- | ⌊SF⌋ | No | 14 | No | 14 |
| Counter sink | -- | ∨ | No | 14 | No | 14 |
| Depth/ Deep | -- | ⩛ | No | 14 | No | 14 |
| Between | -- | ◄──► | No | 14 | No | 14 |
| Statistical tolerance | -- | ⟨ST⟩ | No | 14 | No | 14 |
| Common zone | CZ | -- | No | 13 | No | 13 |
| Continuous feature | -- | ⟨CF⟩ | No | 14 | No | 14 |
| Minor diameter | LD | MINOR∅ | No | 19 | Yes | 19 |
| Major diameter | MD | MAJOR∅ | No | 19 | Yes | 19 |
| Pitch diameter | PD | PD | Yes | - | Yes | - |
| Line element | LE | EACH ELEMENT | No | - | Yes | - |
| Any cross-section | ACS | -- | No | 13 | No | 13 |
| By or number of places | 2x | 2X | No | - | Yes | - |
| Auxiliary dimension[1] | (50) | (50) | Yes | - | Yes | - |
| Angle dimension | X.X° | X.X° | Yes | 25 | No | 25 |
| Dimensions not to scale | 50 | 50 | Yes | - | Yes | - |
| Arc length | ⌒ 10 | ⏜10 | No | 20 | Yes | 20 |
| Cone[2] | ▷ | ▷ | Yes | - | Yes | - |
| Slope | ◁ | ◁ | Yes | - | Yes | - |
| Square | □ | □ | Yes | - | Yes | - |
| Origin indication[3] | ⊕──► | ⊕──► | Yes | - | No | 21 |
| Tolerance frame[4] | ⊕ ∅0.2 A B | ⊕ ∅0.2 A B | Yes | - | Yes | - |
| Simultaneous | Default | Default | No | 22 | Sometimes | - |
| Separate requirement | -- | SEPT REQT | No | 14 | No | 14 |
| Boundary | -- | BOUNDARY | No | 14 | No | 14 |

1 ASME: Reference dimension

2 ASME: Conical taper

3 ASME: Dimension origin

4 ASME: Feature control frame

| Description | ISO | ASME | Same Specification/ Use | Table | Same Interpretation | Table |
|---|---|---|---|---|---|---|
| Datum feature indication[1] | A (boxed) | A (boxed) | Often | 23 | No | 23 |
| LMB datum references | -- | Ⓛ | No | 15 | No | 15 |
| MMB datum references | -- | Ⓜ | No | 15 | No | 15 |
| RMB datum references | | Default | No | 15 | No | 15 |
| Datum target frame[2] | A1 | A1 | Yes | - | Yes | - |
| Datum target point | X | X | Yes | - | Yes | - |
| Datum target line | X—X | — — — — | No | 24 | Yes | 24 |
| Datum target area | (hatched circle) | (hatched circle) | Yes | - | Yes | - |
| Movable datum target | -- | Ø1 / A1 | No | 15 | No | 15 |
| Chain line | -- | — — — | No | 15 | No | 15 |
| Datum translation | -- | ▷ | No | 15 | No | 15 |
| Datum reference frame | -- | Z X Y (axes) | No | - | No | 15 |
| Customized datum reference frame | -- | ⊕ \| 0.2 \| A[x,y,u,v] \| B[z] | No | 15 | No | 15 |

1 ASME: Datum feature symbol
2 ASME: Datum target symbol

# APPENDIX F - BOUNDARY INTERPRETATION

## Why the Boundary Interpretation is Common with Geometric Tolerances Specified at MMC

When the MMC modifier is used in a geometric tolerance, this text uses a VC acceptance boundary interpretation. There are many reasons for using the VC acceptance boundary:

1. The Y14.5, paragraph 7.3.3.1, states that where a geometric tolerance is specified at MMC, there are two interpretations a VC acceptance boundary interpretation and an axis/center plane interpretation. Where there is a conflict between these interpretations, the surface (VC acceptance boundary) interpretation takes precedence. Figure 7-4 in the Y14.5 standard illustrates this concept.

2. When the MMC modifier is specified, the function is often assembly. In these applications the surface of the toleranced feature is functional. The surface of the feature of size must fit within (or around) its virtual condition acceptance boundary that represents the worst case for assembly.

3. When the MMC modifier is used, a functional gage is often used to verify the tolerance requirement. A functional gage verifies that a feature of size fits within (or around) its VC boundary.

4. When the MMC modifier is used, a bonus tolerance is permitted. When using the VC acceptance boundary interpretation, the bonus is automatically accounted for. There is no need to determine how much of the tolerance deviation is permitted from size (bonus) and how much tolerance deviation is permitted from other geometric attributes.

5. When using the axis/center plane interpretation, the bonus tolerance must be calculated. How to determine the amount of bonus tolerance when using the axis method is controversial. Is the bonus based on one or more actual local sizes of the feature of size or is the bonus based on the size of the related actual mating envelope? Does each cross-section element of a feature of size have a different amount of bonus based on its local size? Or is the amount of bonus available based on the smallest actual local size of the feature of size?

6. The boundary interpretation can also reduce inspection time / cost when using a CMM. The acceptance boundary is more repeatable and the deviation of a feature of size can be verified by collecting a set of surface points and comparing the points to the virtual condition boundary.

Based on this list, the simplest and most clear approach is to use the VC acceptance boundary method.

In many cases, the acceptance boundary interpretation and the axis / center plane interpretation are equivalent. The figure on the opposite page shows an example using the axis interpretation and the acceptance boundary interpretation.

**Author's Comment**
There is a case where the boundary interpretation does not work. This occurs on internal features of size when the position tolerance is so large that the virtual condition boundary would be less than zero. In this case, the axis interpretation must be used.

## Drawing

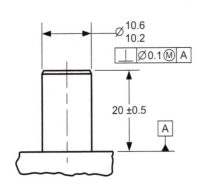

## Axis Interpretation

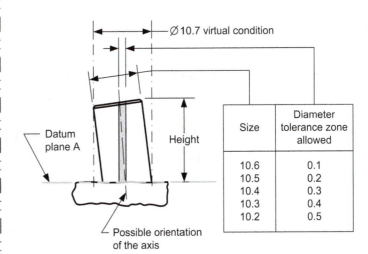

| Size | Diameter tolerance zone allowed |
|------|---------------------------------|
| 10.6 | 0.1 |
| 10.5 | 0.2 |
| 10.4 | 0.3 |
| 10.3 | 0.4 |
| 10.2 | 0.5 |

When the pin diameter is at MMC (10.6), the perpendicularity tolerance zone is 0.1 diameter. When the pin diameter departs from its MMC size, an increase in the perpendicularity tolerance zone is allowed equal to the amount of such departure. This is the bonus tolerance. The axis of the unrelated actual mating envelope pin must be within the tolerance zone allowed based on the size of the pin.

## Acceptance Boundary Interpretation

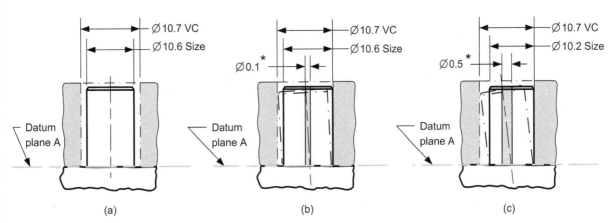

(a)  (b)  (c)

(a) The MMC (10.6) pin with perfect orientation is shown in an acceptance boundary of 10.7 diameter

(b) When the pin is at maximum diameter (10.6), the acceptance boundary will accept the part with up to 0.1 deviation in perpendicularity

(c) When the pin is at minimum diameter (10.2), the acceptance boundary will accept the part with up to 0.5 deviation in perpendicularity

* Equivalent axis tolerance zone

# APPENDIX G - BIBLIOGRAPHY

1.    American Society of Mechanical Engineers. Dimensioning and Tolerancing: ASME Y14.5M-1994 [Revision of ANSI Y14.5M-1982 (R1988)]. NY: ASME, 1995.

2.    American Society of Mechanical Engineers. Dimensioning and Tolerancing: ASME Y14.5-2009. NY: ASME, 2009.

3.    Krulikowski, Alex. Advanced Concepts of GD&T. Westland, MI: Effective Training, 1998.

4.    Krulikowski, Alex. Alex Krulikowski's ISO Geometrical Tolerancing Guide. Westland, MI: Effective Training, 2010.

5.    Krulikowski, Alex. Geometric Dimensioning and Tolerancing Self-Study Workbook. Westland, MI: Effective Training, 1990.

6.    Krulikowski, Alex. Geometric Tolerancing Applications Workbook. Westland, MI: Effective Training, 1994.

7.    Krulikowski, Alex. Tolerance Stacks: A Self-Study Course. Westland, MI: Effective Training, 1992.

# APPENDIX H - GLOSSARY

*Actual Local Size* - The measured value of any individual distance at any cross section of a regular feature of size.

*Actual Mating Envelope* - A similar perfect feature counterpart of the smallest size that can be contracted about an external feature of size, or the largest size that can be expanded within an internal feature of size, so that it coincides with the surfaces at the highest points. There are two types of actual mating envelopes: related and unrelated.

*Actual Minimum Material Envelope* - This envelope is within the material. A similar perfect feature(s) counterpart of largest size that can be expanded within an external feature(s) or smallest size that can be contracted about an internal feature(s) so that it coincides with the surface(s) at the lowest points. There are two types of actual mating envelopes: unrelated and related.

*All-Around Symbol* - The symbolic means of indicating that a profile tolerance zone applies to all surfaces, all around the true profile in the view shown.

*All-Over Symbol* - The symbolic means of indicating that a profile tolerance zone applies to all surfaces, all over the three-dimensional profile of a part.

*Angularity* - The condition where a surface, axis, or center plane is at the specified angle to a datum plane or datum axis.

*Angularity Tolerance* - A geometric tolerance that limits the amount a surface, axis, or center plane is permitted to deviate from a basic angle relative to a datum reference frame.

*Annotation* - Dimensions, tolerances, notes, text, or symbols visible without any manual or external manipulation.

*Annotation Plane* - A conceptual plane containing annotation that either perpendicularly intersects or is coincident with one or more surfaces of a feature.

*Arc* - A segment of a curve.

*ASME Y14.41-2003* - American Society of Mechanical Engineers Standard on Digital Product Definition Data Practices.

*ASME Y14.5-2009* - The standard for dimensioning and tolerancing. At a minimum, an engineering drawing should specify this standard. An engineering drawing will often invoke several additional standards.

*ASME Y14.5M-1994* - The national standard for dimensioning and tolerancing in the United States. ASME stands for American Society of Mechanical Engineers. The Y14.5 is the standard number. "M" is to indicate the standard is metric, and 1994 is the date the standard was officially approved.

*Assembly* - A number of parts, or combination thereof, that are joined together to perform a specific function and subject to disassembly without degradation of any of the parts (e.g., power shovel-front, fan assembly, audio frequency amplifier).

*Attribute Gage* - The family of receiver gages used to collect attributes data; for example, GO and functional gages.

*Average Diameter* - The average of several diametric measurements across a circular or cylindrical feature.

*AVG* - Placed near the qualified size dimension where a form tolerance (such as circularity) is specified in a free state. It is then used to ensure that the actual diameter of the feature can be restrained to the desired shape at assembly.

*Axis* - A theoretical straight line about which a geometric object rotates or may be imagined to rotate. Where the term "axis" is used in this text, unless otherwise indicated, it refers to the axis of the unrelated actual mating envelope of a cylindrical feature of size.

*Axis/Center Plane Interpretation* - The axis (or center plane) of the unrelated actual mating envelope of the feature of size must be within the tolerance zone.

*Axis Offset* - The parallel distance between two axes; for example, the distance between a part feature of size axis and a datum axis.

*Axis Theory* - When the axis (or center plane) of the feature of size must be within the tolerance zone.

*Basic Angle* - A type of basic dimension with an angular value used to define the orientation of a part feature or datum target.

*Basic Dimension* - A numerical value used to describe the theoretically exact size, true profile, orientation, or true position of a feature, feature of size, or gage information (e.g., datum targets).

*Between Symbol* - The symbolic means of indicating that a profile tolerance zone applies to all surfaces between two points.

*Bidirectional Position Tolerance* - Where the location tolerance of a hole is defined with different tolerance values in two directions.

***Bilateral Equally Disposed Tolerance Zone*** - A tolerance zone that is equally divided bilaterally to both sides of the true profile. Where an equally disposed (centered) bilateral tolerance zone is intended, the profile tolerance has a leader line directed to the surface or an extension line of the surface.

***Bilateral Unequally Disposed Tolerance Zone*** - An unequally disposed tolerance zone is indicated with an unequally disposed profile symbol in the feature control frame following the tolerance value. A second value after the symbol indicates the tolerance in the direction that adds material to the true profile.

***Bilateral Tolerance*** - A type of plus-minus tolerance that allows a dimension to vary in both the plus and minus directions.

***Bilateral Tolerance Zone*** - A tolerance zone that is divided bilaterally to both sides of the true profile

***Bonus Tolerance*** - A potential additional tolerance for a geometric tolerance.

***"By" Symbol*** - A symbol that indicates a relationship between two coordinate dimensions.

***Center Plane*** - A theoretical plane about which a geometric object is equally disposed. Where the term "center plane" is used, unless otherwise indicated, it refers to the center plane of the unrelated actual mating envelope of a width feature of size.

***Circular Runout*** - A geometric tolerance that affects the circularity and coaxiality of circular elements of a surface of revolution, or affects the wobble of circular elements of a plane surface perpendicular to a datum axis. It applies independently at each circular measuring position.

***Circular Runout Tolerance*** - A geometric tolerance that limits the high to low point deviation (runout) of the circular elements of any surface of revolution or the circular elements of a planar surface that is perpendicular to and intersects the datum axis. The tolerance zone (runout) applies independently at each circular element of the toleranced surface (i.e., at each cross section perpendicular to the datum axis).

***Circularity*** - The condition of a surface where: a) For a feature other than a sphere (e.g., cylinder, cone, etc.), all points of the surface intersected by any plane perpendicular to an axis or spine (curved line) are equidistant from that axis or spine (curved line). b) For a sphere, all points of the surface intersected by any plane passing through a common center are equidistant from that center.

***Circularity Tolerance*** - A geometric tolerance that limits the circularity deviation (radial deviation between highest and lowest points) on individual circular elements

***Coaxial Datum Features*** - Coaxial features of size used to create a single datum axis.

***Coaxiality*** - The condition where the axes of the unrelated actual mating envelopes, axis of the unrelated minimum material envelope, or median points (as applicable) of one or more surfaces of revolution are coincident.

***Complex Feature*** - A single surface of compound curvature or a collection of other features that constrains up to six degrees of freedom.

***Composite Position Tolerance*** - A feature control frame that contains multiple segments with a single entry of a position tolerance that is applicable to all horizontal segments of the feature control frame.

***Composite Profile Tolerance*** - A feature control frame that contains a single entry of a profile tolerance that is applicable to all horizontal segments of the feature control frame.

***Concentricity*** - The condition where the median points of all diametrically opposed elements of a surface of revolution (or the median points of correspondingly located elements of two or more radially disposed features) are congruent with the datum axis (or center point).

***Concentricity Tolerance*** - A geometric tolerance that defines the permissible deviation that all median points of a surface of revolution (or correspondingly located elements of two or more radially disposed features) are permitted to vary from a datum axis or center point.

***Conical*** - A surface of revolution that has a taper symmetrical about its axis. See conical taper.

***Conical Taper*** - The ratio of the difference in the diameters of two sections (perpendicular to the axis) of a cone to the distance between these sections. See ASME Y14.5-2009 paragraph 2.13 for acceptable methods of dimensioning.

***Constraint*** - A limit to one or more degrees of freedom.

***Continuous Feature Modifier*** - A symbol that indicates where a group of two or more features of size are to be treated geometrically as a single feature or feature of size.

***Controlled Radius*** - A radius where the part surface must have a smooth curve within the tolerance zone. In a controlled radius, no flats or reversals are permitted on the arc surface.

***Coordinate Measuring Machine (CMM)*** - A precision three-axis machine used to collect and analyze surface points from a part to verify the geometric requirements (size, location, orientation, and form) of part features.

**Coordinate Tolerancing** - A dimensioning system where the coordinates (X,Y,Z) of the centers of features of size and surfaces are located (or defined) by means of linear dimensions with plus-minus tolerances.

**Coplanar Datum Features** - When two or more coplanar surfaces are designated as a single datum plane, simulated by coinciding with the datum feature simulator that simultaneously contacts the high points of two surfaces.

**Coplanarity** - The condition of two or more surfaces having all elements on one plane.

**Counterbore** - A flat-bottomed diameter.

**Counterbore Symbol** - Symbolic means of indicating a counterbore.

**Countersink** - A conical shape added to the beginning of a hole. The large diameter of the cone and the included angle are specified along with the countersink symbol.

**Countersink Symbol** - Symbolic means of denoting the shape of geometry on a part.

**Cylindricity** - The condition of a surface of revolution in which all points of the surface are equidistant from a common axis.

**Cylindricity Gage** - An instrument used to measure cylindricity deviation.

**Cylindricity Tolerance** - A geometric tolerance that defines the deviation permitted on a cylindrical surface. A cylindricity tolerance applies along the entire cylindrical surface simultaneously.

**Datum** - A theoretically exact plane, point or axis from which a dimensional measurement is made.

**Datum Axis** - The axis of a datum feature simulator established from the datum feature.

**Datum Center Plane** - The center plane of a datum feature simulator established from the datum feature.

**Datum Center Point** - The center plane of a datum feature simulator established from the datum feature.

**Datum Feature** - A feature that is identified with either a datum feature symbol or a datum target symbol.

**Datum Feature Reference** - The desired order of precedence is indicated by entering the appropriate datum feature reference letter, from left to right, in the feature control frame.

**Datum Feature Shift** - The allowable movement or looseness between the part datum feature and the datum feature simulator.

**Datum Feature Simulator (Physical)** - The physical boundary used to establish a simulated datum from a specified datum feature.

**Datum Feature Simulator (Theoretical)** - The theoretically perfect boundary used to establish a datum from a specified datum feature.

**Datum Plane** - A datum established from the datum feature simulator of a nominally flat datum feature. It constrains three degrees of freedom (one translation and two rotations).

**Datum Reference Frame** - A set of three mutually perpendicular datum planes. The datum reference frame provides direction as well as an origin for dimensional measurements.

**Datum Reference Frame Symbol** - A symbol used to label axes of a datum reference frame. The X, Y, Z labels represent the translational degrees of freedom.

**Datum System** - A set of symbols and rules on how to constrain a part to establish a relationship between the part and geometric tolerance zones.

**Datum Targets** - A set of symbols that describe the shape, size, and location of datum feature simulators that are used to establish datum planes, axes, or points.

**Datum Target Area** - A designated area used in establishing a datum.

**Datum Target Identification System** - A means of designating a datum target on a drawing. The symbol is paced outside the part outline with a radial line directed to the target.

**Datum Target Line** - A designated line used in establishing a datum.

**Datum Target Point** - A designated point used in establishing a datum.

**Datum Target Symbol** - A circular symbol with a horizontal line dividing the symbol into two halves. The bottom half denotes the datum letter and the target's number. The top half denotes the datum feature simulator size, when applicable.

**Degrees of Freedom (DOF)** - The movement of a part in space. A rigid part has six degrees of freedom: three translational degrees of freedom and three rotational degrees of freedom.

**Depth** - Refers to how deep a feature is on a part.

**Depth Symbol** - Indicates the depth of a hole, counterbore, or spotface.

**Derived Median Line** - An imperfect line formed by the center points of all cross sections of a feature of size.

**Derived Median Plane** - An imperfect plane formed by the center points of all line segments bounded by the feature of size.

**Dial Indicator** - An instrument used to accurately measure small linear distances, and is frequently used in inspection of parts.

**Diameter** - A symbol that describes the shape, size, and location of gage elements that are used to establish datum planes or axes.

**Diameter Symbol** - Used two different ways on engineering drawings: a) Indicates that a dimension applies to a circular or cylindrical shape. b) Indicates a cylindrical-shaped tolerance zone.

**Dimension** - A numerical value expressed in appropriate units of measure and used to define the size, location, orientation, form, or other geometric characteristics of a part.

**Dimension Origin Symbol** - A symbol that consists of a circle in place of an arrowhead on one end of a dimension line, denoting that a dimension between two surfaces originates from a plane established from one surface and not the other.

**Drawing** - An engineering document or digital data file(s) that discloses (either directly or by reference) by means of graphic or textural presentations, or by combinations of both, the physical or functional requirements of an item.

**Elongated Hole** - A hole with a different length and width (commonly referred to as a slot).

**Engineering Drawing** - A document that communicates a precise description of a part. This description consists of pictures, words, numbers, and symbols.

**Equal Bilateral Plus-Minus Tolerance** - A tolerance where the allowable variation from the nominal value is the same in both directions.

**Feature** - A physical portion of a part, such as a surface, pin, hole, or slot, or its representation on drawings, models, or digital data files.

**Feature Axis** - The axis of the unrelated actual mating envelope of a feature.

**Feature Center Plane** - The center plane of the unrelated actual mating envelope of a feature.

**Feature Control Frame** - A rectangular box that is divided into compartments within which the geometric characteristic symbol, tolerance value, modifiers, and datum references are placed.

**Feature of Size** - A general term that is used to refer to either a regular feature of size or an irregular feature of size.

**Fixed Fastener Assembly** - An assembly of two or more parts where a fastener is held in place in (threaded or pressed into) one of the components of the assembly.

**Fixed Fastener Formula (General)** - (for equally distributed tolerances) is:
$$H = F + 2T \text{ or } T = \frac{H - F}{2}$$
Where:   T = total position tolerance for assembly
H = MMC of the clearance hole
F = MMC of the fastener

**Fixed Fastener Formula (Modified)** - (for unequally distributed tolerances) is:
$$H - F = T_1 + T_2$$

Where:   H = MMC of the clearance hole

F = MMC of the fastener

$T_1$ = tolerance for the threaded hole

$T_2$ = tolerance for the clearance hole

**Fixed-Limit Gage** - A device of defined geometric form and size used to assess the conformance of a feature of size of a workpiece to a dimensional specification.

**Flag Notes** - Notes that are located with the general notes but apply only at specific areas or points on the drawing.

**Flatness** - The condition of a surface or a derived median plane having all of its elements in one plane.

**Flatness Tolerance** - Specifies a tolerance zone of two parallel planes within which a surface or derived median plane must lie.

**Floating Fastener Assembly** - An assembly of two or more parts where two (or more) components are held together with fasteners (such as bolts and nuts), and both components have clearance holes for the fasteners.

**Floating Fastener Formula** - A formula used to calculate positional tolerance values in a floating fastener assembly.
$$T = H - F$$
Where:
T = position tolerance zone diameter (for each part)
H = MMC of the clearance hole
F= MMC of the fastener

**Form Tolerance** - A geometric tolerance that defines the allowable flatness deviation permitted on a planar surface or derived median plane.

**Free State** - The condition of a part free of applied forces.

*Free State Variation* - The distortion of a part after removal of forces applied during manufacture.

*Full Indicator Movement* - The total movement of an indicator where appropriately applied to a surface to measure its variations.

*Functional Dimensioning* - A dimensioning approach that defines a part based on the product requirements.

*Functional Gage* - A fixed-limit gage used to verify virtual condition acceptance boundaries.

*Fundamental Dimensioning Rules* - A set of general rules that apply to dimensioning and interpreting engineering drawings.

*Gage* - A device used to measure a part characteristic.

*Gage Element* - A physical portion of a gage.

*General Notes* - Notes that apply to the entire drawing; they are always numbered as a single-numbered list in the notes area of the drawing.

*Geometric Characteristic Symbols* - A set of 14 symbols used in the language of GD&T to describe the geometry attributes of a part.

*Geometric Dimensioning and Tolerancing (GD&T)* - An international language used on engineering drawings that consists of well-defined of symbols, rules, definitions and conventions, used on engineering drawings to accurately describe a part. GD&T is a precise mathematical language that can be used to describe the size, form, orientation, and location of part features. GD&T is also a design philosophy on how to design and dimension parts.

*Geometric Tolerance* - The general term applied to the category of tolerances used to control form, profile, orientation, location, and runout.

*Geometry Attribute* - A characteristic of a feature or feature of size.

*GO Gage* - A a fixed-limit gage that checks a feature of size for acceptance within MMC perfect form boundary.

*Great Myth of GD&T* - The misconception that geometric tolerancing raises product costs.

*Height Gage* - A measuring device used for measuring a vertical distance of a feature from a reference surface.

*Implied 90° Angle* - Where center lines and lines depicting features are shown on a 2D orthographic drawing at right angles, and no angle is specified.

*Implied 90° Basic Angle* - A 90° basic angle that applies where center lines of features in a pattern (or surfaces shown at right angles on a drawing) are located and defined by basic dimensions, and no angle is specified.

*Implied Basic Zero Dimension* - Where a center line or center plane of a feature of size is shown in line with a datum axis or center plane. The distance between the center lines or center planes is an implied basic zero.

*Implied Datum* - An assumed plane, axis, or point from which a dimensional measurement is made.

*Implied Parallelism* - Where two surfaces are shown parallel on a drawing and the size dimension controls the parallelism between the surfaces.

*Implied Self-Datum* - When a geometric tolerance implies that a feature is to be inspected relative to its perfect counterpart. Example: a position tolerance specified without a datum reference creates a relationship between the two co-axial features.

*Independency Concept* - Size and form are independent; Rule #1 does not apply

*Indicator Zero Block* - A gage detail that has two parallel surfaces set at a set distance apart.

*Individual* - Notation used along with an indication of the number of places a datum feature and either a single feature of size or a pattern of features of size apply on an individual basis.

*Inner Boundary* (of an internal feature of size) - A worst-case boundary generated by the smallest feature of size (MMC) minus the effects of the applicable geometric tolerance and any additional tolerance (bonus) that may apply.

*Inner Boundary* (of an external feature of size) - A worst-case boundary generated by the smallest feature of size (LMC) minus the effects of the applicable geometric tolerance and additional tolerances (bonus) that may apply.

*Irregular Feature of Size* - A general term for two types of irregular features of size:

    Type A: A directly toleranced feature or collection of features that may contain or be contained by an actual mating envelope: that is a sphere, cylinder, or pair of parallel planes.

    Type B: A directly toleranced feature or collection of features that may contain or be contained by an actual mating envelope: other than a sphere, cylinder, or pair of parallel planes.

*Least Material Boundary (LMB)* - The limit defined by a tolerance or combination of tolerances that exists on or inside the material of a feature(s).

*Least Material Condition (LMC)* - The condition in which a feature of size contains the least amount of material within the stated limits of size.

*Limit Dimensioning* - Where a dimension has its high and low limits stated. In a limit tolerance, the high value is placed on top, and the low value is placed on the bottom.

*Limits of Size* - The specified maximum and minimum sizes.

*Local Notes* - Notes that are located at the specific area or point of application on the drawing.

*MAX Dimension* - The drafting symbol for a maximum dimension is the letters "MAX" following the dimension value; where a maximum dimension is indicated, the opposite end of the specification is usually zero.

*Maximum Material Boundary (MMB)* - The boundary established by the collective effects of the MMC of a datum feature and any applicable tolerances.

*Maximum Material Condition (MMC)* - The condition in which a feature of size contains the maximum amount of material within the stated limits of size.

*Median Point* - The mid-point of a two-point measurement.

*MIN Dimension* - The drafting symbol for a minimum dimension is the letters "MIN" following the dimension value; where a minimum dimension is indicated, the opposite end of the specification is as large as the dimension could be and still fit on the part

*Minimum Radial Separation (MRS)* - The radial distance between two concentric circles which just contain the measured polar profile as a minimum.

*Model* - A combination of design model, annotation, and attributes that describes a product.

*Model Coordinate System* - A representation of a Cartesian coordinate system shown on a CAD model or on an engineering drawing.

*Modifiers* - Symbols or keywords that communicate additional information about the tolerancing of a part.

*Movable Datum Target* - A datum target with a movable datum target datum feature simulator.

*Movable Datum Target Symbol* - Indicates that the datum feature simulator is movable.

*Multiple Single-Segment* - When two (or more) single-segment feature control frames are stacked together. Each segment is an independent requirement.

*Multiple Single-Segment Position Tolerance* - Two or more single-segment position tolerances applied to the same feature of size or pattern of features of size.

*Multiple Single-Segment Profile Tolerance* - Two or more profile tolerances are applied to a surface (or collection of surfaces) relative to different datums.

*NOGO Gage* - A fixed-limit gage that checks a feature of size for violation of its LMC actual local size.

*Nominal Size* - The designation used for purposes of general identification. On a drawing, it is the specified value of a dimension.

*Non-Opposed* - Two planar surfaces are non-opposed if none of the rays projected from each surface intersect the other surface.

*Non-Uniform Tolerance Zone* - A maximum material boundary and least material boundary, of a unique shape, that encompasses the true profile. The method to indicate a non-uniform tolerance zone is use the term "NON-UNIFORM" to replace the tolerance value in the feature control frame.

*Number of Places Symbol* - Symbolic means of indicating how many places a dimension or shape apply on a drawing.

*Opposed* - All rays normal from each planar surface intersects the other surface.

*Optical Comparator* - (Often called a comparator or profile projector) is a device that applies the use of optics to measure part geometry.

*Orientation* - The orientation of a feature or a feature of size is its angular relationship to a datum. Orientation tolerances control parallel, perpendicular, and all other angular relationships.

*Orientation Tolerance* - A geometric tolerance that limits the amount of orientation error a surface, axis, or center plane is allowed.

*Outer Boundary (OB)* - A worst-case boundary generated by the largest feature of size (MMC) plus the effects of the applicable geometric tolerance.

*Parallel* - The condition of being an equal distance apart at every point.

*Parallelism* - The condition where a surface, axis, or center plane is equidistant at all points from a datum plane or axis.

*Parallelism Tolerance* - A geometric tolerance that limits the amount a surface, axis, or center plane is permitted to deviate from parallelism, relative to a datum reference frame.

**Part** - One item, or two or more items joined together, not normally subject to disassembly without destruction or impairment of designed use.

**Partial Datum Reference Frame** - Where only one or two datum planes of the datum reference frame are used.

**Partially Opposed** - Two planar surfaces are partially opposed if some of the rays projected normal from each planar surface intersect the other surface..

**Pattern** - Two or more features or features of size to which a locational geometric tolerance is applied and is grouped by one of the following methods: nX, INDICATED, n CO-AXIAL HOLES, ALL OVER, ALL AROUND, A←→B (between symbol), n Surfaces, simultaneous requirement

**Perpendicularity** - The condition where a surface, axis, or center plane is exactly 90° to a datum plane or axis.

**Perpendicularity Tolerance** - A geometric tolerance that limits the amount a surface axis, or center plane is permitted to deviate from being perpendicular to a datum. A perpendicularity tolerance can only constrain rotational degrees of freedom relative to the datum referenced.

**Planar Feature of Size** - A feature of size that contains two features: the two parallel plane surfaces (such as a width or a thickness). A planar feature of size is one type of a regular feature of size; capable of producing a center plane.

**Plus and Minus Tolerancing** - Where the nominal of a dimension is given first, followed by a plus-minus expression of a tolerance.

**Position Tolerance** - A geometric tolerance that limits the amount the center point, axis, or center plane of a feature of size is permitted to deviate from true position.

**Precision Square** - A gage block that has a precise 90° angle between the bottom surface and one of the sides of the block.

**Primary Datum** - The first datum that the part contacts in a dimensional measurement.

**Profile** - An outline of a surface, a shape made up of one or more features, or a two-dimensional element of one or more features.

**Profile Tolerance** - A geometric tolerance that is used to define a tolerance zone to control form or combinations of size, form, orientation, and location of a feature relative to a true profile.

**Profile of a Line Tolerance** - A geometric tolerance that establishes a two-dimensional tolerance zone that is normal to the true profile at each line element.

**Profile of a Surface Tolerance** - A geometric tolerance that establishes a three-dimensional tolerance zone that extends along the length and width (or circumference) of the considered feature or features.

**Projected Tolerance Zone** - A tolerance zone that is projected (moved) outside of the feature of size.

**Radius** - A straight line extending from the center to the periphery of a circle or sphere.

**Reference Dimension** - a) A repeat of a dimension b) A dimension derived from other values shown on the drawing or on related drawings. One method for identifying a reference dimension on drawings is to enclose the dimension within parentheses.

**Regardless of Feature Size** - The term that indicates a geometric tolerance or datum reference applies at any increment of size of the actual mating envelope of the feature of size.

**Regardless of Material Boundary (RMB)** - Indicates that a datum feature simulator is adjustable (or movable) from MMB towards LMB until it makes maximum contact with the extremities of the datum feature(s).

**Regular Feature of Size** - One cylindrical or spherical surface, a circular element, a set of two opposed (or partially opposed) parallel elements, or opposed (or partially opposed) parallel surfaces, each of which is associated with a directly toleranced dimension (size dimension).

**Related Actual Mating Envelope** - A similar perfect feature counterpart expanded within an internal feature of size, or contracted about an external feature of size, while constrained in either orientation or location (or both) to any applicable datums.

**Related Actual Minimum Material Envelope** - A similar perfect feature counterpart contracted about an internal feature(s) or expanded within an external feature(s) while constrained within orientation or location (or both) to the applicable datum(s).

**Restraint** - The application of force(s) to a part to simulate its assembly or functional condition resulting in possible distortion of a part from its free-state condition.

**Reversal** - A change of direction (or interruption) of the surface of the arc (i.e., a tool mark).

**Right-Hand Rule** - The right-hand rule works as follows: if you hold the back of your right hand against a drawing and hold the thumb and index finger at 90° to each other:
- The thumb pointing to the right represents the X-axis positive direction.
- The index finger pointing to the top of the drawing represents the Y-axis positive direction.

The middle finger pointing at you represents the Z-axis positive direction.

**Rotation** - Angular movement about an axis.

**Roundness (Circularity) Gage** - A precision spindle machine used to measure the out of roundness (circularity deviations) of a surface of revolution.

**Rule #1** - The form of an individual regular feature of size is controlled by its limits of size.

**Rule #2** - RFS applies with respect to the individual tolerance, and RMB applies, with respect to the individual datum feature reference, where no modifying symbol is specified.

**Runout** - The high to low point deviation of the surface elements relative to a datum axis.

**Secondary Datum** - The second datum that the part contacts in a dimensional measurement.

**Separate Requirement** - This tolerance is verified as a separate inspection item. The part must meet this requirement to be accepted.

**Simulated Datum** - A datum established from a physical datum feature simulator.

**Simultaneous Requirement** - Where two or more geometric tolerances have identical datum references and are considered a pattern.

**Sine Bar** - A precision inspection tool used to orient parts at any angle, usually between zero and sixty degrees. It is so named because the sine of the angle determines the height of a stack of gage blocks used to set the angle.

**Size Dimension** - A dimension across two opposed (or partially opposed) surfaces, line elements, or points.

**Spherical Diameter** - The length of a straight line that passes through the center of a sphere and touches two points on its edge.

**Spherical Radius** - A radius that generates a segment of a sphere.

**Spine** - A simple (non self-intersecting) curve.

**Spotface** - The min size of the flat on a flat-bottomed diameter.

**Square Symbol** - Indicates that a dimension applies to the length and width of a square shape.

**Straightness (Axis or Center Plane)** - The condition where an element of a surface or a derived median line is a straight line.

**Straightness of a Line Element** - The condition where each line element (or axis or center plane) is a straight line.

**Straightness Tolerance** - A geometric tolerance that, when directed to a surface, limits the amount of straightness error allowed in each surface line element. A straightness tolerance is applied in the view where the elements to be controlled are represented by a straight line.

**Surface Interpretation** - The tolerance zone is a virtual condition (acceptance boundary), located at true position, which the surface(s) of the tolerance features(s) of size must not violate.

**Symmetrical Relationship** - Symmetrical relationships may be controlled using position, profile, or symmetry tolerances.

**Symmetry** - The condition where all the median points of all opposed correspondingly located elements of two or more feature surfaces are congruent with a datum axis or center plane.

**Symmetry Tolerance** - A geometric tolerance that defines the permissible deviation that median points of all opposed correspondingly located elements of two or more feature surfaces are permitted to vary from a datum axis or center plane.

**Surface Plate** - A solid, flat plate used as the main horizontal reference plane (datum feature simulator, see Chapter 13) for precision inspection and tooling setup.

**Tangent Plane** - A plane that is tangent to the high points of the surface.

**Tangent Plane Modifier** - Indicates that a geometric tolerance only applies to the tangent plane of the toleranced surface.

**Tertiary Datum** - The third datum that the part contacts in a dimensional measurement.

**Theoretical Datum Feature** - The theoretically perfect boundary used to establish a datum from a specified datum feature.

**3-2-1 Rule** - A part will have a minimum of three points of contact with its primary datum plane, two points minimum contact with its secondary datum plane, and a minimum of one point of contact with its tertiary datum plane.

***Tolerance*** - The total amount a specific dimension is permitted to vary. The tolerance is the difference between the maximum and minimum limits.

***Tolerance Accumulation*** - A condition where the tolerances from several dimensions combine to result in a greater amount of tolerance for a part distance.

***Tolerance Analysis Chart*** - A chart that graphically displays the available tolerances of a feature of size as defined by the drawing specifications.

***Tolerance of Position Tolerance*** - A geometric tolerance that defines the location tolerance of a feature of size from its true position.

***Tolerance Stack*** - A calculation that determines the theoretical maximum or minimum distance between two features on a part or in an assembly.

***Tolerance Zone*** - The area (zone) that represents the total amount that part features are permitted to vary from their specified dimension.

***Total Runout*** - The area (zone) that represents the total amount that part features are permitted to vary from their specified dimension.

***Total Runout Tolerance*** - A geometric tolerance that limits the high to low point deviation of all surface elements of a cylindrical surface, or all surface elements of a planar surface that is perpendicular to and intersects the datum axis.

***Translation*** - The ability to move without rotating.

***True Position*** - The theoretically exact location of a feature of size as established by basic dimensions.

***True Profile*** - A profile located and defined by basic radii, basic angular dimensions, basic coordinate dimensions, basic size dimensions, profile to drawings, formulas, or mathematical data, including design models.

***Unequal Bilateral Plus-Minus Tolerance*** - The allowable variation is from the nominal value, and the variation is not the same in both directions.

***Unequally Disposed Profile Symbol*** - Indicates a unilateral or unequally disposed profile tolerance.

***Unilateral - In the Direction That Adds Material*** - A uniform boundary tolerance zone that is offset to the outside of the true profile.

***Unilateral - In the Direction That Removes Material*** - A uniform boundary tolerance zone that is offset to the inside of the true profile.

***Unilateral Tolerance*** - Where the allowable variation from the nominal value is all in one direction and zero in the other direction.

***Unilateral Tolerance Zone*** - A tolerance zone that is offset to one side of the material, either "all outside" or "all inside" of the material.

***Unrelated Actual Mating Envelope*** - A similar perfect feature counterpart expanded within an internal feature of size or contracted about an external feature of size and not constrained to any datums.

***Unrelated Actual Mating Envelope - External*** - A similar perfect feature(s) counterpart of smallest size contracted about an external feature(s) and not constrained to any datum(s).

***Unrelated Actual Mating Envelope - Internal*** - A similar perfect feature(s) counterpart of largest size that can be expanded within an internal feature(s) and not constrained to any datum(s).

***Unrelated Actual Minimum Material Envelope*** - A similar perfect feature counterpart contracted about an internal feature(s) or expanded within an external feature(s) and not constrained to any datum reference frame.

***Variable Gaging*** - Any gaging operation that measures part variation and provides a quantitative value of the variation.

***Virtual Condition (VC)*** - A fixed-size boundary generated by the collective effects of a considered feature of size's specified MMC or LMC and the geometric tolerance for that material condition.

***Worst-Case Boundary (WCB)*** - A general term to refer to the extreme boundary of a feature of size that is the worst case for assembly.

***Zero Tolerance at MMC*** - A dimensioning method that takes the tolerance stated in the feature control frame, combines it with the size tolerance, and states a zero tolerance in the feature control frame.

# INDEX

## A

## B

## C

# ETI PRODUCTS AND SERVICES

Effective Training Inc. (ETI) is an internationally recognized leader in the field of geometric dimensioning and tolerancing (GD&T). Company founder Alex Krulikowski is a noted GD&T expert, educator and author of more than 20 books and articles on the subject. Alex gained more than 30 years of industrial experience as a design supervisor with one of the world's largest manufacturing corporations. He also has a degree in industrial vocational education from Eastern Michigan University. He has developed all of ETI's products and services.

## Onsite Workshops
- Applications of GD&T (ASME Y14.5M-1994)
- ASME Y14.5M-1994-2009 Update
- ASME to ISO Standards Comparison
- Engineering Drawing Requirements - Exclusively from ETI
- Executive Overview of GD&T - Exclusively from ETI
- Functional Gaging and Measurement (Y14.5, Y14.5.1, Y14.43)
- GD&T Advanced Concepts of GD&T (ASME Y14.5M-1994)
- GD&T Fundamentals (ASME Y14.5M-1994)
- GD&T Fundamentals (ASME Y14.5-2009)
- GD&T Fundamentals for Inspectors (ASME Y14.5M-1994)
- GD&T Overview Workshop (ASME Y14.5M-1994) - Exclusively from ETI
- ISO Geometrical Tolerancing
- Solid Model Tolerancing: The ASME Y14.41 Standard - Exclusively from ETI
- Statistical Tolerance Stacks - Exclusively from ETI
- System Approach to Component Tolerancing - Exclusively from ETI
- Tolerance Stacks (concept driven)

## Training Software
GD&T Trainer Professional Edition (Single-User, Multi-User, LAN)

## Web-Based Courses
- ASME Y14.5 Standard Comparison: 1994 to 2009
- Engineering Drawing Requirements
- Fundamentals of GD&T (ASME Y14.5M-1994)

## Self-Study Workbooks
- Fundamentals of GD&T Self-Study Workbook, 2nd Edition
- Tolerance Stacks Self-Study Course

## Video Training
- Fundamentals of GD&T Video Training (ASME Y14.5M-1994) (Individual Module or Full Course - 8-DVD Set)
- Fundamentals of GD&T Video Training Overview Course - 4-DVD Set
- Fundamentals of GD&T 10-Tape VHS Video Set

## Self-Study Workbook
- GD&T Workbook With Engineering Drawings (ASME Y14.5M-1994)

www.etinews.com

CPSIA information can be obtained
at www.ICGtesting.com
Printed in the USA
FFHW012337090919
54846693-60539FF